Paradigm Shift in E-waste Management

Paradigm Shift in E-waste Management

Vision for the Future

Edited by
Abhijit Das
Biswajit Debnath
Potluri Anil Chowdary
Siddhartha Bhattacharyya

CRC Press
Taylor & Francis Group
Boca Raton London New York

CRC Press is an imprint of the
Taylor & Francis Group, an **informa** business

First edition published 2022
by CRC Press
6000 Broken Sound Parkway NW, Suite 300, Boca Raton, FL 33487-2742

and by CRC Press
4 Park Square, Milton Park, Abingdon, Oxon, OX14 4RN

Library of Congress Cataloging-in-Publication Data
A catalog record has been requested for this book

ISBN: 978-0-367-55985-4 (hbk)
ISBN: 978-0-367-55989-2 (pbk)
ISBN: 978-1-00-309597-2 (ebk)

DOI: 10.1201/9781003095972

Typeset in Times
by Newgen Publishing UK

Dedication

Abhijit would like to dedicate this book to his beloved wife Amrita, daughter Abhipsa and son Amritesh.

Biswajit would like to dedicate this book to his lovely and cheerful wife Ankita & Dr. Arabinda Sen, his inspiration towards metal recovery.

Anil would like to dedicate this book to his beloved family, his father Potluri Suresh, Guru K. Kameswara Rao and Green Waves partner K. Pranesh Varma.

Siddhartha would like to dedicate this volume to the memory of his sister, Late Prof. Madhura Dutta, an ardent teacher of Computer Science.

Contents

PART 1 *Global Status of E-waste Recycling and Management*

PART 2 *Benchmark Practices and Case Studies*

PART 3   Technologies for E-waste Valorization and Management

PART 4   Vision for the Future: Towards Resource Efficient E-waste Management

Preface

Growth of the electronics industry has been tremendous in the last two decades, which ensured a lot of cash flow, but it also contributed to the material flow stream of electronic waste. The demand of Electronic and Electrical Equipment (EEE) is ever increasing and the driving force behind this demand is often the technological advancement coupled with short innovation cycles and business strategies, which shortens the lifespan of the equipment. The high rate of obsolescence of electronic items is due not only to the abovementioned reasons, but also to intelligent marketing gimmicks of the electronics industry. Globally, e-waste is the fastest expanding waste stream in the world increasing at an annual rate of 3-5 percent. The global issue of e-waste is about to set out into its pinnacle which is a serious threat to the overall anthropogenosphere. There has been acceleration in research and development of environmentally sound e-waste recycling technologies in the past decade which is laudable. However, an alternative approach is to look at it as anthropogenic stockpiles. This concept is easily extrapolated to link with the core concept of urban mining. Urban mining is a concept which facilitates the recovery of material and energy from urban waste and brings them back into the economy. E-waste is a rich source of metals, glass fibre, polymers etc, which makes it the most potential candidate for urban mining. This huge resource present in the molecules and networks of e-waste will be in vain unless tapped and brought back to the economy. This offers an opportunity for implementing circular economic approaches and moving towards a sustainable future. E-waste management and valorization is a very complex and transdisciplinary field. Upcoming technologies such as the Internet of Things, blockchain technologies, nanotechnology and concepts such as smart city, green computing, green economy, sustainable city, and so forth can be clubbed together to proliferate in real life. Additionally, a cross-disciplinary approach at this rate can be expected to complement the sustainable development goals (SDGs).

Written during 2021, this book aims to provide an overview towards the future of the e-waste management sector. The paradigm shift in e-waste management towards a sustainable future needs to be understood while keeping in mind the targets of the SDGs. In view of that, there are four major parts in this book divided into several chapters. Corresponding authors are denoted by an asterisk in the chapter openers.

Part 1 describes the global status of E-waste Recycling and Management with country specific contributions. It also covers e-waste recycling technologies, supply chain aspects and trans-boundary movement issues.

Part 2 focuses on policy tools such as EPR, ARF etc; policy gaps and the informal sector activities. Additionally, it features case studies and benchmark practices from local and renowned industries around the world.

Part 3 offers detailed information about globally implemented technologies for e-waste valorization and management including the evolving biotechnological advancement.

Part 4 contributes to the visions of the future i.e. towards resource efficient e-waste management amalgamating the sustainable benefits of e-waste plastics recycling and life cycle assessment of e-waste systems.

This book is intended for researchers, academicians and practitioners. It will serve as a ready-made material for the researchers and academicians as it covers a wide range of subject areas belonging to several majors falling under the umbrella of e-waste management. Additionally, it may be treated as a handbook which will meet the needs of policymakers, supply chain managers and technology ninjas. The editors would feel rewarded if the concepts presented in the book add to the social cause.

Biswajit Debnath
Kolkata, India
and
Birmingham, UK

Anil Potluri
Visakhapatnam, India

Abhijit Das
Kolkata, India

Siddhartha Bhattacharyya
Birbhum, India
June, 2021

Acknowledgments

The editors express their gratitude to CRC Press for publishing the book. We sincerely thank the contributing authors and the following persons who helped bring this volume to life.

Dr. Gagandip Singh
Ms. Ankita Das
Prof. Amar Chandra Das
Dr. Saswati Gharami
Ms. Aryama Raychaudhuri
Ms. Moumita Sardar
Late Guru Ramkrishna Goswami

Editors

Dr. Abhijit Das received his B.Tech. (IT) from the University of Kalyani, M.Tech. (IT) from the University of Calcutta and Ph.D. (Engineering) from the Department of CSE Jadavpur University, India.

Dr. Das has over 17 years of teaching and research experience and more than 40 publications and three edited books of international repute. Presently he is serving as an Associate Professor in the Department of IT, RCC Institute of Information Technology, Kolkata, India. He had been the Head of the Department of IT (Jan 2018–Jan 2020) and has convened various committees at an institutional level.

Dr. Das has organized and chaired various international conferences and seminars. He has served as a resource person in various institutes and universities and television channels at state and national levels. Currently, six scholars are working with him on different research topics such as IoT, e-Waste Management, Data Science, Quantum Computing, Object Oriented Categorization, and the like.

Abhijit has published four patents and eight copyrights to date. He serves as a reviewer for many reputed journals and is a professional member of IEEE, IETE (Fellow) and ACM.

Abhijit is a professional singer as well and frequently performs in All India Radio Kolkata, DoorDarshan, and various private television channels. He has more than 10 music albums on his credit.

Biswajit Debnath is a Senior Research Fellow in the Chemical Engineering Department, Jadavpur University. He was a Commonwealth Split-Site Scholar at Aston University, Birmingham, UK (2019–2020). He received his B.Tech and M.E. in Chemical Engineering in 2013 and 2015, respectively. His area of specialization is waste valorization and sustainability with special focus on e-waste and plastic waste. His other research interests include Circular Economy, Climate Change, SDGs, Supply Chain Management, Sustainable Smart City, Environmental Chemical Engineering and so forth. He has worked in UKIERI projects and published nearly 60 articles including conference proceedings, peer-reviewed journals and contributed to 29 book chapters published by Springer, Wiley, Elsevier, IEEE and River Publishers. His h index is 12 with 506 citations. He is a CPCB certified trainer for the six waste management rules and provided training on e-waste & plastic waste rules on invitation. He has won best paper award several times in international conferences, which offered him ten invited lectures including webinars. Since December 2018, he has been one of the most read authors from his department in ResearchGate. He has completed five collaborative (unfunded) projects with colleagues in the USA, Finland, Saudi Arabia and India. He is a reviewer of reputed journals namely *ACS Sustainable Chemistry and Engineering*, *Journal of Material Cycles and Waste Management*, *Journal of Network and Computer Applications*, *Environment, Development and Sustainability*, and *Waste Management* published by Springer, ACS and Elsevier. He also works as sustainability consultant for national and international clients.

Mr. Potluri Anil Chowdary has been the Managing Director of Green Waves Environmental Solutions, the only First Authorized E-waste collection and handling unit in Andhra Pradesh (unit in Visakhapatnam), since April 2015. Mr. Potluri graduated with a Diploma in Environment Resource Management from the Waiariki Institute of Technology (New Zealand). He completed his Master in Science (M.Sc) in Environmental Science branch from GITAM Institute of Science, GITAM University (India). He did his Engineering in Chemical Technology from Chaitanya Bharathi Institute of Technology, Osmania University (India). He has worked in several projects in New Zealand including Plantation of Native Flora and Pest Management (2014); 'Love Your Water' program organized by Sustainable Coastline on cleaning of fresh and marine water bodies (2014); Composting and Vermi Composting for the Linton Park Community Centre under the guidance of Mr. Rick Mansell (Centre Coordinator, Linton Park Community Centre, Rotorua, New Zealand) in 2014. Under his leadership, Green Waves Environmental Solutions has won multiple national and international awards such as National Awards for its excellence in e-Waste Recycling at Indian Industry Session (at 8th Regional 3R Forum in Asia and the Pacific); Seva Puraskar award by Andhra Pradesh Pollution Control Board for contributions toward sensitizing people on e-waste management and for effective recycling of e-waste on World Environmental Day 2018; IconSWM Award for excellence in E-Waste recycling at 8th International Conference on Sustainable Waste Management, 22 November 2018, Acharya Nagajurna University, Guntur, Andhra Pradesh. He has delivered invited lectures and participated in workshops on waste management, e-waste, upcycling and nature conservation.

Dr. Siddhartha Bhattacharyya did his Bachelors in Physics, Bachelors in Optics and Optoelectronics and Masters in Optics and Optoelectronics from University of Calcutta, India in 1995, 1998 and 2000, respectively. He completed his Ph.D in Computer Science and Engineering from Jadavpur University, India in 2008. He is the recipient of the University Gold Medal from the University of Calcutta for his Masters. He also received several coveted awards including the Distinguished HoD Award and

Distinguished Professor Award conferred by Computer Society of India, Mumbai Chapter, India in 2017, the Honorary Doctorate Award (D. Litt.) from The University of South America and the South East Asian Regional Computing Confederation (SEARCC) International Digital Award ICT Educator of the Year in 2017. He was appointed the ACM Distinguished Speaker for 2018–2020. He was been inducted into the People of ACM hall of fame by ACM, USA in 2020. He was named the IEEE Computer Society Distinguished Visitor for the tenure 2021–2023. He was elected as a full foreign member of the Russian Academy of Natural Sciences and a full fellow of The Royal Society for Arts, Manufacturers and Commerce (RSA), London, UK.

Dr. Bhattacharyya currently serves as the Principal of Rajnagar Mahavidyalaya, Rajnagar, Birbhum. He has served as a Professor in the Department of Computer Science and Engineering of Christ University, Bangalore and as the Principal of RCC Institute of Information Technology, Kolkata, India during 2017–2019. He has also served as a Senior Research Scientist in the Faculty of Electrical Engineering and Computer Science of VSB Technical University of Ostrava, Czech Republic (2018–2019), prior to which, he was the Professor of Information Technology of RCC Institute of Information Technology, Kolkata, India. He served as the Head of the Department from March, 2014 to December, 2016 following tenure as an Associate Professor of Information Technology of RCC Institute of Information Technology, Kolkata, India from 2011–2014. Before that, he served as an Assistant Professor in Computer Science and Information Technology of University Institute of Technology, The University of Burdwan, India from 2005–2011. He was a Lecturer in Information Technology of Kalyani Government Engineering College, India during 2001–2005.

Dr. Bhattacharyya is a co-author of six books and the co-editor of 78 books and has more than 300 research publications in international journals and conference proceedings to his credit. He has two PCTs and 19 patents to his credit. He has been the member of the organizing and technical program committees of several national and international conferences. He is the founding Chair of ICCICN 2014, ICRCICN (2015, 2016, 2017, 2018), ISSIP (2017, 2018) (Kolkata, India). He was the General Chair of several international conferences like WCNSSP 2016 (Chiang Mai, Thailand), ICACCP (2017, 2019) (Sikkim, India) and (ICICC 2018 (New Delhi, India) and ICICC 2019 (Ostrava, Czech Republic).

He is the Associate Editor of several reputed journals including *Applied Soft Computing, IEEE Access, Evolutionary Intelligence* and *IET Quantum Communications*. He is the editor of *International Journal of Pattern Recognition Research* and the founding Editor-in-Chief of *International Journal of Hybrid Intelligence, Inderscience*. He has guest edited several issues in international journals and serves as the Series Editor of IGI Global Book Series Advances in Information Quality and Management (AIQM), De Gruyter Book Series Frontiers in Computational Intelligence (FCI), CRC Press Book Series(s) Computational Intelligence and Applications & Quantum Machine Intelligence, Wiley Book Series Intelligent Signal and Data Processing, Elsevier Book Series Hybrid Computational Intelligence for Pattern Analysis and Understanding and Springer Tracts on Human Centered Computing.

His research interests include hybrid intelligence, pattern recognition, multimedia data processing, social networks and quantum computing.

A life fellow of Optical Society of India (OSI) and of International Society of Research and Development (ISRD), UK, Dr. Bhattacharyya is also a fellow of Institution of Engineering and Technology (IET), UK, of Institute of Electronics and Telecommunication Engineers (IETE), India and of Institution of Engineers (IEI), India. He is also a senior member of Institute of Electrical and Electronics Engineers (IEEE), USA, International Institute of Engineering and Technology (IETI), Hong Kong and Association for Computing Machinery (ACM), USA.

Further, Dr. Bhattacharyya is a life member of Cryptology Research Society of India (CRSI), Computer Society of India (CSI), Indian Society for Technical Education (ISTE), Indian Unit for Pattern Recognition and Artificial Intelligence (IUPRAI), Center for Education Growth and Research (CEGR), Integrated Chambers of Commerce and Industry (ICCI), and Association of Leaders and Industries (ALI). He is a member of Institution of Engineering and Technology (IET), UK, International Rough Set Society, International Association for Engineers (IAENG), Hong Kong, Computer Science Teachers Association (CSTA), USA, International Association of Academicians, Scholars, Scientists and Engineers (IAASSE), USA, Institute of Doctors Engineers and Scientists (IDES), India, The International Society of Service Innovation Professionals (ISSIP) and The Society of Digital Information and Wireless Communications (SDIWC). He is also a certified Chartered Engineer of Institution of Engineers (IEI), India. He is on the Board of Directors of International Institute of Engineering and Technology (IETI), Hong Kong.

Contributors

Biswaranjan Acharya
School of Computer Engineering
KIIT Deemed to be University
Bhubaneswar, Odisha, India

Yasanthi Alakahoon
Department of Business Administration
University of Sri Jayewardenepura
Nugegoda, Sri Lanka

M. Tanvir Alam
Department of Chemical Engineering
Monash University
Clayton, Victoria, Australia

Anan Ashrabi Ananno
Department of Management and
    Engineering
Linköping University
Linköping, Sweden

Siddhartha Bhattacharyya
Rajnagar Mahavidyalaya
Birbhum, India

Ariel L. Cappelletti
Centro Experimental de la Vivienda
    Económica
Córdoba, Argentina

Sandip Chatterjee
Director
Ministry of Electronics and
    Information Technology
    (Meity)
New Delhi, India

Abhijit Das
RCC Institute of Information
    Technology
Kolkata, India

Peter Dabnichki
School of Engineering
RMIT University
Melbourne, Victoria, Australia

Biswajit Debnath
Chemical Engineering Department
Jadavpur University
Kolkata, India
And
Department of Mathematics
ASTUTE, Aston University
Birmingham, UK

Satarupa Dey
Shyampur Siddheswari Mahavidyalaya
University of Calcutta
India

Valerio Elia
Department of Innovation Engineering
University of Salento
Lecce, Italy

Chukwunonye Ezeah
Department of Civil Engineering
Faculty of Engineering and Technology
Alex Ekwueme Federal University
Ndufu-Alike, Ikwo, Ebonyi State, Nigeria

Fariborz Faraji
The Robert M. Buchan Department
    of Mining
Queen's University
Kingston, Ontario, Canada

Marlia M. Hanafiah
Department of Earth Sciences and
    Environment
Faculty of Science and Technology
Universiti Kebangsaan Malaysia
Bangi, Selangor, Malaysia

Haikal Ismail
School of Technology Management and
   Logistics
College of Business
Universiti Utara Malaysia
Sintok, Kedah, Malaysia

Manya Khanna
Jindal School of Government and
   Public Policy
O.P. Jindal Global University
Sonipat, Haryana, India

Rosana Gaggino
Centro Experimental de la Vivienda
   Económica
Córdoba, Argentina

Julián González Laría
Centro Experimental de la Vivienda
   Económica
Córdoba, Argentina

Rabeeh Golmohammadzadeh
Department of Chemical Engineering
Monash University, Clayton, Victoria,
   Australia

Melina Gómez
Centro Experimental de la Vivienda
   Económica
Córdoba, Argentina

Maria Grazia Gnoni
Department of Innovation Engineering
University of Salento
Lecce, Italy

Nuwan Gunarathne
Department of Accounting
University of Sri Jayewardenepura
Nugegoda, Sri Lanka
And
Department of Business Strategy and
   Innovations

Griffith University
Southport, Australia

Mosarrat Mahjabeen
Shaheed Suhrawardy
   Medical College
Sher-e-Bangla Nagor, Bangladesh

Mahadi Hasan Masud
School of Engineering
RMIT University
Melbourne, Victoria, Australia
And
Department of Mechanical Engineering
Rajshahi University of Engineering and
   Technology
Rajshahi, Bangladesh

Christia Meidiana
Department of Regional and Urban
   Planning
Brawijaya University
Malang, Indonesia

Florin-Constantin Mihai
Environmental Research Center
   (CERNESIM)
Interdisciplinary Research Institute
"Alexandru Ioan Cuza" University
   of Iasi
Romania

Monjur Mourshed
School of Engineering
RMIT University
Melbourne, Victoria, Australia
And
Department of Mechanical
   Engineering,
Rajshahi University of Engineering and
   Technology
Rajshahi, Bangladesh

Lucas E. Peisino
Centro Experimental de la Vivienda
  Económica
Córdoba, Argentina

Anil Potluri
Managing Director
Green Waves Environmental Solutions
www.greenwavesrecyclers.in

Bárbara B. Raggiotti
Centro de Investigación
Desarrollo y Transferencia de
  Materiales y Calidad (CINTEMAC),
  UTN–FRC
Córdoba, Argentina

Petra Schneider
Magdeburg-Stendal University of
  Applied Sciences
Magdeburg, Germany

M. Shahabuddin
Carbon Technology Research Centre
School of Engineering, Information
  Technology and Physical Sciences
Federation University
Gippsland, Victoria, Australia

1 Introduction

Biswajit Debnath*, Anil Potluri, and Abhijit Das

CONTENTS

1.1 BRIEF COMMENTS ON CURRENT SITUATION OF E-WASTE MANAGEMENT

E-waste is the fastest growing waste stream worldwide growing with an alarming rate of 3–5 percent per year. According to the global e-waste monitor 2020, an outstanding 53.6 million metric tons (Mt) of e-waste was generated around the globe, which corresponds to an average of 7.3 kg per capita (Forti et al. 2020). In the year 2014, nearly 42 Mt of e-waste was generated globally which increased to 44.7 Mt in 2016 (Baldé et al. 2015; Baldé et al. 2017). The global e-waste generation has increased by 9.2 Mt since 2014 and is expected to become almost double by 2030 (Forti et al. 2020). In other words, the global e-waste generation in 2019 is equivalent to 5516 Eiffel towers compared to e-waste generation in 2016 i.e. 4600 Eiffel towers (Baldé et al. 2017; Forti et al. 2020). The global e-waste generation is expected to reach 74.7 million metric tons by 2030, and 120 million metric tons by 2050. This global mushrooming of e-waste generation is primarily because of the electronics industry. The growth of the electronics industry has been stupendous in the past two decades. It is expected to reach $400 billion in 2022 from $69.6 billion in 2012 (Corporate Catalyst (India) Pvt. ltd. 2015). But we cannot blame the industry alone. The electronics industry is demand driven and we, the consumers are to blame. There are two faces of this – a) The consumers' ability to buy because of economic development and b) the industry's intelligent marketing gimmicks. There is a direct relationship between the electronics item consumption and extensive international economic development. Electronic items have become an essential commodity in modern and evolving society where the grade and version of electronics is considered as a benchmark of living standards. Additionally, high disposable income, urbanization and better industrialization are auxiliary factors that drive the enormous amount of electronics items (Forti et al. 2020). But this doesn't end here. The industry employs short innovations and minor upgrades based on the latest technological proliferations coupled with intelligent marketing strategies that lure the users to discard their old electronics and buy a new product. In a sustainable society, this is very unsustainable practice as it doesn't maximize the resource efficiency. Moreover, the product is not utilized till its full lifespan,

DOI: 10.1201/9781003095972-1

1

which leads to product obsolescence and thereby contributing to the e-waste stream (Debnath 2020).

E-waste management has become very important in order to ensure resource efficiency and material circularity (Debnath et al. 2021). The global e-waste management market was worth nearly USD 42 billion in 2019, which is expected to grow at a CAGR of 14.1 percent between 2020 and 2027 (Nair 2021). The ongoing COVID-19 has a huge impact on the e-waste recycling industry. As more offices, both in government and private sectors are opting to work from home, there has been an increase in laptop and PC demand, which are potential future e-waste. At the same time, the desktop PCs and other IT equipment in office areas are no longer required. In 2020, nearly 29 percent of the desktop computers were abandoned and more than 23 percent of these computers are going to sit idle in 2021. In 2020, the estimated post-Covid e-waste management market was USD 47.5 billion which is expected to reach USD 119.94 billion by 2027. This estimation is higher than the pre-Covid scenario estimation (Nair 2021).

The ongoing COVID-19 pandemic has not only disrupted the supply chains in major sectors but also affected the waste management industry. Proper collection of e-waste has been hampered due to lockdowns and unavailability of proper logistics. At the same time, logistics cost has increased while the copper price has been skyrocketing since the end of February 2021 (Paben 2021). E-waste needs to be recycled with better efficiency and more sustainability. E-waste is a secondary source of resources as well which has enormous potential to enhance urban mining and help to establish a circular economy. Material recovered from e-waste could be a feedstock for several other allied industries which can also bring up industrial symbiotic models. Hence, for the future to be greener the urban mining of e-waste is not an option, but rather a necessity. The majority of the e-waste recyclers around the globe perform mechanical recycling. The resulting fractions are metals, plastics, glass and other materials, which are potentially recycled by third party recyclers (Debnath 2020). Due to its hazardous nature, it is often shipped to other developing and underdeveloped nations. Sometimes e-waste scraps are mixed with other metal scraps and shipped to middle income or developing countries (Shittu et al. 2020). In many developed countries the burden of e-waste is ignored. China banned the import of waste materials including plastic waste and electronic waste from other developed countries. Philippines declared war over the issue of e-waste against Canada (The Guardian 2019). As a result, waste disposal has become a problem for these countries. There are policy gaps as well which allow this illicit trade even though the countries involved in this practice are signatory to Basel conventions. Hence, even there is great potential in the e-waste management sector, the whole assay needs to be incorporated into and streamlined with policy, better management practices, supply chain optimization and, most importantly, an inclusive attitude of e-waste recyclers. Next, we look into the industry perspective to e-waste management with a focus on India.

1.2   LATEST SITUATION FROM THE INDUSTRY'S PERSPECTIVE

E-waste has become an important waste stream in terms of its volume and toxicity. It is a complex category of hazardous waste. E-waste contains a wide variety of elements including common metals, rare earth metals, polymers, glass, glass fiber, rubber, concrete and ceramics etc. Hence e-waste recycling is important for a sustainable future, and to ensure essential raw materials do not run out (Ottoni et al. 2020). The main driving force behind e-waste recycling is recovery of metals. Metal recovery from e-waste is now technologically feasible, yet the sustainability of the business is a matter of concern as the electronics are becoming lighter. This is due to the percentage of metals in e-waste decreasing and the plastics increasing (Debnath 2020). The recovery and reuse of the plastic part of e-waste is comparatively a less discussed topic in contemporary literature as well as the conferences. Utilization of this huge source is imperative to maintain business sustainability. The issues arise due to improper segregation as plastics containing halogens cannot be recycled in an environment friendly way via extrusion. Additionally, to recycle a specific category of waste plastic, it is essential to avoid contamination with plastics containing Halogenated Flame Retardants (HFRs). Sometimes heavy metals from e-waste migrates to the plastics where they are used to make secondary products (Mao et al. 2020).

In India, nearly 3.2 million metric tons of e-waste was generated in 2019 (Forti et al. 2020). To tackle this huge e-waste, the government of India has published the E-waste Rules in 2016 with some later amendments in 2018. Currently, there are 400 e-waste dismantlers and recyclers with 1068542.72 metric tons per annum capacity in the country (CPCB 2021a). But today Indian E-waste recyclers are able to collect 30–35 percent of e-waste material to the facility. At present, the scrap metals rates are low in the market. For instance, due to COVID-19, the effect on automobile industry scrap metal rates like iron, aluminum and steel was low. Currently, the selling rate of scrap metals India is quite low – Iron is 16–18 INR (0.22–0.25 USD) per kg; Copper is 340 INR (4.70) per kg; Aluminum is 80–90 INR (1.11–1.24 USD) per kg; Steel is 30 INR (0.41 USD) per kg and Brass is 240 INR (3.32 USD) per kg. But still e-waste buying rates are constant and also GST being 18 percent also impact profit.

E-waste recyclers in India use both state-of-art and indigenous machineries for recycling (Figure 1.1a, b). Most of the e-waste recyclers in India are struggling to see 25 percent profit margin. As a result, they are trying their hands on other activities such as upcycling (Figure 1.2a, b). In the last couple of years, there has been an increase in the number of Producer Responsibility Organizations (PRO) in India. In 2019, there were 31 PROs that have increased to 74 as of February 2022 (CPCB 2022).

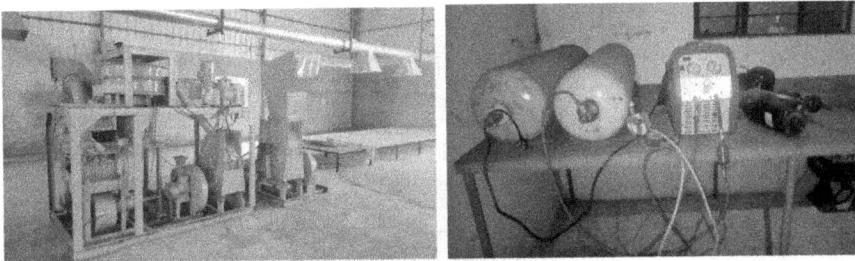

FIGURE 1.1 (a) State-of-art shredder and separator machine and (b) indigenous machine for degassing and storage of CFC from compressors.

FIGURE 1.2   (a) Capacitors upcycled as jewelry and (b) incandescent bulb upcycled to show piece.

PROs are responsible for smooth channelization of e-waste from point of generation to recyclers so as to ensure Extended Producer Responsibility (EPR). They are also attributed with the responsibility to create symbiosis with formal and informal sector. The PROs make more money than the standalone e-waste recyclers.

On the other hand, refurbishment of IT equipment like laptop, tablet and smart mobile phone saw increased demand in the market due to digital education. Additionally, the demand of peripherals and accessories increased. Both in India and abroad, there was a surge in electronics prices as the supply was disrupted due to the pandemic. With a fairly middle-class natives, there was a sudden demand in the second hand market as well. As a result, the e-waste under these categories saw an increase in price range. But market trends still feel this sudden hike in demand on this following IT equipment is temporary and dependent on COVID conditions. In the global scenario, the trend seems to be better than Indian markets due to the difference in policy in procuring e-waste.

Circular economy in e-waste management gives a good result in handling this waste but over the long run. Recently, companies like DELL, APPLE etc. started using secondary metals extracted from e-waste to manufacture its products. This type of initiative helps e-waste recycling industries to grow in future. EPR policy in India for e-waste management still didn't seem to see the desire result. And e-waste recyclers seem to be third party after bulk producer and PRO. This EPR policy can help in reaching capacity target, but financially it's not so promising, thus this is the reason behind the recent increase of PRO organizations compared to recyclers. Even some recyclers have converted into PROs as well. However, there should be a paradigm shift in terms of technological proliferation as well as in the policy level to ensure better collection and business sustainability.

1.3   PARADIGM SHIFT TOWARDS ICT FOR CIRCULAR E-WASTE MANAGEMENT

A paradigm shift in e-waste management is necessary and it can be achieved with evolving Information and Communication Technologies (ICT). Digital technologies such as cloud computing, fog computing, Internet-of-Things (IoT) etc. can be well implemented to improve e-waste management. Recently, augmented reality has been employed for managing and monitoring e-waste. Machine learning techniques are also being researched for suitable application in e-waste supply chain management. Green computing is another branch which is exploring harmless and energy efficient alternatives for e-waste at the usage phase. IoT applications in e-waste management are engaging citizens and cities alike in the project of making our waste practices more sustainable. Optimizing collection routes based on actual disposal unit fill levels – as measured by fill level sensors – is one such application that's proving to be quite impactful. Ultimately, truly transforming e-waste management will require deeper collaboration between public and private stakeholders. Sensor-enabled and internet-connected garbage bins can collect information on fill level, temperature, location, or whatever data types the sensors gather and the sanitation department finds useful. When the bins are full, they notify the collectors and recyclers. With the use of the Internet of Things (IoT) and cloud computing as a backup, we can manage, detect and monitor the electronic waste which is generated. It helps in managing electronic waste in smart cities. The role of Artificial Intelligence (AI) in e-waste management begins with intelligent garbage bins. E-waste management companies can take advantage of Internet of Things (IoT) sensors to monitor the fullness of smart bins throughout the city. This will allow the collectors and PROs to optimize e-waste collection routes, times, and frequencies. Making proper authority for the disposal of electronic waste, initiation of proper recycling locations and severe authorization of laws on electronic waste which can help to solve the rate of growth of electronic waste which can result in managing electronic waste in a safe process and also in a sustainable manner. Blockchain allows real-time tracking of waste management. It helps to track the amount of waste collected, who collected it, and where it is being moved for recycling or disposal. Blockchain helps by rewarding the waste segregation, real-time tracking of waste and securing data transactions. Blockchain is the technology that enables us to write smart contracts. Smart contracts are self-executing computer codes that take specified actions when certain conditions are met in the real world. E-waste management using smart contracts will bring more coordination among producers, importers, retailers and recyclers of electronic items. It will enable the government to regulate e-waste collection and recycling. It will also reduce the imbalance between the organized and unorganized sectors, which will lead to increased transparency throughout the process.

1.4 CONCLUSION

In this introductory chapter, we have outlined the current scenario of e-waste management and e-waste market. The effect of the ongoing COVID-19 pandemic has been highlighted by the latest statistics. The on-ground status of e-waste recycling has been depicted from the industry perspective with a focus on India. It was identified that a paradigm shift is necessary in e-waste business both in technological advancement and policy development. A brief idea on how digital technologies can be the precursor for the paradigm shift was outlined. Nevertheless, in-depth analysis of specific areas of e-waste management will provide a more exemplified picture of the current situation and the paradigm shift towards sustainability.

REFERENCES

Baldé, Cornelis P., Feng Wang, Ruediger Kuehr and Jaco Huisman. (2015). The global e-waste monitor – 2014, United Nations University, IAS – SCYCLE, Bonn, Germany.

Baldé, Cornelis P., Vanessa Forti, Vanessa Gray, Ruediger Kuehr, and Paul Stegmann. (2017). *The global e-waste monitor 2017: Quantities, flows and resources.* United Nations University, International Telecommunication Union, and International Solid Waste Association.

Corporate Catalyst (India) Pvt. ltd. (2015). A brief report on Electronics Industry in India. Available from: www.cci.in/pdfs/surveys-reports/Electronics-Industry-in-India.pdf. (Accessed 4 Jan 2016).

CPCB. (2021a). "List of authorised e-waste dismantler/recycler.". https://cpcb.nic.in/uploads/Projects/E-Waste/List_of_E-waste_Recycler.pdf (Accessed 29 May 2021).

CPCB. (2022). "Producer responsibility organisation (PRO) registered with CPCB". https://cpcb.nic.in/list-of-registered-pro/. (Accessed 15 February 2022).

Debnath, Biswajit. (2020). "Towards Sustainable E-Waste Management Through Industrial Symbiosis: A Supply Chain Perspective." In *Industrial Symbiosis for the Circular Economy*, pp. 87–102. Springer, Cham.

Forti, Vanessa, Cornelis P. Balde, Ruediger Kuehr, and Garam Bel. (2020). "The Global E-waste Monitor 2020: Quantities, flows and the circular economy potential." United Nations University (UNU)/United Nations Institute for Training and Research (UNITAR) – co-hosted SCYCLE Programme, International Telecommunication Union (ITU) & International Solid Waste Association (ISWA), Bonn/Geneva/Rotterdam.

Mao, Shaohua, Weihua Gu, Jianfeng Bai, Bin Dong, Qing Huang, Jing Zhao, Xuning Zhuang, Chenglong Zhang, Wenyi Yuan, and Jingwei Wang. (2020). "Migration of heavy metal in electronic waste plastics during simulated recycling on a laboratory scale." *Chemosphere* 245: 125645.

Nair, Abhijith. (2021). "E-Waste Management Market Size, Share and Industry Analysis | 2027"| 2027". www.alliedmarketresearch.com/e-waste-management-market. (Accessed 29 May 2021).

Ottoni, Marianna, Pablo Dias, and Lúcia Helena Xavier. (2020). "A circular approach to the e-waste valorization through urban mining in Rio de Janeiro, Brazil." *Journal of Cleaner Production* 261: 120990.

Paben, Jared. (2021). "Copper Price Climbs To Recent Record – E-Scrap News". *E-Scrap News*. https://resource-recycling.com/e-scrap/2021/02/25/copper-price-climbs-to-recent-record/?utm_medium=email&utm_source=internal&utm_campaign=Feb+25+ESN. (Accessed 29 May 2021).

Shittu, Olanrewaju S., Ian D. Williams, and Peter J. Shaw. (2021). "Global E-waste management: Can WEEE make a difference? A review of e-waste trends, legislation, contemporary issues and future challenges." *Waste Management* 120: 549–563.

The Guardian. (2021). "Trash Talk: Philippine President To 'Declare War' On Canada In Waste Dispute". *The Guardian*. www.theguardian.com/world/2019/apr/24/philippine-president-rodrigo-duterte-to-declare-war-on-canada-in-waste-dispute. (Accessed 29 May 2021).

Part 1

Global Status of E-waste
Recycling and Management

2 Global Electronic Waste Management

Current Status and Way Forward

*Mahadi Hasan Masud, Monjur Mourshed,
Mosarrat Mahjabeen, Anan Ashrabi Ananno,
and Peter Dabnichki*

CONTENTS

DOI: 10.1201/9781003095972-3

9

2.1 INTRODUCTION

E-waste is a summary term for electrical and electronic instruments, consumables, and subassemblies used for generating, transferring, or measuring signals/responses (electrical and magnetic) in their service life and discarded as unwanted or end-of-life (EoL) disposable or obsolete products (Ranasinghe and Athapattu 2020). SteP, an international organization developing sustainable solutions for e-waste management, defines e-waste as the electrical and electronic equipment (EEE) or parts thrown away by the end-user without any intention to reuse (StEP 2014). StEP also identified a higher growth rate of e-waste in comparison to municipal solid waste (MSW) generation and reported about 46.4 million tonnes (Mt) of e-waste generation in 2015 (Baldé et al. 2017). Rapid technological development and the industrial revolution undoubtedly made our lives easier; however, this widespread use of technology has created a considerable amount of e-waste. Moreover, the EEE markets are continuously updated with ever more appealing appearance, features, performance, and shorter life cycles that allure customers to buy newer products and discard older ones (Kumar, Holuszko, and Espinosa 2017). A report by the United Nations predicts that it will globally increase to 52.2 Mt in 2021 (Baldé et al. 2017). However, in reality, according to the study of Forti et al., the global generated e-waste in 2019 was already 53.6 Mt. According to the United States Environment Protection Agency (USEPA), the global annual generation of E-waste is rising at a rate of 5 to 10 percent, whereas the recovery rate is just around 10 percent of total waste generation (Forti et al. 2020).

When e-waste is not recycled appropriately, it is placed in landfills, a detrimental practice to the environment, and economically inept due to the loss of recoverable materials (Masud et al. 2020). Landfill sites pose higher risks in terms of social, economic, and environmental perspectives (Ananno et al. 2020). The mounting global e-waste generation has a higher economic value than the combined annual GDP of over 120 nations (Ryder and Zhao 2019). International Criminal Police Organisation (INTERPOL) reported the value of e-waste products of nearly 20.5 to 25 billion dollars/ year (500 dollars per tonne) (Rucevska I. 2015). Therefore, countries are being deprived of the potential economic benefits of e-waste due to a lack of recycling and consumer awareness. Moreover, after collecting and separating metal and plastic, scrap parts (wire cover, plastic circuit boards, plastic cases, etc.) are dumped for landfilling or directly burned in open spaces (Wang, Guo, and Wang 2016). Besides, nitrates, acids, and aqua regia from lead-acid batteries are left in open space after collecting the valuable materials without maintaining any safeguard for human health or the environment (Srivastava and Pathak 2020). Unfortunately, most people only see social and economic benefits when burning and dismantling e-waste without being conscious of health hazards and negative impacts on the environment (Borthakur and Govind 2018). Furthermore, thyroid, lungs, neural system, and human fertility are susceptible to the harmful substances in e-waste (Grant et al., 2013). Hence, appropriate e-waste management is of paramount concern as around 80 percent of the global e-waste is recycled informally (Ryder, 2019). Lack of awareness about environmental pollution made informal processing predominant in e-waste management and recycling in the developing world (Borthakur, 2015). Although e-waste is a source of income to many people engaged in informal sectors for their livelihood; however, due to the unsafe practices, concerns remain regarding health hazards and the environment (Heacock et al. 2016). In the developed world, special legislations have been enacted to control the e-waste drift. Integrating sustainable techniques into the management of e-waste is the way forward to retain safety in business and support their efforts to keep the world cleaner as they work to earn a living. In addition, sustainable solutions are required to ameliorate the crisis of exponential e-waste generation. Therefore, in order to fulfill the goal, authors of this chapter will analyse the global e-waste generation scenario and highlight the international e-waste management policies. Moreover, the overall management scenario of e-waste in different continents around the world will be summarized in this chapter and based on the current level effort, a sustainable solution is proposed.

The chapter is organized into several sections. In the first section, the global e-waste generation scenario for different continents is illustrated along with annual generation trends. Section 2.3 discusses the implications of global policies and frameworks of the United Nations (UN) and other international bodies on e-waste management. The e-waste management efforts of different continents and global e-waste management responsibility are critically discussed in Section 2.4. Finally, plausible future research directions are presented in Section 2.5.

2.2   E-WASTE GENERATION SCENARIO

EEE covers all the household and official items from essential to luxurious gadgets in today's day to day life. From Figure 2.1, it can be seen that, though Europe, Oceania, America have the highest per capita e-waste generation but lower total volume of e-waste production; which can be attributed to the level of income, available recycling and recovery option, trans-border shipment of e-waste from developed to developing countries, etc.

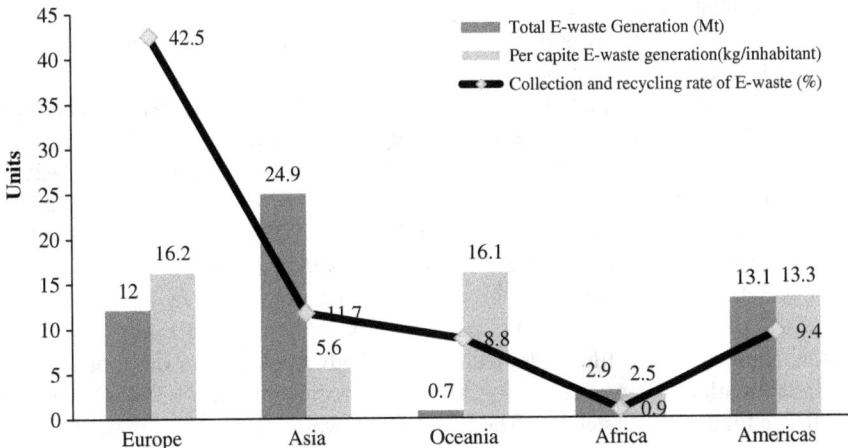

FIGURE 2.1   Continent wise e-waste generation and recycling trends. (Adapted from Forti et al. 2020.)

The figure also shows that Asia produces the highest amount of e-waste in the world – 24.9 Mt in 2019 (5.6 kg/capita/year), and only 11.7 percent (2.9 Mt) of this goes through the recovery and recycling process (Forti et al. 2020). Eastern Asia has the highest generation and recycling rate in Asia which is 13.7 Mt and 20 percent respectively; however, Western, Central, South-Eastern and Southern region have a recycling rate of 6 percent, 5 percent, 0 percent, and 0.9 percent respectively against their generation of 2.6 Mt, 0.2 Mt, 3.5 Mt, and 4.8 Mt respectively (Forti et al. 2020).

In 2019, North America generated 7.7 Mt of e-waste (20 kg/capita/year), but only 15 percent of their total scrap was officially processed (Baldé et al. 2017, Kumar and Holuszko 2016, Forti et al. 2020). Whereas Latin America generates 5.4 Mt of e-waste in 2019 from its 20 member countries, which is 31 percent (4.2 Mt) higher than in 2016, and the average per capita production rate is 8.4 kg (Baldé et al. 2017, Forti et al. 2020). In this continent, Brazil (2,143 kt), Mexico (1,220 kt), Argentina (465 kt), and Colombia (318 kt) are at the highest position, and from Table 2.1, it is also evident that their GDP (about 69% used in household), technical advancement, and population acts as the main factors behind this scenario (Forti et al. 2020, Heacock et al. 2016, Kaza et al. 2018, Kiddee, Naidu, and Wong 2013). In Europe, per capita e-waste generation is 16.2 kg, which is the second-highest among the continents, and the total volume is more than a quarter (12 Mt) of the world's total e-waste generation in 2019 (Forti et al. 2020).

However, in some cases, the EEE consumption volume is much higher in developing countries than developed ones. It is estimated that by 2030, 400–700 million outdated computers will be discarded from developing countries compared to 200–300 million by the developed countries (Srivastava and Pathak 2020). United Nations Environment Program (UNEP) and United Nations University (UNU) jointly estimated that by 2020, India would face 500 percent growth of e-waste derived from computers and from obsolete cell phones, an increase of about 18 times and 7 times will be faced by India and China respectively (Kumar and Rawat 2018, Lu et al. 2015). However, in developed economies like the EU, manufacturers are responsible for disposing of e-waste through take-back system (TBS) and other special facilities. About 75 percent of EEE are manufactured in developing countries like China and India, and consequently, these or other developing countries have to manage this e-scrap after EoL (Wang, Zhang, and Guan 2016).

Figure 2.2 shows the global e-waste generation and predicted e-waste generation scenarios from 2014 to 2030. Baldé et al. reported around 48.2 Mt (6.1 kg/capita/year) in 2016, with the largest contribution from Asia (~41%), then USA (~29%) and European Union (EU) (~27%) (Baldé et al. 2017, Srivastava and Pathak 2020). Moreover, a recent report, "Global E-waste Monitor 2020 (GEM)" shows a striking growth of e-waste to a total volume of 55.5 Mt (7.5 kg/capita/year) (Forti et al. 2020), whereas Baldé et al. predicted 57.4 Mt by 2021 with a growth rate of 4 percent (Baldé et al. 2017). Higher production and consumption rate, industrialization and urbanization, shorter product life cycle, technological advancement are the root cause of this inflammation of e-waste, and by 2030, it is predicted to be 74.7 Mt (Forti et al. 2020) (see Figure 2.2 for details). The increasing e-waste production can be attributed to the technological advancement, industrial automation, progress in information and

TABLE 2.1
Effect of Population and GDP on E-Waste Generation and Management

Region	Countries in this region	Population (Million)	GDP (Per capita in USD)	Per capita e-waste Generation (Kg/year)	Total e-waste Generation (Mt)	Collection and Recycling (Mt)	Value of Raw Materials (Billion USD)	GHG emission in CO_2 equivalent (Mt)	Mercury (kt)	BFR Plastics (kt)
Africa	53	1308.06	1930	2.5	2.9	0.03 (0.9%)	3.2	9.4	0.01	5.6
America	35	1014.72	57800	13.3	13.1	1.2 (9.4%)	14.2	26.3	0.01	18
Asia	49	4601.37	7350	5.6	24.9	2.9 (11.7%)	26.4	60.8	0.04	35.3
Europe	40	747.18	29410	16.2	12	5.1 (42.5%)	12.9	12.7	0.01	11.4
Oceania	13	42.13	53220	16.1	0.7	0.06 (8.8%)	0.7	1.0	0.001	1.1

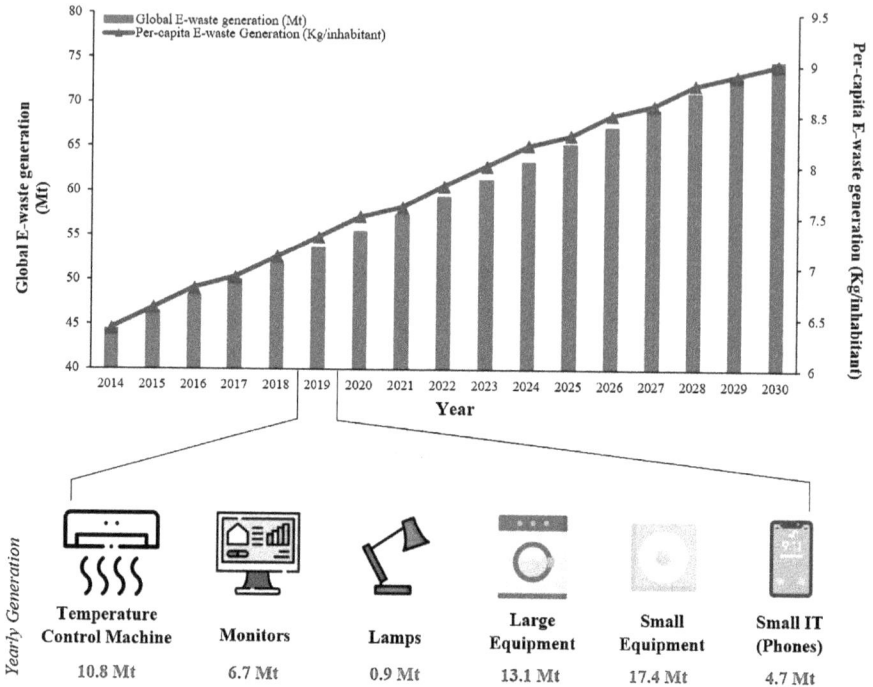

FIGURE 2.2   Global e-waste generation trend. (Adapted from Forti et al. 2020.)

communication technologies (ICT), economic level growth rate, and competition for the versatility of EEE that makes the price of EEE downwards (Borthakur and Govind 2017, Kumar, Holuszko, and Espinosa 2017, Borthakur, Govind, and Singh 2019, Li et al. 2015). Moreover, this generation pattern is also closely related to income level, social status, and geographical location as the e-waste growth rate is also influenced by advanced materials and manufacturing technologies, fast market penetration, stable economy, etc. (Tansel 2017).

In most developing countries, e-waste growth rate is closely related to the gross domestic product (GDP) rather than on population (Kumar, Holuszko, and Espinosa 2017), as GDP covers the material flow, purchasing power parity (PPP), economic elasticity of any country (Kusch and Hills 2017). It is estimated that with an increment of 100 GDP per capita in purchasing power standards (PPS) is responsible for every 0.27 kg of e-waste collection and 0.2 kg of e-waste ready for recovery and recycle (Awasthi et al. 2018). Table 2.1 illustrates the effect of population and GDP on e-waste generation in 2019 across continents (Forti et al. 2020, Olimid and Olimid 2019, IMF 2019). Asia has the highest volume of e-waste generation and an inferior e-waste management system due to its large population and low GDP. On the other hand, Oceania with the lowest population generates as high per capita volume as Europe due to its high GDP per capita, which is not being used to recycle it effectively.

Europe keeps balance with waste generation and management as their collection efficiency is extraordinary (35%). However, Africa has the lowest e-waste generation, and collection efficiency is below 1 percent (4-kilotonne out of 2.2 Mt). America has the highest GDP per person, and North America contributes maximum e-waste generation (7 Mt) compared to Central (1.2 Mt) and South (3 Mt) American countries.

Besides, it is also worth mentioning that e-waste collected solely in 2019 has produced 71 kt of Brominated Flame Retardants (BFR) plastics, 50 t of mercury, and 98 Mt of greenhouse gases (CO_2 equivalent GHG) (Forti et al., 2020). Moreover, the recovery of e-waste acts as a secondary resource for potential precious materials collection, and the material flow at different stages of the recovery process has a significant impact on the successful e-waste management system. Researchers already investigated the technological opportunities for the recovery and recycling of e-waste, and it has been found that both formal and informal recovery approaches focus on the material value of e-waste rather than the environmental and health aspect of the partially disposed substance (Zeng et al. 2016). Hence, a holistic approach is necessary to develop a sustainable strategy to recover entire products rather than specific content. However, the Association of Plastics Manufacturers in Europe (APME) also described e-waste as a combination of multifarious materials like ferrous, non-ferrous, plastic, and ceramic (Srivastava and Pathak 2020, Masud et al. 2019, Mourshed et al. 2017, Newaj and Masud 2014). EEE contains metals like platinum, ruthenium, indium, gold along with cobalt, copper, which are not only precious (USD 57 billion only for the raw materials), but also have minimal availability in earth crest (Baldé et al. 2017, Forti et al. 2020). Since an average of 15 percent of e-waste was recovered globally (Heacock et al. 2016) and only 17.4 percent or 9.3 Mt of the total e-waste has undergone a formal recycle process in 2019, which is more than fivefold the formal recycling portion (1.8 Mt) in 2014 (Forti et al. 2020). However, about 82.6 percent or 44.3 Mt e-waste data is not available (Heacock et al. 2016, Baldé et al. 2017, Ilankoon et al. 2018, Forti et al. 2020). A small portion, about 8 percent of this unrecorded e-waste has been used for landfills and incineration with common household wastes, and 7 to 20 percent has been exported from rich to poor countries as scrap or second-hand goods (Forti et al. 2020). In addition, about 70-75 percent of e-waste generated in USA and India remains unaltered due to the lack of proper management (Borthakur and Govind 2017), and 0.6 Mt in Europe is disposed in waste bins (Forti et al. 2020). Chemicals like lead, chromium, mercury, and cadmium in e-waste also negatively impact this planet (Kumar, Holuszko, and Espinosa 2017, Kumar and Holuszko 2016, Masud et al. 2019, Heacock et al. 2016). Therefore, proper knowledge about the metal recovery and control over heavy metal mixing with new products at recycling must be needed to avoid the spread of ecotoxicology of e-waste.

2.3 INTERNATIONAL E-WASTE MANAGEMENT POLICIES AND INITIATIVES

2.3.1 AT A GLANCE

There are different policies developed, adopted, and evaluated by different countries as well as different regional and international bodies to combat e-waste related problems. Some of these policies discuss the fate of e-waste being the manufacturer's responsibility; others recommend the collection of advanced Waste Electrical and

Electronic Equipment (WEEE) processing fees from the consumers while selling to cover the cost of recycling e-waste (Herat 2009). However, all the policies are concerned with the hazards and effects of e-waste on health and environment and implemented to mitigate those effects. In this section, e-waste management policies of International bodies like United Nations, Regional bodies like the European Union, and several developed countries of different continents are discussed. E-waste management systems in low-income countries are embryonic, with only 60 percent of low-income countries have some regulations for solid waste (SW) management, and for high-income countries, it is about 96 percent (Kaza et al. 2018). Specifically, for e-waste management, about 71 percent of the total population of the world in 2019 was covered by legislation, which was 66 percent and 44 percent in 2016 and 2014, respectively (Heacock et al. 2016, Baldé et al. 2017, Ilankoon et al. 2018, Forti et al. 2020).

2.3.2 Policies of International Body (United Nations)

2.3.2.1 Basel Convention

Early in the 1970s and 1980s, some developed countries exported 75 to 80 percent of hazardous e-waste in Africa, Asia, and other continents for dumping and recycling (Golev et al. 2016). To remedy this problem, the United Nations Environment Programme (UNEP) collaborating with the United Nations introduced the Basel Convention in 1992 (Basel Convention on the Control of Transboundary Movements of Hazardous Wastes and their Disposal) (Shinkuma and Huong 2009).

The Basel Convention is one of the first and most comprehensive global agreements which aims to reduce hazardous waste generation, promote environment-friendly waste management irrespective of the place of disposal and maintain a regulatory as well as legislative system while applying to the cases where the transboundary movements of this agreement are allowable (Convention 1992). Some countries have not entered into the Basel Convention yet, whereas the developed countries are looking for dumping sites and recycling through cheap labour; hence, in 1995, representatives from developed countries passed a Ban Amendment to completely restrict any hazardous junk shipment from developed to developing countries (Tansel 2017). Including "Ban Amendment 1995" (proposed in 1995, came into force on December 5, 2019), which includes a new preambular clause which provides a new prohibition to the transfer of hazardous wastes in developing countries for their final disposal. The Basel Convention has several other amendments from time to time, especially noting two amendments. One important amendment in 1998 further elaborates on the convention's waste management regulation. At the fourteenth meeting in 2019, plastic waste inclusion was amended and will come into force in 2021 (Convention 2021). Along with these amendments, 28 other decisions were agreed upon at this meeting; the most noteworthy one is the interim adoption of the technical guidance on the management of e-waste and usage of electrical and electronic equipment (Convention 2019). However, since its sixth meeting in 2002, the Basel Convention has been focused on e-waste management, strategic plans, and innovative solutions for environment-friendly management (Herat 2009).

2.3.2.2 Solving the E-waste Problem (StEP) Initiative

In 2004, The UN solving e-waste problem initiative was introduced as an independent platform consisting of research institutes, international organizations, government agencies, and Non-Governmental Organizations (NGOs) dedicated to design strategies to advance the development and management of global e-waste (StEP 2014). StEP works for finding scientific, environment-friendly, ethically, and economically sound salient solutions to worldwide e-waste challenges. In 2009 they arranged the first StEP e-waste Summer School, and in 2012, they published the first E-waste Academy Management Edition (EWAM) in Ghana. Moreover, they published StEP E-waste World Map in 2013, and they were declared as a standalone entity in 2019 (StEP 2014).

2.3.2.3 Others

International Environmental Technology Centre (IETC) of UNEP has been working on e-waste management since 2007 (IETC 2020). They have made several e-waste foresight reports, created advisory councils to advise different states on making e-waste policy, and participated in several international platforms who work to manage e-waste. Also, they created training courses that provide recent and updated knowledge on e-waste management (UN 2020).

Global e-Sustainability Initiative (GeSI) is a universal organization of Information Communication and Technology companies, industry associations, and NGOs who are working collectively to address global problems and achieve sustainable objectives by innovative technologies (GeSI 2020).

Sustainable Cycles (SCYCLE) is a program by United Nations Universities that aims to develop sustainable environment-friendly production, usages, disposal, and recycling of hazardous substances with a particular focus on electronic and electrical equipment. SCYCLE works in a leading position to conduct the global e-waste discussion and helps the states advance sustainable e-waste management strategies.

United States Environmental Protection Agency (USEPA) and Taiwan Environmental Protection Administration (Taiwan EPA) have collaborated and made an international platform naming International E-waste Management Network (IEMN) (EPA 2020) to ensure sound management of e-waste since 2011. On September 24–29, 2018, the 8th workshop was arranged by 11 government officials from different countries, and they have discussed e-waste markets, new technological possibilities, and environment-friendly management policies.

2.3.3 POLICIES OF REGIONAL BODIES

2.3.3.1 European Union

Recently, the EU has launched several new initiatives to address the negative effect on environment and human health due to waste and hazardous substances. These policies regulate the e-waste management of 27 European countries and help decision-makers, manufacturers, and consumers worldwide. There are four recent initiatives of EU, described in Table 2.2:

TABLE 2.2

Policies of the European Union Regarding E-waste Management

Name of the policy	Year of implementation	Major Content	Recent meetings and remarks
Directive on waste electrical and electronic equipment (e-waste Directive)	Introduced in February 2003, in force since 2012, effective since February 14, 2014 (Comission 2020a).	This directive's main objectives are to recycle, reuse, and reduce the e-waste, thus minimizing the generation rate along with safe disposal. It also motivates the manufacturers to design environmentally-friendly equipment (Selin and VanDeveer 2006).	On April 18, 2017, EU adopted "e-waste package", and the recent adoption was about "Implementing Decision (EU) 2019/2193" (Comission 2020a).
Directive on the restriction of the use of certain hazardous substances in electrical and electronic equipment (RoHS Directive)	In force since February 2003, effective since January 3, 2013 (Comission 2020a).	This legislation suggests safe alternatives to hazardous substances, including flame retardants such as polybrominated diphenyl ethers (PBDE) or polybrominated biphenyls (PBB) and heavy metals like mercury, lead, cadmium, and hexavalent chromium (Comission 2020a).	EU developed RoHS 2 in 2012 and continued a study on reviewing the directives in 2018 (Comission 2020a). Besides, two draft amendments were proposed on July 22, 2019 (Cusack and Perrett 2006, Rivera 2019).
EU directive on Energy-using-Products (EuP) and Energy-related Products (ErP)	Introduced on July 6, 2005 (Herat 2009) and revised in 2009 (EuP 2020).	This directive comprises some Eco-design requirements for products (consumer electronic devices, water heaters, electric motor systems, lighting, office equipment etc.) (Herat 2009). This directive aims to reduce the energy in every step of the supply chain, including production, transport, packaging, and usage (Comission 2020b).	Has not been amended recently

TABLE 2.2 (Continued)
Policies of the European Union Regarding E-waste Management

Name of the policy	Year of implementation	Major Content	Recent meetings and remarks
Regulation on the Registration, Evaluation, and Authorization of Chemicals (REACH)	Issued on December 30, 2006, in force since June 1, 2007 (Herat 2009).	The regulation is based on making a database of hazardous and chemical substances in the EU member states and controlling, regulating, testing, estimating the risk of using hazardous materials, and motivating the manufacturers to find an alternative to those harmful substances (Herat 2009).	The regulation was amended on April 26, 2018, including details about nanomaterials (EUON 2020).

2.3.3.2   G8 Countries

G8 countries came to an agreement to promote the initiative called 3Rs: Reduce, Reuse and Recycle in April 2005 with a dream to create a "sound-material-cycle society" (Ilankoon et al. 2018). With the collaboration of United Nations Centre for Regional Development (UNCRD) in 2009 (Herat and Agamuthu 2012), the regional 3R forum has been formed in Asia to arrange the meetings of member countries, advise the Government about environment friendly policy-making, encourage multi-stakeholder collaboration in the waste sector, and reduce the obstacles between the e-waste flow from manufacturers to recycling (UNCRD 2020). At a recent meeting of the forum held on 4–6 March 2019 emphasizing circular economic utilization of waste, proper flow of material from manufacturer to recycling, using technology for clean energy and proper e-waste management for the green industry, leadership of policymakers, and sufficiency of the economy have been explicitly discussed (3R 2020).

Other organizations like the NORDIC co-operation council, BRICS, and ASEAN do not have structured law like European Union or G8 countries. NORDIC countries follow the legislation of their own countries along with the European Union. BRICS and ASEAN are yet to affirm a common E-waste management regulation for their own organization. ASEAN countries ratify Basal Convention (Ibitz 2012), and the RoHS Directive and WEEE Directive have influence on BRICS countries along with their own countr's legislations (Borthakur 2020).

2.3.4   POLICIES OF COUNTRIES OF DIFFERENT CONTINENTS

Over 90 jurisdictions and well above 2000 sets of legislation are found in different countries around the world to safeguard e-waste management (Veit 2014). Some developed economies (EU, Japan, South Korea) have implemented both regional (EuP Directives, ROHS Directives) and international (Basel Convention) for the proper management of e-waste while USA, Canada, China, Brazil are still developing relevant regulations and laws (Li et al. 2015, Zeng et al. 2016). Some countries of different continents are selected randomly, and their national policies, acts, and initiatives are represented in Table 2.3.

2.4   GLOBAL E-WASTE MANAGEMENT EFFORTS

2.4.1   OVERVIEW

Sustainable e-waste management is a vital challenge because of its volume and the subsequent impact on the environment. Regulations described in the previous section focus on government legislation to minimize e-waste disposal volume by maximizing reuse, recovery, and recycling approaches. In addition to the 4R strategies, global and regional legislation also come into effect and cover up to 66 percent of the total world population under the legal framework for discarded e-waste (Baldé et al. 2017). However, a good number of developing and underdeveloped countries are still lagging behind in setting up a sound e-waste management guideline. EU directives (2003) also mandate that its 27 member countries recycle a minimum of 85 percent of their e-waste goods by 2019 (Ilankoon et al. 2018, Garlapati 2016). Canada has developed an industry-grade e-waste recycling system for its own e-waste; whereas, the USA does not have a federal policy, but half of its states follow a range of legislations regarding e-waste recycling processes. The 'National Television and Computer Recycling Scheme' developed by Australia combines government and industry actions to ease the collection and post-consumption processing of e-waste. In the Asian territory, Taiwan, Japan, and South Korea recycle 82 percent, 75 percent, and 75 percent of their e-waste, respectively, while China, India, Pakistan, and Bangladesh has some sets of rules but they are not yet implemented. Latin American countries have e-waste rules, while African countries have mixed experience with their laws to import and reuse EEE. About 85 percent of the imported EEE in Africa is second-hand, and productive use of this EEE is two to three times higher than in developed countries (Garlapati 2016). Due to the lack of strict legislation, formal e-scraps collection and recycling activities have been seized by unauthorized and illegal sectors that make this e-waste management much more challenging. Figure 2.1 shows a snapshot of formal legislative components for waste electrical and electronic equipment management (WEEM).

In this way, EPR and ARF have been considered the backbone of the organized WEEEM system, as Switzerland is the pioneer to adopt this system globally. Developed countries like USA and EU countries have started developing standardized disposal, collection, and recycling systems to form an effective income stream. EPR is considered a powerful tool for e-waste management that encourages value recovery and recycle before disposal. It has set policies for applying the principles of completing and extending product useful lifetime and effective collection and processing for recycling and recovery after the products' EoL. The core content of EPR is the identification of active TBS or design for the environment (DfE)

TABLE 2.3
Country-wise Policies on E-waste Management of Different Continents

	Country name	Country status	Policies	Remarks
Asia	India	Developing	• E-waste (Management) Amendment Rules 2018 (Baidya et al. 2020)	E-waste management rules are amended in 2018 which included the targets of EPR (Extended Producer Responsibility) along with guidelines for the new companies who have just started the business. This law will effectively handle the sound management of E-waste in India (Borthakur 2016, 2020, Borthakur and Govind 2017).
	Bangladesh	Developing	• 'Proposed Hazardous E-waste Management Rules (Drafted in 2019) (Ananno et al. 2021)	This new proposed draft has included definition of e-waste, guideline of recycling e-waste, responsibilities of the consumers and manufacturers along with government. This new law is a milestone to solve E-waste related problems of Bangladesh.
	China	Developing	• Administration Regulation for the Collection and Treatment of Waste Electrical and Electronic Products naming China e-waste came into force on January 1, 2011.	This regulation makes the recycling of e-waste mandatory and establish special fun for recycling. It also gives importance to the implementation of the Extended Producer Responsibility (EPR) and proper certification of second-hand products and recycling Enterprises (Yu et al. 2010).
			• Administrative Measures for the Prevention and Control of Environmental Pollution by Electronic Waste (2007), which came into force in 2006 (Li et al. 2006).	This policy gives importance to prevent environmental pollution, which can be caused during any recycling stage. Also, it specifies the responsibilities of manufacturers and discloses the licensing scheme for e-waste (Yu et al. 2010).

(continued)

TABLE 2.3 (Continued)
Country-wise Policies on E-waste Management of Different Continents

	Country name	Country status	Policies	Remarks
	Japan	Developed	• Technical Policy on Pollution Prevention and Control of Waste Electrical and Electronic Products (2006)	This policy prohibits the import of hazardous waste, including e-waste. It describes the framework of eco-design, environment-friendly management systems such as production, recycling, reusing, and disposal of e-waste (Yu et al. 2010).
			• Law for the Promotion of Effective Utilisation of Resources (LPUR), which came into force in April 2001	This law covers electronics like small size batteries as well as personal computers. On 1 July 2006, an amendment was granted in the Law for the effective utilization of resources (Pariatamby and Victor 2013).
			• Law for Recycling Specified Kinds of Home Appliances (LRHA) 2001	LHRA covers refrigerators, televisions, air conditioners, washing machines, cloth dryers, and others. LRHA imposes obligations on manufacturers for making proper recycling facilities and consumers for paying the cost of recycling and transportation of the waste (Pariatamby and Victor 2013).
Americas	USA	Developed	• National Strategy for Electronics Stewardship (NSES) 2011 (EPA 2020).	It recommends the federal governments, businessmen, and consumers to safe and effective management and handling of used electronics and e-waste (EPA 2020).
	Canada	Developed	• Canadian Environmental Assessment Act, 2012 (CEAA 2012)	There are two regulations made under this act • Cost Recovery Regulations (SOR/2012-146) Physical Activities, Regulations Designating (SOR/2012-147) (Laws 1999)
			• Canada's Toxic Substances Management Policy 1995	The policy directs the decision-makers about federal programs based on the scientific management framework of toxic substances and e-waste (Canada 1994).

	Country		Regulation	Description
	Columbia	Developing	• Política Nacional (Colombia): Gestión Integral de Residuos de Aparatos Eléctricos y Electrónicos (Columbian National Regulations of Electronic and Electrical waste) 2017 (MADS 2017).	Columbia was the first country in Latin America, which has launched a separate National e-waste policy (Forum 2017).
Africa	South Africa	Developing	• South African e-waste Association (Ecroignard 2006)	This association works in a system that focuses on the reduction of e-waste with proper treatment and monitoring.
	Nigeria	Developing	• National Environmental Regulations (Electronics Sector) (Benebo 2011)	The regulation bans unused electrical goods explicitly (Sthiannopkao and Wong 2013a).
			• Harmful Wastes Act 1988 (ICLG 2008)	This act defines hazardous materials by the effects caused by them. This act also mentions that trading such banned materials, especially causing death or incurable impairment, can be a reason for life imprisonment punishment.
Oceania	Australia	Developed	• National Waste Policy 2018 (DAWE 2020)	Australian laws are lack of proper and updated e-waste management policies
	New Zealand	Developed	• The New Zealand Waste Strategy 2010 (MFE 2010)	This strategy focuses on reducing the harmful effects of different kinds of waste and the proper and effective use of the resources present.
			• Waste Minimization Act of 2008 (MFE 2008)	Three sets of regulations are present under this Act. • Waste Minimization (Calculation and Payment of Waste Disposal Levy) Regulations 2009 • Waste Minimization (Microbeads) Regulations 2017 • Waste Minimization (Plastic Shopping Bags) Regulations 2018

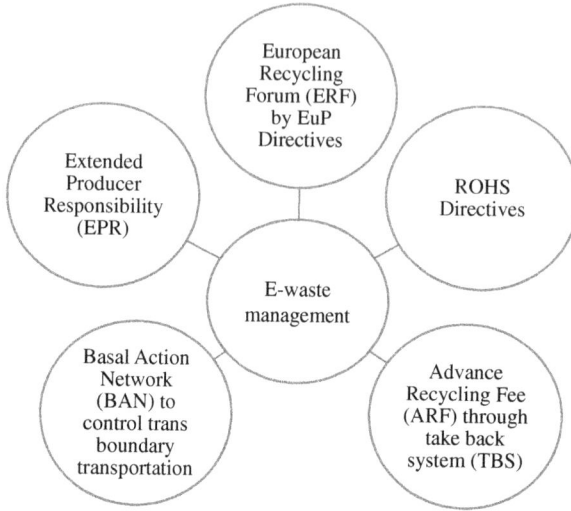

FIGURE 2.3 Legislative components for e-waste management. (Pathak and Srivastava 2019.)

components for safe e-waste disposal (Palmeira, Guarda, and Kitajima 2018). Conversely, due to this strict environmental legislation, first world countries using developing and under-developed countries like India, China, African countries as dumping sites simulating donation or take back strategies (Kumar, Holuszko, and Espinosa 2017). Based on socio-economic conditions, infrastructure for e-waste management (formal and informal), government support (financial and legal), some discrepancies have been identified while the same EPR schemes are applied in both developed and developing countries (Sepúlveda et al. 2010). Hence, Herat and Pariatamby (Herat and Agamuthu 2012) suggested some modifications of EPR for developing countries (Herat and Agamuthu 2012). Moreover, in low and middle-income countries, informal sectors are mainly engaged due to the lack of strict e-waste management legislation, and this practice acts as the main reason behind improper e-waste management (Srivastava and Pathak 2020).

The collection, processing through recovery and recycling, and disposal by following existing rules and regulations are the factors in designing a global e-waste management framework (Srivastava and Pathak 2020). By adopting a protocol for WEEEM throughout the world, setting rules and legislations to regulate export and import of e-waste, fund allocation for infrastructure development, employing effective collection and recycling mechanisms, developing consciousness among the consumers can help to handle this e-waste flow in a better way (Baldé et al. 2017, Borthakur 2016, Kumar, Holuszko, and Espinosa 2017, Kumar and Rawat 2018). Researchers and administrators are also focusing on developing an e-waste management flow system to trace EEE's life cycle from manufacturing to disposal and recycling stages that will ultimately help create a circular economy with the least e-waste consumption (Wang, Zhang, and Guan 2016). They also suggest the 3R (reuse, recycle and reduce) and 4R concept (reuse, recycle, reduce, and replace) to

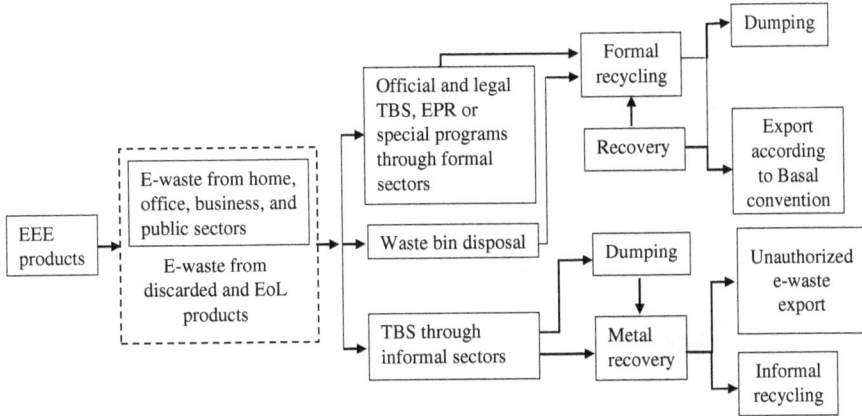

FIGURE 2.4 Flow diagram of a typical e-waste management system. (Pathak, Srivastava, and Ojasvi 2017, Srivastava and Pathak 2020, Sthiannopkao and Wong 2013b.)

develop sustainable e-waste management goals (Garlapati 2016, Mourshed et al. 2017). Also, safe disposal and collection of e-waste through eco-friendly devices, proper recovery, control WEEE flow to developing countries, and above all raise the awareness among the consumers and manufacturers have a vital role in achieving the 4R strategy (Kiddee, Naidu, and Wong 2013, Masud et al. 2019, Mourshed et al. 2017). Figure 2.4 illustrates the state of the global WEEEM system practised by most of the countries (developed and developing).

E-waste management technologies, level of income, population, geographic location dictates the nature and fate of WEEEM policies in developing and developed countries. Global flows of e-waste have been traced from North to South in the form of waste or resource, and in this meantime, it imparts environmental and health hazards (Daum, Stoler, and Grant 2017, Sugimura and Murakami 2016). Though the e-waste generation rate for rich countries is significantly higher (19.6 kg/capita/year) compared to poor countries (0.06 kg/capita/year), but this status is changing rapidly (Srivastava and Pathak 2020, Borthakur, Govind, and Singh 2019).

Reusing the exported or domestically collected EEE in full or partially is another approach to control WEEE in developing and under-developed countries. A study on recycling of computers demonstrated that USA charged 20 USD for recycling a single computer while it is ten times less if done in India (Borthakur 2016). Now-a-days, along with extracting optimum financial benefits from e-scraps, consumers from China, Thailand, Vietnam, and South American countries like Brazil and Mexico also develop voluntary program like donation or pay for e-waste management (Kaya 2016). From an ideological point of view, developed countries in North America, EU, and Australia occupy the highest level with respect to investment in the development of collection and recycling systems (Borthakur and Govind 2017). However, developing and underdeveloping countries are still struggling with their crude style of e-waste management of their own and the paraphernalia imported through middlemen and unauthorized dealers (Tansel 2017, Borthakur, Govind, and Singh 2019). Thus,

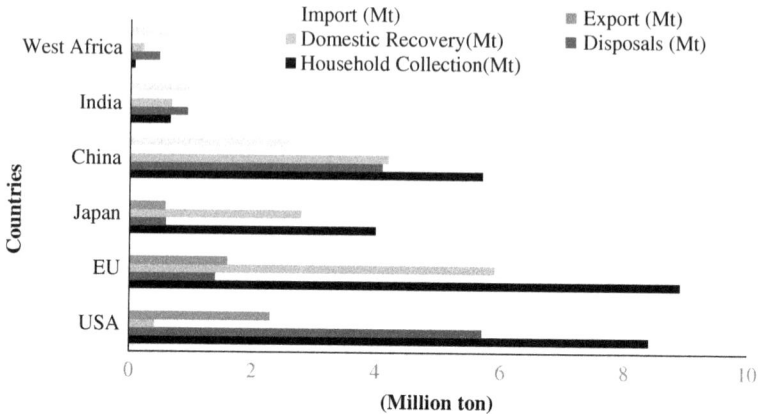

Import (Mt)   ■ Export (Mt)
Domestic Recovery(Mt) ■ Disposals (Mt)
■ Household Collection(Mt)

FIGURE 2.5 Breakdown of WEEEM strategies practised by selected developing and developed countries. (Adapted from Garlapati 2016.)

rigorous legislation should be employed to control WEEE flow in poor countries to attain EPR objective and promote the local economy by considering the benefits for all stakeholders (Nations U 2020).

In developed countries, official TBS, ERP, or special programs are launched by the respective authority, responsible for the e-waste collection points (curb-side, primary municipal or commercial dumping sites) are the most common practices. Apart from these, individual dealers or vendors are also engaged in e-waste collection and recovery processes like metal, non-metal collection, and shipment preparation. Due to the lack of active small metal recovery units, open landfill opportunities in developing countries, more than 50 percent of their WEEE has been exported to China, India, Mexico, Brazil, Vietnam, Pakistan, Bangladesh, Philippines, Pakistan, and Ghana (Cucchiella et al. 2015, Golev et al. 2016). Whereas in developing and under-developed countries, door-to-door collection through self-employed staff or individual waste pickers (called 'Tokai' in Bangladesh and 'Kawariwala' in India) mainly accomplished the collection process. Table 2.2 represents the WEEE generation, collection strategies and corresponding legislations that are practised by the formal and informal sectors of some selected countries (Borthakur and Govind 2017).

Global e-waste management is basically accomplished by three major ways: i) Global recovery (in developing countries of Asia, Africa, South Africa), ii) Regional recovery and dumping (practised in Japan, EU), and iii) Export and dumping (practised in North America, Australia) (Borthakur and Govind 2017). A pictorial view of the e-waste collection, disposal through landfills and incineration, export, and import scenario is given in Figure 5. Considering WEEE importing, about 93 percent of WEEE exported to Asia, where China and India are in the leading position. China mainly imports from South-East Asia, Korea, Japan, Western states of USA, Australia, while India covers Central and part of South-East Asia, EU, Eastern states of USA (Borthakur 2016).

2.4.2 NORTH AMERICA

Two of the most developed nations in North America, USA, and Canada have pro-vincial e-waste management laws rather than national legislation. Until 2010 they did not endorse the Basel Convention for the transboarder shipment of hazardous e-waste, used TVs, laptops, monitors, cellular phones to China, Latin America, and India, which was 8.5 percent (26.5 kt) in 2010 (Baldé et al. 2017). Currently, 84 per-cent of USA residents are under the control of e-waste legislation, and since 2011, they have implemented the "Responsible Electronics Recycling Act," banning the transborder shipment of WEEE that contains toxic elements like lead, cadmium, mer-cury, chromium, etc. USA residents used to keep 70 percemt of their EEE close to EoL in storage up to 3–5 years, as disposal was by the responsible state authority (Kang and Schoenung 2006, Wagner 2009). As a result, heavy metals (about 70%) in USA landfills originated from WEEE due to the improper management of e-waste as well as the lack of adequate processing units compared to the volume generated (Dwivedy and Mittal 2012). However, the lack of strict federal law and low con-sciousness are the potential drivers behind this alarming growth of e-waste (Saphores et al. 2006). Moreover, drop-off locations always prefer domestic electronics rather than heavy power tools, and people living at a distance (> 5 miles) from drop-off points are not usually interested in delivering their obsolete e-products to that point. It is also interesting that consumers are willingly paying for processing their discarded e-waste; hence, extended consumer responsibility (ECR) is endorsed rather than EPR. According to the Environmental Protection Agency (EPA), around 80 percent of the consumers are ready to pay less than USD 5 for e-waste management under the ECR program (Kang and Schoenung 2006). In 2003, California first implemented ECR to collect fees from the consumer for e-waste management operations (Li 2011), and in 2004, Maine engaged municipal authority along with producer and consumer to contribute financially for e-waste management (Wagner 2009).

Canada accounted for 6.2 percent of North American regional and 1.2 percent of world e-waste generation in 2014 despite having a lower population density and higher GDP, as mentioned in Table 2.4. Moreover, about 0.14 Mt of WEEE is dumped for landfills (Kumar and Holuszko 2016). Limited and complicated collection systems make the consumers and citizens drop off their e-waste as general waste or to store in the house. Another key challenge is the limited availability of recycling and recovery facilities. Almost all the provinces offer segregation and subsequent dismantling opportunities, but recycling and recovery options are not available. Therefore, the transportation of those products requires extra shipment costs and time. Canadian e-waste recovery plans depend on consumer financing them through visible or hidden fees. Along with ECR, "Electronics Product Stewardship Canada (EPSC)" has been introduced to provide product stewardship from public and/or environmental funds without demanding post-consumption responsibility from producers (Lepawsky 2012, Borthakur and Govind 2017). Moreover, Alberta collects "Advanced Disposal Surcharge (ADS)" under "Electronics Recycling Administrative Policy" for spe-cific and eligible EEE products as a management cost after EoL (Lepawsky 2012). EPR programs are also practised at the provincial level, which is mainly launched by brand owners or producers with consumers contributing through eco-fees. Moreover,

TABLE 2.4
E-waste Management Practice Scenario by Different Countries

Country	GDP (Per capita in USD)	Total e-waste Generation (kt)	Per capita e-waste generation (Kg/year)	Collection and Recycling (kt)	E-waste used as		Dominating Sectors for e-waste Management		Having e-waste Management Legislation
					Waste	Valuables	Formal	Informal	
Asia									
China	10 261.7	10129	7.2	1546 (2018)		Yes		Yes	Yes
Japan	40246.9	2569	20.4	570 (2017)	Yes		Yes		Yes
South Korea	31762.0	818	15.8	292 (2017)	Yes		Yes		Yes
Thailand	7808.2	621	9.2	Not Available		Yes		Yes	No
Vietnam	2715.3	48	15	Not Available		Yes		Yes	No
Iran	5520.3	790	9.5	Not Available		Yes		Yes	No
Iraq	5955.1	278	7.1	Not Available		Yes		Yes	No
Malaysia	11414.8	364	11.1	Not Available		Yes		Yes	No
Kuwait	32032.0	74	15.8	Not Available		Yes		Yes	No
India	2104.1	3230	2.4	30 (2016)	Yes	Yes		Yes	Yes
Saudi Arabia	23139.8	595	17.6	Not Available	Yes			Yes	No
Pakistan	1284.7	433	2.1	Not Available		Yes		Yes	No
Bangladesh	1855.7	199	12	Not Available		Yes		Yes	No
Sri Lanka	3853.1	138	6.3	Not Available					No
Turkey	9042.5	847	10.2	125 (2015)		Yes		Yes	Yes
Europe									
Switzerland	81993.7	201	23.4	123 (2017)	Yes		Yes		Yes
Spain	29613.7	888	19.0	287 (2017)	Yes		Yes		Yes
United Kingdom	42300.3	1598	23.9	871 (2017)	Yes		Yes		Yes
France	40493.9	1362	21.0	742 (2017)	Yes		Yes		Yes
Italy	33189.6	1063	17.5	369 (2016)	Yes		Yes		Yes

Region	Country								
	Belgium	46116.7	234	20.4	128 (2016)	Yes		Yes	Yes
	Denmark	59822.1	130	22.4	70 (2017)	Yes		Yes	Yes
	Germany	46258.9	1607	19.4	837 (2017)	Yes		Yes	Yes
	Poland	15595.2	443	11.7	246 (2017)	Yes		Yes	Yes
Africa	Nigeria	2229.9	461	2.3	Not Available	Yes	Yes		No
	Ghana	2202.1	53	1.8	Not Available	Yes	Yes		No
	Libya	7683.8	76	11.5	Not Available	Yes	Yes		No
	Egypt	3020.0	586	5.9	Not Available	Yes	Yes		No
	South Africa	6001.4	416	7.1	18 (2015)	Yes		Yes	No
North America	United States of America	65118.4	6918	21.0	1020	Yes		Yes	Yes
	Canada	46194.7	757	20.2	101 (2016)	Yes		Yes	Yes
	Mexico	9863.1	1220	9.7	36 (2014)	Yes	Yes		No
South America	Brazil	8717.2	2143	10.2	0.14 (2012)	Yes	Yes		Yes
	Ecuador	6183.8	99	5.7	0.005 (2017)	Yes		Yes	Yes
	Australia	54907.1	554	21.7	58 (2018)	Yes		Yes	Yes
Australia and Oceania	Fiji	6220.0	5.4	6.1	Not Available	Yes		Yes	No
	New Zealand	42084.4	96	19.2	Not Available	Yes		Yes	No

Source: Borthakur and Govind 2017, Jang 2010, Kahhat et al. 2008, Kumar and Holuszko 2016, Lu et al. 2015, Masud et al. 2019, Menikpura, Santo, and Hotta 2014, Pariatamby and Victor 2013, Prakash et al. 2010, Queiruga, Benito, and Lannelongue 2012, Ranasinghe and Athapattu 2020, Saldaña-Durán et al. 2020, Souza 2020, Van Eygen et al. 2016, Vanegas et al. 2020, Veenstra et al. 2010, Widmer et al. 2005, Ilyas et al. 2020, Baldé et al. 2017, Forti et al. 2020, GDP 2019.

"Canadian Environmental Protection Act (CEPA)" works to provide regulations for organizations dealing with toxic elements, and "Electronics Reuse and Refurbishing Program (ERRP)" issues certificates for the companies engaged with reuse and recycling of e-waste (Kumar and Holuszko 2016). Some other organizations run various campaigns and programs to collect specific e-waste under the government support and finance; such as, "Electronic Product Recycling Association (EPRA)" for collecting monitors, laptops, TVs through return program, "Canadian Electrical Stewardship Association (CESA)" for collecting small home appliances through Electro Recycle program, "Canadian Wireless Telecommunication Association (CWTA)" for collecting cell phones under "Recycle My Cell," Cell2Recycle campaign for batteries collection and recycling (Kumar and Holuszko 2016).

2.4.3 SOUTH AMERICA

Almost 30 percent of Latin American countries have been abiding by the e-waste legislation, but an integrated and collective approach to the application of a legal framework for WEEEM through formal (municipal authority) and informal sectors (scrap dealers, collectors, etc.) is still lacking (Baldé et al. 2017, Forti et al. 2020). Bolivia, Chile, Colombia, Costa Rica, Ecuador, Mexico, and Peru have national level e-waste management legislation, whereas Argentina, Brazil, Panama, and Uruguay are trying to develop their e-waste acts. Moreover, Mexico is better positioned with its formal collection efficiency, implementing regulations, and recycling rates compared to other leading countries of South America like Brazil and Argentina (Baldé et al. 2017). Their waste generation rate per person was 7 kg in 2016 (Baldé et al. 2017, Kumar and Holuszko 2016), and WEEEM mostly depends on the informal sector due to their socio-economic condition. Restriction on e-waste free disposal and scarcity of formal drop-off points helps to grow informal channels.

Moreover, e-waste collects from the consumers' doorstep brings some positive exchange value (money/ fancy items) (Rodríguez-Bello and Estupiñán-Escalante 2020). More than 700 private and informal institutions are working with the WEEE recycling process, and the Government tries to bring them under a certification process (Baldé et al. 2017). For example, a few years back, there were only three companies with R2 certification (companies that recycle electronics in an environmentally friendly manner) in Mexico, and now there are 15 more companies. In this manner, Colombia and Peru, already reached a steady collection rate and improved its skills. However, this informal e-waste management sector is also addressing Latin America's labor market and getting higher social imperative as it creates job opportunities for some marginalized groups (Vanegas et al. 2020). Hence, it is essential to acknowledge and regulate their role as like that practiced in Brazil and Chile. Moreover, the OECD countries Mexico, Chile, and Brazil have implemented EPR schemes, and Bolivia, Costa Rica, Ecuador, Guatemala, Panama, Peru, and Uruguay are joining this effort (Rodríguez-Bello and Estupiñán-Escalante 2020). After completing the collection, dismantling and sorting of valuables has been done by some intermediate recyclers and sent for recycling to large companies with R2 certification in Brazil, Mexico, Ecuador, Costa Rica, Chile, and Colombia (Vanegas et al. 2020,

Baldé et al. 2017). In order to implement an adequate, up-to-date WEEEM model, it is important to update waste generation data. Latin America also lags behind in providing up to date information about their generated and imported waste volume (Forti et al. 2020).

2.4.4 EUROPE

Tables 2.1 and 2.4 indicate that GDP and technological advancements are very high in Europe, enabling methodological e-waste management plans reaching a collection and recycling rate of almost 42.5 percent (Baldé et al. 2017, Forti et al. 2020). In this region, UK, Italy, Germany, and France had the highest e-waste consumption in 2017, respectively 1598 kt, 1063 kt, 1607 kt, and 1362 kt against their respective per capita GDP (in USD), 42300.3, 33189.6, 46258.9, and 40439.9 (Borthakur and Govind 2017, Jang 2010, Kahhat et al. 2008, Kumar and Holuszko 2016, Lu et al. 2015, Masud et al. 2019, Menikpura, Santo, and Hotta 2014, Pariatamby and Victor 2013, Prakash et al. 2010, Queiruga, Benito, and Lannelongue 2012, Ranasinghe and Athapattu 2020, Saldaña-Durán et al. 2020, Souza 2020, Van Eygen et al. 2016, Vanegas et al. 2020, Veenstra et al. 2010, Widmer et al. 2005, Ilyas et al. 2020, Baldé et al. 2017, Forti et al. 2020, GDP 2019). Moreover, Northern and Western Europe have the highest per capita e-waste consumption of 22.4 and 20.3 kg, respectively as well as the maximum collection and recycling efficiency of 59 percent and 54 percent respectively (Forti et al. 2020). Most of the WEEEM legalities for collection, recycling, recovery, and reuse are based on the e-waste Directive (2012/19/EU), which is followed by EU member and Balkan countries. EU identified e-waste as a hazard for humans and the environment and passed necessary legislation for the formal processing of e-waste in 2004 (Kiddee, Naidu, and Wong 2013). EU's most significant contribution is the development of legislation for EPR based on e-waste Directive 2002/96/EC, which is now a model e-waste management system all over the world (Borthakur, Govind, and Singh 2019, Gupt and Sahay 2015, Kaza et al. 2018, Lundgren 2012). According to article 7 of this e-waste Directive, member countries (Bulgaria, Czech Republic, Latvia, Lithuania, Hungary, Malta, Poland, Romania, Slovenia, and Slovakia) have the provision to exclude themselves from this campaign by 2021 due to their very low e-waste generation (Van Eygen et al. 2016, Forti et al. 2020). Also, Article 10 of the EU e-waste directive states that the member countries can export e-waste to other member countries and third countries to achieve their least recovery targets, but this article does not give any clear recommendation about the shipment of reused products (Forti et al. 2020). Also, the EU's RoHS Directive acts as controlling authority for implementing legislation for six hazardous chemicals, namely: hexavalent chromium, lead, polybrominated biphenyls, cadmium, mercury, and polybrominated diphenyl ethers to be used in EEE (Borthakur, Govind, and Singh 2019). EU managed the collection and recovery of recyclable components through private organizations and municipalities. This well-developed WEEEM system follows the e-waste legislation set up in 2003 for safe and eco-friendly dumping of WEEE. Though the recycling target is set to 65 percent, a still significant amount of recovery is performed outside of the compliant recycling services as well as mixed recycle of non-compliantly parts with metals. As an example,

Netherlands exports almost 8 percent of their used IT items, laptops, and household EEE to Africa for second-hand usage. The Public Waste Agency of Flanders regulates WEEEM for both consumers and producers, whereas Norway makes it compulsory for producers and importers to be a member of TBS company (Pathak and Srivastava 2019). Belarus has the most advanced e-waste management facilities that offer numerous municipal collection places, and e-waste has been taken from repair and scrap shops providing proper incentives to the collectors/dealers (Wang, Zhang, and Guan 2016). Swiss, WEEEM system, based on EPR is the pioneer in the EU due to its effective implementation in both operational and legal frameworks as well as eco-friendly treatment and disposal (Sinha-Khetriwal, Kraeuchi, and Schwaninger 2005). The consumers play an important role by discarding their e-waste at the designated collection points and also pay through Advanced Recycling Fee (ARF) to support TBS system (Hischier, Wäger, and Gauglhofer 2005). Two mother organizations, Sustainable Energy Solutions (SENS) and The Swiss Association of Information, Communication, and Organizational Technology (SWICO), work on behalf of the producer's responsibility to take back and processing (more than 90%) of e-waste (Savi, Kasser, and Ott 2013). Also, the Southern European countries Spain, Italy, Greece have increased their recycling rate to 34 percent, for example, Spain's was 25 percent before adopting EPR (Queiruga, Benito, and Lannelongue 2012, Forti et al. 2020). Royal Decree (RD) 208/2005 was aligned with EU e-waste directives and mandated the producer to finance the collection and appropriate consumption with no charge to consumers (Queiruga, Benito, and Lannelongue 2012). In Germany, consumers are charged if e-waste is collected from the doorstep, and with the help of German Elektro Geraete Act, EAR project (Elektro-Altgerate Register) works as a clearinghouse for both producer and municipal authority for collecting, monitoring, recovery, and disposal of WEEE (Widmer et al. 2005). However, research indicates that in the UK about 88 percent of large households, 26 percent of office and ITs are recycled, whereas about 10 percent of e-waste is exported to non-OECD countries (mainly Asia and Africa) (Cui and Roven 2011). After transposing the UK's e-waste Regulations 2006 into EU e-waste Directives, the sellers (producers, retailers, etc.) are responsible for the safe disposal of old items against which the seller sells a new item to the customer (Borthakur and Govind 2017). The Eastern European region also developed its infrastructure for e-waste processing. Russia and Ukraine have enterprises for managing e-waste in an environmentally friendly manner, but mixed recycling and open landfilling are common practices. EU countries, however, do not update data on exported reusable products. Although EU countries offer the highest recycling rate, the small e-waste is still discarded as household waste, and in some cases, mixed recycling occurs (Forti et al. 2020, Van Eygen et al. 2016).

2.4.5 ASIA

Many regions of this continent have become recycling, recovery, and dumping zones for the wealthiest countries. EEEs from developed countries are exported to poor countries of Asia for second-hand use, reusable goods, and registered as e-waste generation by the recipients (Borthakur 2016, Cucchiella et al. 2015, Dwivedy and Mittal

2012, Heacock et al. 2016, Jain 2010, Jang 2010, Kumar, Holuszko, and Espinosa 2017, Menikpura, Santo, and Hotta 2014, Mourshed et al. 2017, Pariatamby and Victor 2013, Pathak and Srivastava 2019).

Asia controls the supply of reusable components (26.4 billion USD in 2019) throughout the world by engaging their informal sectors, but this increases environmental and human health hazards due to unwise dumping of WEEE. Economic conditions in Southern and part of Eastern and South-Eastern Asian countries allow multiple users or prolonged effective use life cycle of EEE, which helps to decrease EEE consumption and management costs (Borthakur and Govind 2017, Forti et al. 2020). Moreover, almost 95 percent of WEEE in most of Asia is being collected and recycled through informal dealers, as customers sell their used EEE or e-waste (Borthakur and Govind 2017).

China also employs legislation to control and regulate the 14 types of e-waste, including almost all households and office appliances, whereas Japan and South Korea play a leading role by implementing their advanced WEEEM regulation. Japan is one of the pioneer countries who applied EPR, and under Home Appliance Recycling Law (HARL-2001), retailer shops or municipalities to take back e-waste from consumers and again return it to stockyards for further processing as 70 percent of home appliances get recycled (Zoeteman, Krikke, and Venselaar 2010, Menikpura, Santo, and Hotta 2014). However, to avoid payment for recycling, consumers also sell half of their old EEE to the exporter that ultimately weakens local recycling organizations (Wang, Zhang, and Guan 2016). To discourage this outflow, increase recycling capacity, and restrict landfill opportunities, Japan implemented HARL and "Small Appliance Recycling Law" (Kumar, Holuszko, and Espinosa 2017, Pathak, Srivastava, and Ojasvi 2017).

In South Korea, consumers have to pay the authority while disposing of their e-waste, and they can also avoid the fees for obsolete EEE collection if they buy a replacement from sellers (Kahhat et al. 2008, Jang 2010). Under the Producer Recycling system (PRS-2003), e-waste (telecommunication devices, computers, printers, photocopiers), recycling responsibility extended to the manufacturer ends. Usually, e-waste has been collected from some designated points or sometimes directly from consumers and then sent to the recycling centre supported by "National Recycling Network" (Pathak and Srivastava 2019) as well as stored for export (Jang 2010). This total collection, recycling, storage, and import operations are regulated by "The Korea Environment and Resource Corporation (ENVICO)" (Pathak and Srivastava 2019). However, Western and Central Asia are still in the developing stage to implement e-waste legislation, whereas both rich and poor countries are part of this region. In the Middle East, migrant workers use and reuse refurbished EEE that restricts e-waste generation along with the investment of United Arab Emirates for a 100 kt recycling facility in Dubai.

In South Asia, only India has e-waste management rules from 2011 to develop formal collection and dismantling locations for e-waste with an annual capacity of 800 kt and licensing 312 dealers. India also employed the Producer Responsibility Organization (PRO) act in 2016 and engaged 31 certified PROs to facilitate e-waste collection, raise awareness, maintain collection and processing standards while drafting legislation for safe recovery and shipment of secondary materials (Forti et al. 2020). Pakistan updated their import policy for EEE and banned importing large household appliances like refrigerators, air conditioners, monitors, etc. (Pathak and Srivastava 2019). Malaysia passed the National Solid Waste Management Policy, but having a limited budget for WEEEM turns it into being ineffective in practice (Kaza et al. 2018). Taiwan demonstrates outstanding e-waste management under the Recycling Fund Management Board (RFMB), where 20 recycling facilities have recycled 64 percent of collected e-waste as its recycling capacity is more compared to their own WEEE generation volume. China stands as the highest WEEE consumers (10.1 Mt) in the world, which roughly contributes one-fifth of the world's and half of Asia's e-waste. Large population, low manufacturing and refurbishment cost, high local and international demand are the factors for this colossal e-waste pile. China claims a 40 percent recycling rate for five of their e-waste products, and after considering the 54 products listed by (UNU-Keys), this rate declines to 15 percent (Forti et al. 2020, Lu et al. 2015, Wang, Zhang, and Guan 2016). Although China faces challenges to raise e-waste management funding, they set a target to recover 20 percent for new EEE and achieve a 50 percent recycling rate by 2025 (Forti et al. 2020, Wang, Zhang, and Guan 2016).

2.4.6 AUSTRALIA AND OCEANIA

Australia imports most of its EEE and exports a small volume of old reusable products. Through the adoption of the 'National Television and Computer Recycling Scheme' (NTCRS), the Government has reduced open dumping of e-waste for landfills (Dias, Bernardes, and Huda 2019). However, due to inadequate downstream recycling structure, a major part of the collected e-waste is exported overseas for recycling (Golev et al. 2016, Lane, Gumley, and Santos 2015). Although exports of e-waste have gradually increased over time, information regarding destination countries and their processing methods is not well documented (Golev et al. 2016, Lane, Gumley, and Santos 2015). Currently, e-waste collected through NTCRS initiative is processed in two stages:

1. Domestic and initial recycling according to Australian regulation. This includes shredding and dismantling of products (AU 2011)
2. Domestic and international downstream recycling that mainly covers separation, product-specific treatment, and various refining process (Dias et al. 2018)

First stage recyclers (FSR) in Australia receive funding from liable parties to initiate the recycling process. Once the e-waste is collected, FSR sort and dismantle waste into different components. Afterwards, FSR may perform minimal waste treatment (e.g., battery removal) and export major portion of the waste overseas for dismantling (Dias, Bernardes, and Huda 2019). FSR then sells the dismantled parts to domestic or international downstream recyclers. E-waste from Australia is primarily exported to Indonesia, Malaysia, China, South Korea, and Japan (Dias et al. 2018). Automated domestic recycling centres of Australia, on average, recycles 3.84 kt of e-waste/year.

Manual dismantling facilities recycle, on average, 1361 tons of e-waste each year. First stage domestic recycling in Australia utilizes 39.7 kW energy per ton of e-waste processing (Dias, Bernardes, and Huda 2019). Australia's current e-waste management practice focuses on international downstream recycling due to inexpensive international freights and manpower. Moreover, for many electronic components, downstream recycling is virtually non-existent in Australia. While the export of e-waste may be a viable option, for now, the country is losing potentially 120 million USD worth of recycled metals (Golev and Corder 2017). Furthermore, the Australian Government can create more jobs by investing in the domestic downstream recycling industry.

Every year, in New Zealand, an estimated 97000 tonnes of e-waste is dumped into landfill, which negatively impacts the public health and environment (Blake, Farrelly, and Hannon 2019). Lack of waste generation data hinders policymakers from drafting a sustainable e-waste management policy for the country (MFE 2014, 2017). Blake et al. performed a survey in the Whangarei district, and an estimated 98.2 percent of total household WEEE can end up in landfills (Blake, Farrelly, and Hannon 2019). They opined that the lack of strict waste management policy and inadequate recycling centres as a leading cause of this crisis in New Zealand. Global E-waste Monitor reported that e-waste generation in New Zealand increased by 1.1 percent from 2015 to 2017 (Baldé et al. 2017, David C. Wilson 2015). Although the Government of New Zealand acknowledges the environmental concern, there are no policies for e-waste management. There are insufficient data resources in terms of public awareness and concepts such as 'willingness to pay.' Moreover, there Ministry for the Environment (MFE), New Zealand, failed to declare WEEE as a 'priority' in waste stream under Section 9 of the Waste Minimization Act 2008 (Blake, Farrelly, and Hannon 2019). This situation can be ameliorated by the application of a national scale mandatory WEEE stewardship scheme.

2.4.7 AFRICA

Many South African countries are engaged in the refurbishment business both in a formal and informal mode, which helps them to use EEE at a comparatively low price. Interestingly a survey in Ghana shows that male consumers are more concerned about the environmental aspects of e-waste than females (Borthakur and Govind 2017). Apart from this, consumers' consciousness and attitude towards e-waste collection and recycling process, level of income, educational qualification, ideology, convenience, age, gender (though some researchers also opposed this issue) etc. also actively drive the total e-waste management process. EEE consumption rate is higher in developing countries than developed nations. Although Basel Act is to regulate transboundary shipment, the least developed African nations are often chosen as the destination for e-waste, although they cannot manage the recycling process in an environmentally sound manner. For example, about 70 percent of second-hand EEE products exported to Ghana in 2009 are not functional or close to their EoL (Heacock et al. 2016).

According to international sources, in 2010, West Africa generated 210 kt of e-waste and imported 0.61 million tons (Asante, Amoyaw-Osei, and Agusa 2019). In the same year, West African nations disposed of 450 kt of e-waste through landfills and incineration (Zoeteman, Krikke, and Venselaar 2010). This indicates that nations in this region are importing more than their capacity for safe recycling. Per capita e-waste in Africa ranges between 4.6 kg (Liberia) and 44 kg (Nigeria). Although Nigeria is a least developing country with a small ICT sector, it has a per capita waste generation similar to the developed countries (Schluep et al. 2012). This happens because Nigeria imports (or developed countries exports) excess amount of WEEE from the developed countries. In order to address this issue, in 1991, 51 representatives of the African Union held a convention in Bamako, Mali. Through this Bamako convention (effective since 1998), African nations have restricted bulk import of e-waste.

Currently, most of WEEE in Africa is managed informally through manual collection and dismantling and recovery metals through incineration. The remaining waste is dumped in landfills. In some countries like Liberia and Benin individuals perform informal waste treatment for low material throughput. At the same time, countries such as Ghana, Cote d'Ivoire, and Nigeria have structured informal sectors for retrieving metals like aluminium, steel, and copper in high volume (Manhart et al. 2011). In most countries, manual labour and tools such as hammers, chisels are used to dismantle and separate e-waste products. Reusable parts are normally sold to second-hand markets or repair shops. Open burning is the most common procedure in all African nations for recovering valuable materials. In Nigeria, advanced refining techniques are used for leaching gold from printed wiring boards. In most African nations, the residual e-waste parts like plastics and CRT-glasses are dumped in the open areas (Schluep et al. 2012).

2.5 WAY FORWARD

2.5.1 Reverse Logistics

Most of the existing legislations cover economic aspect of e-waste with respect to recovery and recycling mode. However, less priority is given to the health hazard of living beings and environmental aspects. Reverse logistics emphasize the minimization of ecotoxicology and environmental hazard by maintaining sustainability in developed as well as developing countries (Zeng et al. 2016). Along with this, the types of materials recovered, and their subsequent recycling can be updated through the adoption of proper disassembling processes, upgrading the existing metallurgical recovery steps, and refining the recovered items to segregate chemicals, valuable, hazardous, and heavy materials. As the developing countries already face a great challenge with their own as well as exported e-scraps, developed countries should extend a helping hand by implementing pragmatic steps to solve this crisis.

2.5.2 Developing Waste Flow Model

E-waste management through formal sectors is the utmost priority to positively change the economy, environment, and human being. After assessing the current scenario of global e-waste practices, we detect a shift from informal to formal sectors as informal sectors taking the lead in most areas. However, making informal sectors more accountable and building a bridging network between formal and informal WEEE processing sectors is a feasible option to redefine e-waste flow. In this sense, encouraging, 4-R, 3-R approaches along with the employment of scientific waste flow model for the collection, small scale mobile processing units, and dumping of e-waste could be a sustainable option (Pathak, Srivastava, and Ojasvi 2017, Srivastava and Pathak 2020).

2.5.3 Circular Economy

Circular economy ensures the maximum EEE utilization by introducing new features, modifying the business plan, energy flows through money and products, improving the recycling and recovery systems, smarter product design, etc. In another way, the multi-R approach, including reuse, recycle, reduce, recovery, refuse, rethinking, has been considered to achieve the objectives of UNIDO 2015 (Srivastava and Pathak 2020). Informal sectors, private sectors, financial support, infrastructure development for collection and recycling processes, etc. have been considered as the key character of the circular economy. Efficient metal recovery from e-waste is intricate and costly, but it would decrease mining for new metals as about 25 percent of recovery can be achieved from WEEE (Heacock et al. 2016). Another recent catchphrase is purchasing specific items from the dematerialization of e-waste or refurbishing EEE rather than taking a new product to attain a downstream e-waste generation rate. Thus, supporting this sector financially and technical know-how can contribute to economic development as well as to balance environmental harmony. To make this transition from linear to circular economy, many models such as TBS, cradle to cradle, refurbishment, repair, reverse logistics, cascade system have been proposed to cover not only EoL of EEE products, but also the different stages of its life cycle (Rodríguez-Bello and Estupiñán-Escalante 2020). By successfully employing this circular economy strategy through greater multi-R, recycle, recover, reuse, refurbish approaches, the product lifecycle can be tied into a close loop and, in the meantime, excess consumption of virgin materials, wastage can be cut down to a minimum level.

2.5.4 Global and Domestic Protocol

As already discussed, developed countries export their obsolete EEE to developing countries through authorized and illegal dealers. Moreover, some of this WEEE is used as refurbished products in developing countries, and the scraps are dumped without maintaining any environmental legislation after necessary recovery (Garlapati 2016). These practices maximize economic benefits from e-waste, followed by a serious threat to global health and environment. Hence, a concerted global effort is needed to control transboundary movement as well as social, economic, political, ideological, human health, and environmental challenges of recovery, recycling, reuse of obsolete EEE (Heacock et al. 2016). Till now, the global protocol abiding by the Basel Convention is only implemented to regulate the trading of toxic products, which needs to be revised and must include a second hand, obsolete, close to EoL products.

Strict legislation must curb the trading of toxic and obsolete products from rich countries to poor countries. Also, domestic legislation of developing countries should be strengthened by imposing a complete ban on e-scrap exportation as well as increasing custom duties.

Moreover, international organizations like UN, WB must come forward to develop some standard protocols to ease the collection, recycling, recovery, trading, tracking of e-waste flow throughout the world. Besides, the developed countries should extend their hands through industrial collaboration between developed and developing countries to share advanced and environmentally friendly recycling and recovery technology. Also, financial compensation package (setting up a new production plant, certified recycling units etc.) should be provided to those countries who are using dumping zone and facing challenges to fight against this e-waste; so that their economy and environment become stabilized in the long run (Wang, Zhang, and Guan 2016).

2.5.5 CONSUMERS – MANUFACTURER'S RESPONSIBILITY

The tendency of replacing EEE products is increasing at an alarming rate due to frequent up-grades of EEE, shorter life span of products, low-cost, and product aesthetic appearance. Manufacturers should focus on developing sustainable products, refurbishing second-hand EEE, and EoL treatments that maximize the utilization of e-waste through recycling and recovery before final disposal. Identifying recyclable components can be a viable option to differentiate between toxic and safe components and ensure quality recovery. This manufacturing strategy has two-fold advantages; it will increase environmental safety by lowering the rate of WEEE disposal and improve economic stability of recycling as well as recovery sectors. Consumers are also responsible for the safe disposable of their products, and in this way, TBS, EPR, and other specific programs help to smooth this disposal and collection system. Some countries provide special collection bins (like Japan) while some countries apply deposition fees so that customers can get back their money after returning their EEE through authorized collectors or dealers (Wang, Zhang, and Guan 2016). In contrast, applying higher charge and taxation to luxurious EEE goods can be a viable approach to reduce environmental hazards with a bit negative impact on the market-based economy (Borthakur 2016).

2.5.6 EXTENDED PRODUCER RESPONSIBILITY POLICY

EPR is a TBS based policy introduced by EU to ensure the utilization of product lifetime at its full, proper reuse, low waste, and disposal by adopting sustainable and eco-friendly EEE design. EPR defines producer's responsibility to develop sustainable products and extended responsibilities to facilitate TBS, recovery, recycling, and then finally discarded as waste. Organization for Economic Cooperation and Development (OECD) clarified EPR as a program that starts from the manufacturing stages and extended to the disposal or post-consumption stage (Widmer et al. 2005). Followed by EU, Brazil, India, China, Thailand, and those countries which are environmentally vigilant, trying to impose EPR for attaining a better product lifecycle and extending the responsibility of the manufacturers towards the betterment of society (Borthakur 2016). Extended EPR works together with circular economy and emphasis on the circulatory movement of materials and economy. This e-waste management tool engages both the producers' physical and financial contributions and needs substantial compensation to keep this reuse, recovery, and recycling practices in motion (Rodríguez-Bello and Estupiñán-Escalante 2020). Thus, implementing extended EPR is one of the pragmatic approaches to managing WEEE stream by adopting an improved collection system (TBS, deposition system, etc.) (Manomaivibool 2009, Borthakur 2016).

2.5.7 Developing Relationship between Formal and Informal Sectors

Developing both formal and informal institutions, as well as their integrated approach is required to take the challenge of e-waste management. Development of formal sectors is a continuous process through the adoption of most recent technologies or best available technologies (BAT) for recycling and recovery, setting up new standards to cope with eco-friendly standards, developing infrastructure, and for these processes, adequate support from respective authorities as well as financial back up are the crucial factors. In this manner, informal sectors are the most deprived ones; they have no technical knowledge about the collection, relubrication of second-hand goods, item-wise segregation of e-waste, recovery of valuable materials from toxic items, recycling processes, and the critical challenges with the dumping of scraps. Hence, collaboration and knowledge sharing between informal and formal sectors can be a feasible option for systematically using informal sectors and ensuring the maximum practical management of WEEE.

For effective WEEEM, public awareness, or in a broad sense, consumer behavior and attitude towards control, disposal, recycling of e-waste and their harmful effect on environment must be addressed accordingly. A graphical summary of the proposed methods for achieving a sustainable WEEEM system is represented in Figure 2.6. The effectiveness of each solution would vary depending on the economy, geopolitical stability, consumer psychology, and environmental awareness. Further research is required for selecting a waste management model for a particular country.

Developing and least developed countries contribute significantly to the production of electronic waste. Developing countries such as Bangladesh produces 199 kt e-waste per year (2016–2017), which is higher than developed countries such as Denmark, New Zealand and Kuwait (see Table 2.4). Failing to adequately manage the e-waste, these countries are facing different types of environmental pollution. In this regard, following the established models of the developed countries can be proven useful. These models focus on 4-R and 3-R, develops relationship between formal and informal sectors. In most developing and least developed countries, e-waste is managed in informal ways, which is not only less efficient but also don't promote economic growth. Concepts such as reverse logistics, circular economy and different global protocols encourage a transition from informal to formal waste management sector. Developing countries could also benefit for increased consumer awareness and effective e-waste recycling practices. Methods like willingness to pay (WTP), financial incentives and improved consumer-manufacture responsibility

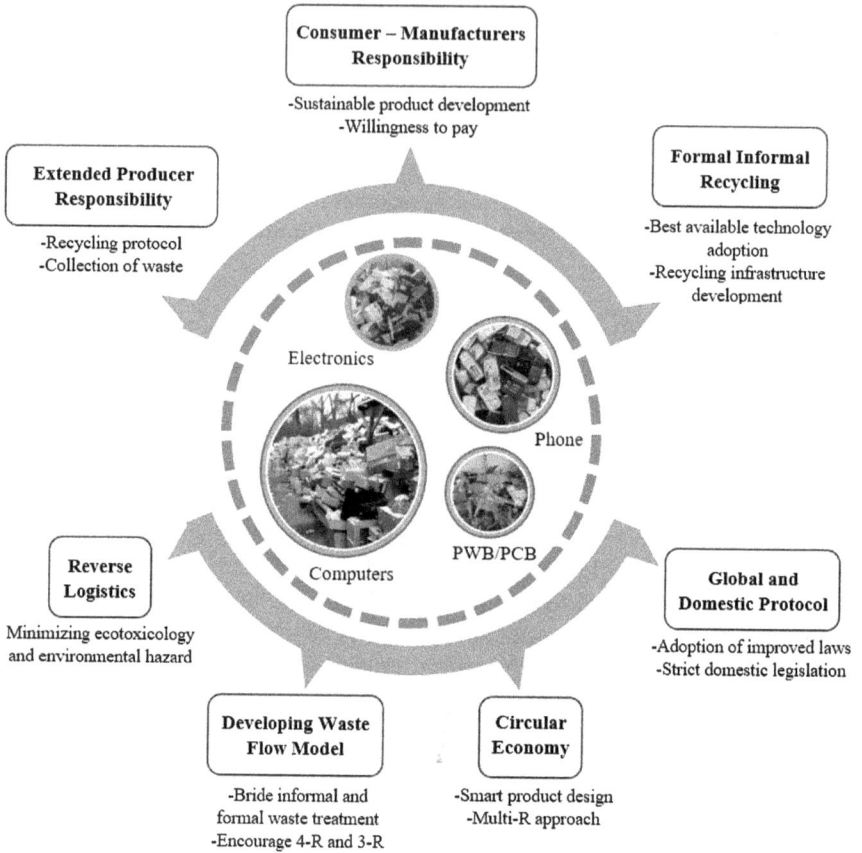

FIGURE 2.6 Graphical summary of the proposed methods for sustainable e-waste management.

can be good options in this regard. Original equipment manufacturers (OEM) from developed countries comply with extended producer responsibility (EPR), which increases the life cycle of electronic products. Small and medium-sized enterprises in the electronic industries of developing countries should also adhere to some form of EPR. Moreover, least developed countries could potentially enforce strict regulation on illegal importing of e-waste following the domestic protocols of developed countries.

2.6 CONCLUSION

The EEE industry is one of the largest consumers of different valuable metals, including gold, iron, copper, and aluminium. Therefore, improper handling of these products can account for significant economic losses at EOL. Proper handling and adoption of the most recent recycling technologies can turn this e-waste into an interesting source for recovering these valuable materials. This is not only beneficial for economy, environment, and human health together. This chapter sheds light on the global e-waste generation scenario and different international policies and legislation to reduce its harmful impact. The chapter further describes the e-waste management strategies of various countries on different continents. A critical discussion is presented to highlight the reasons for inadequate recycling practices in some developed (Australia) and developing (Africa, South Asia) regions. Finally, this chapter provides an insight of different WEEEM models that can aid the governments of different nations to tackle e-waste generation rate and management approaches. Innovative solutions should be taken to effectively and economically collect and recycle e-waste. Instead of landfilling or open burring, e-waste should be recycled through effective treatment methods to recover precious materials and minimize the hazardous impact on living beings. Moreover, international bodies should track the global e-waste stream and prevent illegal trading in developing countries.

REFERENCES

3R. "The Ninth Regional 3R Forum in Asia and the Pacific." accessed 5 October, 2020. www. env.go.jp/recycle/3r/en/results/09.html.

Ananno, A. A., M. H. Masud, S. A. Chowdhury, P. Dabnichki, N. Ahmed, A. M. E. Arefin, and Consumption. 2020. "Sustainable food waste management model for Bangladesh." *Sustainable Production and Consumption* 27:35–51.

Ananno, A. A., M. H. Masud, P. Dabnichki, M. Mahjabeen, and S. A. Chowdhury. 2021. "Survey and analysis of consumers' behaviour for electronic waste management in Bangladesh." *Journal of Environmental Management* 282:111943. doi: https://doi.org/ 10.1016/j.jenvman.2021.111943.

Asante, K. A., Y. Amoyaw-Osei, and T. Agusa. 2019. "E-waste recycling in Africa: risks and opportunities." *Current Opinion in Green Sustainable Chemistry* 18:109–117.

AU, F. R. o. L. 2011. "Product Stewardship (Televisions and Computers) Regulations 2011." Australian Government, accessed 05 October, 2020. www.legislation.gov.au/Details/ F2011C00912.

Awasthi, A. K., F. Cucchiella, I. D'Adamo, J. Li, P. Rosa, S. Terzi, G. Wei, and X. Zeng. 2018. "Modelling the correlations of e-waste quantity with economic increase." *Sci. Total Environ.* 613–614:46–53. doi: https://doi.org/10.1016/j.scitotenv.2017.08.288.

Baidya, R., B. Debnath, S. K. Ghosh, S.-W. J. W. M. Rhee, and Research. 2020. "Supply chain analysis of e-waste processing plants in developing countries." *Waste Management & Research* 38 (2):173–183.

Baldé, C., V. Forti, V. Gray, R. Kuehr, and P. Stegmann. 2017. The Global E-waste Monitor– 2017, Quantities, Flows, and Resources. Bonn/Geneva/Vienna.: United Nations University (UNU), International Telecommunication Union (ITU) & International Solid Waste Association (ISWA).

Benebo, N. 2011. "E-waste issues and experiences: the Nigerian experience." International Dialogue on Electronics Management, Taipei, Taiwan.

Blake, V., T. Farrelly, and J. Hannon. 2019. "Is Voluntary Product Stewardship for E-Waste Working in New Zealand? A Whangarei Case Study." *Sustainability* 11 (11):3063.

Borthakur, A. 2016. "Policy implications of e-waste in India: a review." *International Journal of Environment Waste Management & Research* 17 (3–4):301–317.

Borthakur, A. 2020. "Policy approaches on E-waste in the emerging economies: A review of the existing governance with special reference to India and South Africa." *Journal of Cleaner Production* 252:119885. doi: https://doi.org/10.1016/j.jclepro.2019.119885.

Borthakur, A., and M. Govind. 2017. "Emerging trends in consumers' E-waste disposal behaviour and awareness: A worldwide overview with special focus on India." *Resources, Conservation and Recycling* 117:102–113. doi: https://doi.org/10.1016/j.resconrec.2016.11.011.

Borthakur, A., and M. Govind. 2018. "Public understandings of E-waste and its disposal in urban India: from a review towards a conceptual framework." *Journal of Cleaner Production* 172:1053–1066.

Borthakur, A., M. Govind, and P. Singh. 2019. "Chapter 2 – Inventorization of E-waste and Its Disposal Practices With Benchmarks for Depollution: The Global Scenario." In *Electronic Waste Management and Treatment Technology*, edited by Majeti Narasimha Vara Prasad and Meththika Vithanage, 35–52. Butterworth-Heinemann.

Canada, G. o. 1994. "Toxic substances management policy." Government of Canada, accessed 05 October, 2020. www.canada.ca/en/environment-climate-change/services/management-toxic-substances/policy.html.

Comission, E. European Comission, accessed 05 October, 2020. https://ec.europa.eu/environment/waste/e-waste/index_en.htm.

Comission, E. "Energy efficient products." accessed 05 October, 2020. https://ec.europa.eu/info/energy-climate-change-environment/standards-tools-and-labels/products-labelling-rules-and-requirements/energy-label-and-ecodesign/energy-efficient-products_en.

Convention, T. B. 1992. "The Basel Convention." accessed 10 January. www.basel.int/TheConvention/Overview/tabid/1271/Default.aspx.

Convention, T. B. 2019. "Fourteenth Meeting of the Conference of the Parties to the Basel Convention." accessed 10 January. www.basel.int/TheConvention/ConferenceoftheParties/Meetings/COP14/tabid/7520/Default.aspx.

Convention, T. B. 2021. "Text of the Convention." accessed 10 January. www.basel.int/TheConvention/Overview/TextoftheConvention/tabid/1275/Default.aspx.

Cucchiella, F., I. D'Adamo, S. L. Koh, and P. Rosa. 2015. "Recycling of WEEEs: An economic assessment of present and future e-waste streams." *Renewable Sustainable Energy Reviews* 51:263–272.

Cui, J., and H. J. Roven. 2011. "Electronic waste." Waste.

Cusack, P., and T. Perrett. 2006. "The EU RoHS Directive and its implications for the plastics industry." *Plastics, Additives and Compounding* 8 (3):46–49.

Daum, K., J. Stoler, and R. Grant. 2017. "Toward a more sustainable trajectory for e-waste policy: a review of a decade of e-waste research in Accra, Ghana." *International Journal of Environmental Research Public Health* 14 (2):135.

David C. Wilson, L. R., P. Modak, R. Soos, A. Carpintero, C. Velis, M. Iyer, O. Simonett. 2015. Global waste management outlook: Summary for decision-makers. Edited by UNEP. Japan.

DAWE, A. G. "National Waste Policy." Department of Agriculture, Water and Environment, accessed 05 October, 2020. www.environment.gov.au/protection/waste-resource-recovery/national-waste-policy.

Dias, P., A. M. Bernardes, and N. Huda. 2019. "Ensuring best E-waste recycling practices in developed countries: An Australian example." *Journal of Cleaner Production* 209:846–854.

Dias, P., A. Machado, N. Huda, and A. M. Bernardes. 2018. "Waste electric and electronic equipment (WEEE) management: A study on the Brazilian recycling routes." *Journal of Cleaner Production* 174:7–16. doi: https://doi.org/10.1016/j.jclepro.2017.10.219.

Dwivedy, M., and R. K. Mittal. 2012. "An investigation into e-waste flows in India." *Journal of Cleaner Production* 37:229–242. doi: https://doi.org/10.1016/j.jclepro.2012.07.017.

Ecroignard, L. 2006. E-waste legislation in South Africa.

EPA. "International E-Waste Management Network (IEMN)." accessed 05 October, 2020. www.epa.gov/international-cooperation/international-e-waste-management-network-iemn.

EPA. "National Strategy for Electronics Stewardship (NSES)." accessed 05 October, 2020. www.epa.gov/smm-electronics/national-strategy-electronics-stewardship-nses.

EUON. Improved clarity on nanomaterials in the EU – Member States vote to amend REACH Annexes. Accessed 05 October, 2020.

EuP. "EuP (Eco-Design Of Energy Using Products)." accessed 05 October, 2020. www.fluorocarbons.org/eup-eco-design-energy-using-products/.

Forti, V., C. P. Balde, R. Kuehr, and G. Bel. 2020. The Global E-waste Monitor 2020: Quantities, flows and the circular economy potential. United Nations University (UNU)/ United Nations Institute for Training and Research (UNITAR) – co-hosted SCYCLE Programme, International Telecommunication Union (ITU) & International Solid Waste Association (ISWA), Bonn/Geneva/Rotterdam.

Forum, W. R. 2017. Colombia First Latin American Country With E-waste Management Policy. Bogota: SRI.

Garlapati, V. K. 2016. "E-waste in India and developed countries: Management, recycling, business and biotechnological initiatives." *Renewable and Sustainable Energy Reviews* 54:874–881. doi: https://doi.org/10.1016/j.rser.2015.10.106.

GDP. 2019. "The World Bank GDP per Capita." The World Bank, accessed 05 October, 2020. https://data.worldbank.org/indicator/NY.GDP.PCAP.CD.

GeSI. accessed 05 October, 2020. http://gesi.org/about.

Golev, A., and G. D. Corder. 2017. "Quantifying metal values in e-waste in Australia: The value chain perspective." *Minerals Engineering* 107:81–87.

Golev, A., D. R. Schmeda-Lopez, S. K. Smart, G. D. Corder, and E. W. McFarland. 2016. "Where next on e-waste in Australia?" *Waste Management* 58:348–358. doi: https://doi.org/10.1016/j.wasman.2016.09.025.

Gupt, Y., and S. Sahay. 2015. "Review of extended producer responsibility: A case study approach." *Waste Management & Research* 33 (7):595–611. doi: 10.1177/0734242X15592275.

Heacock, M., C. B. Kelly, K. A. Asante, L. S. Birnbaum, Å. L. Bergman, M.-N. Bruné, I. Buka, D. O. Carpenter, A. Chen, and X. Huo. 2016. "E-waste and harm to vulnerable populations: a growing global problem." *Environ. Health Perspect.* 124 (5):550–555.

Herat, S. 2009. "International regulations and treaties on electronic waste (e-waste)." *International Journal of Environmental Engineering* 1 (4):335–351.

Herat, S., and P. Agamuthu. 2012. "E-waste: a problem or an opportunity? Review of issues, challenges and solutions in Asian countries." *Waste Management Research* 30 (11):1113–1129.

Hischier, R., P. Wäger, and J. Gauglhofer. 2005. "Does WEEE recycling make sense from an environmental perspective?: The environmental impacts of the Swiss take-back and recycling systems for waste electrical and electronic equipment (WEEE)." *Environmental Impact Assessment Review* 25 (5):525–539.

Ibitz, A. J. A. J. o. S.-E. A. S. 2012. "Environmental policy coordination in ASEAN: the case of waste from electrical and electronic equipment." 5 (1):30–51.

ICLG. 2008. Environmental law: Nigeria. International Comparative Legal Guide Series. Nigeria: ICLG.

IETC. "International Environmental Technology Centre." accessed 05 October, 2020. www.unenvironment.org/ietc/.

Ilankoon, I., Y. Ghorbani, M. N. Chong, G. Herath, T. Moyo, and J. Petersen. 2018. "E-waste in the international context–A review of trade flows, regulations, hazards, waste management strategies and technologies for value recovery." *Waste Management* 82:258–275.

Ilyas, S., R. R. Srivastava, H. Kim, and Z. Abbas. 2020. "Electrical and electronic waste in Pakistan: the management practices and perspectives." In *Handbook of Electronic Waste Management*, 263–281. Elsevier.

IMF. 2019. " International Monetary Fund (IMF)." accessed 05 October, 2020. www.imf.org/external/datamapper/NGDP_RPCH@WEO/OEMDC/ADVEC/WEOWORLD.

Jain, A. 2010. "E-waste management in India: current status, emerging drivers and challenges." Regional Workshop on e-waste/WEEE management, Osaka, Japan.

Jang, Y.-C. 2010. "Waste electrical and electronic equipment (WEEE) management in Korea: generation, collection, and recycling systems." *Journal of Material Cycles Waste Management* 12 (4):283–294.

Kahhat, R., J. Kim, M. Xu, B. Allenby, E. Williams, and P. Zhang. 2008. "Exploring e-waste management systems in the United States." *Resources, Conservation & Recycling* 52 (7):955–964.

Kang, H.-Y., and J. M. Schoenung. 2006. "Estimation of future outflows and infrastructure needed to recycle personal computer systems in California." *Journal of Hazardous Materials* 137 (2):1165–1174.

Kaya, M. 2016. "Recovery of metals and nonmetals from electronic waste by physical and chemical recycling processes." *Waste Management* 57:64–90.

Kaza, S., L. Yao, P. Bhada-Tata, and F. Van Woerden. 2018. *What a waste 2.0: a global snapshot of solid waste management to 2050*: The World Bank.

Kiddee, P., R. Naidu, and M. H. Wong. 2013. "Electronic waste management approaches: An overview." *Waste Management* 33 (5):1237–1250. doi: https://doi.org/10.1016/j.wasman.2013.01.006.

Kumar, A., and M. Holuszko. 2016. "Electronic waste and existing processing routes: A Canadian perspective." *Resources* 5 (4):35.

Kumar, A., M. Holuszko, and D. C. R. Espinosa. 2017. "E-waste: An overview on generation, collection, legislation and recycling practices." *Resources, Conservation & Recycling* 122:32–42. doi: https://doi.org/10.1016/j.resconrec.2017.01.018.

Kumar, S., and S. Rawat. 2018. "Future e-Waste: Standardisation for reliable assessment." *Government Information Quarterly* 35 (4):S33–S42.

Kusch, S., and C. D. Hills. 2017. "The link between e-waste and GDP—new insights from data from the pan-European region." *Resources* 6 (2):15.

Lane, R., W. S. Gumley, and D. E. Santos. 2015. Mapping, characterising and evaluating collection systems and organisations. Melbourne VIC Australia: Monash University.

Laws, J. 1999. "Canadian Environmental Protection Act, 1999." Justice Laws Website, accessed 05 October, 2020. https://laws-lois.justice.gc.ca/eng/acts/c-15.31/page-1.html.

Lepawsky, J. 2012. "Legal geographies of e-waste legislation in Canada and the US: Jurisdiction, responsibility and the taboo of production." *Geoforum* 43 (6):1194–1206.

Li, J. 2011. "Opportunities in action: the case of the US Computer TakeBack Campaign." *Contemporary Politics* 17 (3):335–354.

Li, J., B. Tian, T. Liu, H. Liu, X. Wen, and S. i. Honda. 2006. "Status quo of e-waste management in mainland China." *Journal of Material Cycles Waste Management & Research* 8 (1):13–20.

Li, J., X. Zeng, M. Chen, O. A. Ogunseitan, and A. Stevels. 2015. ""Control-Alt-Delete": rebooting solutions for the e-waste problem." *Environmental Science & Technology* 49 (12):7095–7108.

Lu, C., L. Zhang, Y. Zhong, W. Ren, M. Tobias, Z. Mu, Z. Ma, Y. Geng, and B. Xue. 2015. "An overview of e-waste management in China." *Journal of Material Cycles Waste Management* 17 (1):1–12.

Lundgren, K. 2012. *The global impact of e-waste: addressing the challenge*: International Labour Organization.

MADS, M. d. A. y. D. S. 2017. Política Nacional (Colombia): Gestión Integral de Residuos de Aparatos Eléctricos y Electrónicos. Colombia: MATHIAS-SRI.

Manhart, A., O. Osibanjo, A. Aderinto, and S. Prakash. 2011. "Informal e-waste management in Lagos, Nigeria–socio-economic impacts and feasibility of international recycling co-operations." *Final Report of Component* 3:1–129.

Manomaivibool, P. 2009. "Extended producer responsibility in a non-OECD context: The management of waste electrical and electronic equipment in India." *Resources, Conservation & Recycling* 53 (3):136–144.

Masud, M. H., W. Akram, A. Ahmed, A. A. Ananno, M. Mourshed, M. Hasan, and M. U. H. Joardder. 2019. "Towards the effective E-waste management in Bangladesh: a review." *Environmental Science Pollution Research* 26 (2):1250–1276.

Masud, M. H., Karim, A., Ananno, A. A., & Ahmed, A. (2020). Sustainable Food Drying Techniques in Developing Countries: Prospects and Challenges. Springer International Publishing.

Menikpura, S. N., A. Santo, and Y. Hotta. 2014. "Assessing the climate co-benefits from Waste Electrical and Electronic Equipment (WEEE) recycling in Japan." *Journal of Cleaner Production* 74:183–190.

MFE, M. f. t. E. 2008. About the Waste Minimisation Act. MFE NZ.

MFE, M. f. t. E. 2010. The New Zealand Waste Strategy 2010. MFE NZ.

MFE, M. f. t. E. 2014. Briefing to the Incoming Minister for the Environment. edited by Ministry for the Environment MFE NZ: Department of Corrections.

MFE, M. f. t. E. 2017. Review of the Effectiveness of the Waste Disposal Levy 2017. edited by Ministry for the Environment MFE NZ. New Zealand: Ministry for the Environment MFE NZ.

Mourshed, M., M. H. Masud, F. Rashid, and M. U. H. Joardder. 2017. "Towards the effective plastic waste management in Bangladesh: a review." *Environmental Science and Pollution Research* 24 (35):27021–27046. doi: 10.1007/s11356-017-0429-9.

Nations, U. Accessed July, 2020. "The United Nations: Sustainable development goals, 2018." accessed 05 October, 2020. www.un.org/sustainabledevelopment/sustainable-development-goals/.

Newaj, N., and M. H. Masud. 2014. "Utilization of waste plastic to save the environment." International conference on mechanical, industrial and energy engineering.

Olimid, A. P., and D. A. Olimid. 2019. "Societal Challenges, Population Trends and Human Security: Evidence from the Public Governance within the United Nations Publications (2015–2019)." *Revista de Stiinte Politice* (64).

Palmeira, V. N., G. F. Guarda, and L. F. W. Kitajima. 2018. "Illegal international trade of e-waste-Europe." *Detritus* 1 (1):48.

Pariatamby, A., and D. Victor. 2013. "Policy trends of e-waste management in Asia." *Journal of Material Cycles Waste Management & Research* 15 (4):411–419.

Pathak, P., and R. R. Srivastava. 2019. "Environmental Management of E-waste." In *Electronic Waste Management and Treatment Technology*, 103–132. Elsevier.

Pathak, P., R. R. Srivastava, and Ojasvi. 2017. "Assessment of legislation and practices for the sustainable management of waste electrical and electronic equipment in India." *Renewable and Sustainable Energy Reviews* 78:220–232. doi: https://doi.org/10.1016/j.rser.2017.04.062.

Prakash, S., A. Manhart, Y. Amoyaw-Osei, and O. O. Agyekum. 2010. Socio-economic assessment and feasibility study on sustainable e-waste management in Ghana. edited by Ministry of Housing Green Advocacy Ghana, Spatial Planning the Environment, VROM-Inspectorate. Ghnana: Öko-Institut eV in cooperation with Ghana Environmental Protection Agency.

Queiruga, D., J. G. Benito, and G. Lannelongue. 2012. "Evolution of the electronic waste management system in Spain." *Journal of Cleaner Production* 24:56–65.

Ranasinghe, W. W., and B. C. Athapattu. 2020. "Challenges in E-waste management in Sri Lanka." In *Handbook of Electronic Waste Management*, 283–322. Elsevier.

Rivera, C. 2019. "New RoHS Directive Amendment and Category 11 Product Type." July 26, 2019.

Rodríguez-Bello, L. A., and E. Estupiñán-Escalante. 2020. "The impact of waste of electrical and electronic equipment public police in Latin America: analysis of the physical, economical, and information flow." In *Handbook of Electronic Waste Management*, 397–419. Elsevier.

Rucevska I., N. C., Isarin N., Yang W., Liu N., Yu K., Sandnæs S., Olley K., McCann H., Devia L., Bisschop L., Soesilo D., Schoolmeester T., Henriksen, R., Nilsen, R. 2015. Waste crime – waste risks: Gaps in meeting the global waste challenge. In *A UNEP Rapid Response Assessment, United Nations Environment Programme and GRID-Arendal, Nairobi and Arendal*. Norway: UNEP.

Ryder, G., and H. Zhao. 2019. "The world's e-waste is a huge problem. It's also a golden opportunity." World Economic Forum.

Saldaña-Durán, C. E., G. Bernache-Pérez, S. Ojeda-Benitez, and S. E. Cruz-Sotelo. 2020. "Environmental pollution of E-waste: generation, collection, legislation, and recycling practices in Mexico." In *Handbook of Electronic Waste Management*, 421–442. Elsevier.

Saphores, J.-D. M., H. Nixon, O. A. Ogunseitan, and A. A. Shapiro. 2006. "Household willingness to recycle electronic waste: an application to California." *Environment Behavior* 38 (2):183–208.

Savi, D., U. Kasser, and T. Ott. 2013. "Depollution benchmarks for capacitors, batteries and printed wiring boards from waste electrical and electronic equipment (WEEE)." *Waste Management* 33 (12):2737–2743.

Schluep, M., T. Terekhova, A. Manhart, E. Müller, D. Rochat, and O. Osibanjo. 2012. "Where are WEEE in Africa?" 2012 Electronics Goes Green 2012+, 9–12 Sept. 2012.

Selin, H., and S. D. VanDeveer. 2006. "Raising global standards: hazardous substances and e-waste management in the European Union." *Environment: Science Policy for Sustainable Development* 48 (10):6–18.

Sepúlveda, A., M. Schluep, F. G. Renaud, M. Streicher, R. Kuehr, C. Hagelüken, and A. C. Gerecke. 2010. "A review of the environmental fate and effects of hazardous substances released from electrical and electronic equipments during recycling: Examples from China and India." *Environmental Impact Assessment Review* 30 (1):28–41.

Shinkuma, T., and N. T. M. Huong. 2009. "The flow of E-waste material in the Asian region and a reconsideration of international trade policies on E-waste." *Environmental Impact Assessment Review* 29 (1):25–31.

Sinha-Khetriwal, D., P. Kraeuchi, and M. Schwaninger. 2005. "A comparison of electronic waste recycling in Switzerland and in India." *Environmental Impact Assessment Review* 25 (5):492–504.

Souza, R. G. 2020. "16 – E-waste situation and current practices in Brazil." In *Handbook of Electronic Waste Management*, edited by M.N. V. Prasad, M. Vithanage and A. Borthakur, 377–396. Butterworth-Heinemann.

Srivastava, R. R., and P. Pathak. 2020. "Policy issues for efficient management of E-waste in developing countries." In *Handbook of Electronic Waste Management*, 81–99. Elsevier.

StEP. 2014. "Step Initiative Solving the E-Waste Problem (Step) White Paper: One Global Definition of E-waste." Step, accessed 05 October, 2020. www.step-initiative.org/.

Sthiannopkao, S., and M. H. Wong. 2013a. "Handling e-waste in developed and developing countries: Initiatives, practices, and consequences." *Science of the Total Environment* 463–464:1147–1153. doi: https://doi.org/10.1016/j.scitotenv.2012.06.088.

Sthiannopkao, S., and M. H. Wong. 2013b. "Handling e-waste in developed and developing countries: Initiatives, practices, and consequences." *Science of the Total Environment* 463:1147–1153.

Sugimura, Y., and S. Murakami. 2016. "Problems in Japan's governance system related to end-of-life electrical and electronic equipment trade." *Resources, Conservation & Recycling* 112:93–106. doi: https://doi.org/10.1016/j.resconrec.2016.04.009.

Tansel, B. 2017. "From electronic consumer products to e-wastes: global outlook, waste quantities, recycling challenges." *Environment International* 98:35–45.

UN. "E-Waste Management." accessed 05 October, 2020. www.unenvironment.org/ietc/what-we-do/e-waste-management.

UNCRD. "3R Initiative." accessed 05 October, 2020. www.uncrd.or.jp/index.php?menu=388.

Van Eygen, E., S. De Meester, H. P. Tran, and J. Dewulf. 2016. "Resource savings by urban mining: The case of desktop and laptop computers in Belgium." *Resources, Conservation & Recycling* 107:53–64. doi: https://doi.org/10.1016/j.resconrec.2015.10.032.

Vanegas, P., A. Martínez-Moscoso, D. Sucozhañay, P. Paño, A. Tello, A. Abril, I. Izquierdo, G. Pacheco, and M. Craps. 2020. "E-waste management in Ecuador, current situation and perspectives." In *Handbook of Electronic Waste Management*, 479–515. Elsevier.

Veenstra, A., C. Wang, W. Fan, and Y. Ru. 2010. "An analysis of E-waste flows in China." *The International Journal of Advanced Manufacturing Technology* 47 (5-8):449–459.

Veit, R. 2014. Evolution of producer responsibility and product stewardship. In *StEP EWAS*. Shanghai: StEP EWAS E-waste academy, Shanghai.

Wagner, T. P. 2009. "Shared responsibility for managing electronic waste: A case study of Maine, USA." *Waste Management* 29 (12):3014–3021.

Wang, Z., D. Guo, and X. Wang. 2016. "Determinants of residents' e-waste recycling behaviour intentions: evidence from China." *Journal of Cleaner Production* 137:850–860.

Wang, Z., B. Zhang, and D. Guan. 2016. "Take responsibility for electronic-waste disposal." *Nature* 536 (7614):23–25.

Widmer, R., H. Oswald-Krapf, D. Sinha-Khetriwal, M. Schnellmann, and H. Böni. 2005. "Global perspectives on e-waste." *Environmental Impact Assessment Review* 25 (5):436–458.

Yu, J., E. Williams, M. Ju, and C. Shao. 2010. "Managing e-waste in China: Policies, pilot projects and alternative approaches." *Resources, Conservation & Recycling* 54 (11):991–999.

Zeng, X., X. Xu, H. M. Boezen, and X. Huo. 2016. "Children with health impairments by heavy metals in an e-waste recycling area." *Chemosphere* 148:408–415.

Zoeteman, B. C., H. R. Krikke, and J. Venselaar. 2010. "Handling WEEE waste flows: on the effectiveness of producer responsibility in a globalizing world." *The International Journal of Advanced Manufacturing Technology* 47 (5–8):415–436.

3 A Global Outlook on the Implementation of the Basel Convention and the Transboundary Movement of E-waste

Florin-Constantin Mihai*, Maria Grazia Gnoni, Christia Meidiana, Petra Schneider, Chukwunonye Ezeah, and Valerio Elia

CONTENTS

3.1   INTRODUCTION

Waste management is a complex sector involving various industries, institutions, and municipalities, and monitoring of each waste category is a difficult task for national authorities and international organizations. Hazardous wastes must be carefully managed in an environmentally sound manner to reduce environmental contamination risks and related public health threats. E-waste is one of the fastest waste flows due to technological progress and easier access to electronic goods. Daily activities (at work or home) impose the use of several electronic goods feed by a consumer society under the linear economy paradigm which disregards the natural resources depletion and environmental pollution associated with bad waste management practices. The exacerbate consumption patterns of EEE products, particularly in high-income countries, lead to huge amounts of e-waste items on a per capita basis compared to low-income countries (Mihai et al., 2019). The consumption patterns of EEE products emerge in developing countries and the formal e-waste management systems must cope with such increased flows (Gollakota et al., 2020). Second-hand EEE products or e-waste items are subject to transboundary movement from high income to middle and low-income countries often without a sound e-waste management infrastructure. Improvement of international regulations, such as the Basel Convention, is crucial in fighting environmental pollution associated with e-waste flow (Matemilola and Salami, 2020). Africa and Asian countries are most vulnerable to the import of hazardous e-waste items which are further treated in poor equipped facilities while workers are exposed to critical public health threats (Zhavoronkova, 2020; Wang et al., 2020). The Basel Convention aims to supervise the transboundary movement of hazardous waste flow generated by developed countries and to prevent their improper disposal practices in poorer countries. The high-income countries export a higher percentage (6%) of hazardous waste generated compared to low-income countries where export is below 0.5 percent (Secretariat of Basel Convention, 2018). Therefore, this chapter aims to examine the issues related to the transboundary movement of e-waste flow in each major geographical area across the globe (Europe, Asia-Pacific, Africa, and North and Latin America) pointing out environmental implications and socio-economic disturbances. Multi-criteria decision approaches in the e-waste management sector supported by reliable e-waste flows data could lead to adjuted regulations and policies in each country with their particular challenges (Marinello et al., 2021; Borrirukwisitsak et al., 2021). Urban mining and circular economy mechanisms start to emerge as promising solutions to divert the e-waste flow from environmental pollution and to increase resource recovery in the industry (Wiesmeth, 2021). Thus, alternative options related to better e-waste management practices are further discussed to reduce the transboundary movement of e-waste flow and associated environmental pollution threats under the Basel Convention umbrella.

3.2   GLOBAL E-WASTE MANAGEMENT DEFICIENCIES

Mismanagement of e-waste poses serious environmental and public health issues around the world, particularly in low and middle-income countries, through open burning, illegal dumping, or their disposal in mixed municipal solid waste (MSW) landfills (Mihai and Gnoni, 2016; Gangwar et al., 2019). Rudimentary e-waste dismantling activities are performed in Asian and African countries with serious related health issues (Bimir, 2020). These practices are fed by the transboundary movement of e-waste flow from Europe, the US, and other high-income countries. Hot-spots of environmental pollution related to such e-waste disposal sites are identified in countries like Ghana, Nigeria, China, or India (Wang et al., 2020; Singh et al., 2020). There are significant disparities regarding the generation of hazardous waste on a per capita basis between low-income countries 13 kg.per capita.yr^{-1} to 41-42 in middle-income countries peak to 136 kg.per capita.yr^{-1} in high-income countries while the global average is around 58 kg.per capita.yr^{-1} in 2015 (Secretariat of Basel Convention, 2018). However, the hazardous waste flow that must comply with the Basel Convention requirements comprises other waste categories than the e-waste flow. The updated global e-waste monitor data reveals the current geographical disparities in terms of e-waste generation flows as shown in Figure 3.1. The average e-waste generation has a slight increase at the global level compared to 2016, but there are still huge differences between the largest e-waste contributor (Norway) and the lowest e-waste generation rates (0.5 kg.inhab.yr^{-1}) specific to low-income countries particularly in Africa.

EU countries, the USA, Canada, Japan, South Korea, some Gulf countries, Australia and New Zealand generate most of the e-waste flow on a per-capita basis compared to other geographical areas. Norway generated the largest amounts of e-waste items on per-capita basis (26 kg.inhab.yr^{-1}) at the global level. In fact, all Nordic countries have higher generation rates of e-waste in line with higher living standards compared to Balkan countries and Eastern Europe. Eastern Europe and Latin America comprise countries close to the global average (8–9 kg.inhab.yr^{-1}) whereas China is slightly below (7.2 kg.inhab.yr^{-1}) and much lower in the case of India (2.4 kg.inhab.yr^{-1}).

Figure 3.1 shows also the major gap between the Global North and South in terms of per-capita e-waste generation rates. There are 49 countries that generate over 12 kg.inhab.yr^{-1} of e-waste (of which 14 over 20 kg.inhab.yr^{-1}) compared to 96 countries which generated less than 8 kg.inhab.yr^{-1} of e-waste in Asia, Africa, and Latin America. There are 23 countries which generate less than 1 kg.inhab. yr^{-1}, but such countries could be exposed to transboundary movements of e-waste and environmental pollution. These are located in Africa with some exceptions such as Afghanistan (0.6 kg.inhab.yr^{-1}), Nepal (0.9 kg.inhab.yr^{-1}), Solomon Islands (0.8 kg.inhab.yr^{-1}). On the other hand, small island states and touristic destinations are among the highest e-waste generator states across the globe, surpassing 10 kg. kg.inhab.yr^{-1} (Antigua & Barbuda, Saint Kitts, and Nevis, Bahamas, Barbados, etc). In 2019, 17.4 percent of e-waste generated was formally documented and collected, therefore, most of the e-waste stream is susceptible to being processed by the informal sector, to be disposed of in municipal landfills or, to be illegally disposed into the natural environment (Forti et al., 2020). E-waste legislation coverage at the national level plays a key role in safer e-waste management practices and as a starting point to reduce illegal shipments of e-waste to less developed countries. The first e-waste Global Monitor (Baldé et al., 2015) revealed that even high-income countries like the USA, Canada, Israel, Singapore, and New Zealand did not have a special regulation of e-waste stream. This situation has improved since 2013, but there are still high-income countries without a e-waste legislation in force which generate large amounts

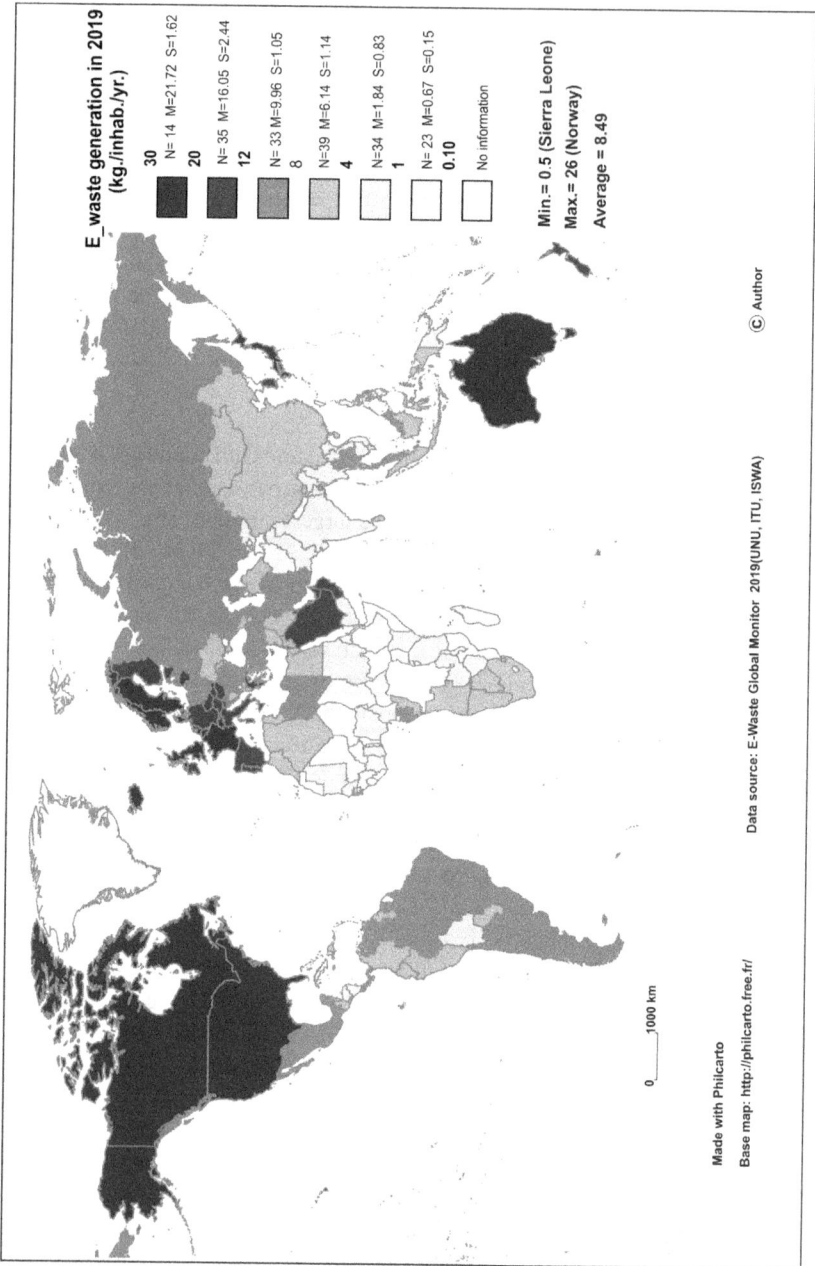

FIGURE 3.1 Geographical Disparities in Terms of e-Waste Generation Rates in 2019. (Author: Mihai F.C. based on E-waste Global Monitor – data for 2019.)

of e-waste, like Bahrain (15.9 kg.inhab.yr[-1]), United Arab Emirates (15 kg.inhab.yr[-1]), Qatar (13.6 kg.inhab.yr[-1]), Brunei (19.7 kg.inhab.yr[-1]) or New Zealand (19.2 kg.inhab. yr[-1]).On the other hand, some countries had e-waste legislation according to previous global monitoring reports but not in 2020 (Russia Federation, Bhutan, Vietnam) or there are not available data (e.g. Cuba) which may suggest some inconsistencies in the reporting system. A widespread e-waste legislation coverage is just a step forward which must be further supported by better law enforcement efforts at regional and local levels and on the other hand, urban and rural communities to be served by adequate e-waste management facilities. In the last years, several countries around the world adopted special e-waste regulations that were in force in 2019 like Argentina (South America), Egypt, Ghana, Rwanda, Zambia, Tanzania (Africa), Iran, Malaysia, Thailand, Mongolia, and Sri Lanka (Asia). In Europe, Montenegro, Belarus, and Moldova are among the latest countries that adopted an e-waste legislation framework. However, Armenia and Azerbaijan are not yet covered by such legislation. Also, small island states and touristic destinations with high per-capita e-waste generation rates (Bahamas, Barbados, Grenada, Antigua, and Barbuda) or large countries like Brazil (> 10 kg.inhab.yr[-1]) have not adopted special regulations regarding the e-waste flows. In such circumstances, high-income countries and emerging economies without a proper e-waste legislation framework are more susceptible to transboundary movement of e-waste to less developed countries or poor e-waste management practices (e.g. open burning/dumping, crude processing of e-waste, disposal in municipal landfills). However, e-waste legislation must be updated worldwide taking into account the feedback received from stakeholders in the industry.

3.3 BASEL CONVENTION: STRENGTHS AND LIMITATIONS

Most low-income and middle-income countries do not have a comprehensive database regarding waste flows at national and subnational levels, therefore, the monitoring process of all waste categories (including e-waste) is difficult to achieve at the international level. Eurostat, UNEP, and OECD statistics provided some reliable data at the national level regarding municipal waste flow. Global analysis of waste management status recognized the gaps in waste statistics data (Kaza et al., 2018), and the need to compile a reliable database regarding various waste flows and basic indicators (e.g. D-Waste Atlas). In this regard, the project of *E-waste global monitor* with three editions (Balde et al., 2015; Balde et al., 2017; Forti et al., 2020) is a crucial forward step to providing a global snapshot of complex interactions regarding e-waste flows. However, the knowledge gap remains a key issue concerning the estimation of e-waste flows at national and subnational levels in many parts of the world. The ratio of e-waste/ second-hand EEE transboundary movements is assumed to be 7–20 percent of e-waste generated (Forti et al., 2020). Hazardous and non-hazardous waste categories are predisposed to be transboundary, moved from the country of origin to be recycled, recovered, treated, or disposed of in another country. Despite the advanced waste management systems, developed countries often exported the e-waste flow to low and middle-income countries that were facing severe environmental pollution threats. To counteract such practices and to offer a specific legal framework and guidelines regarding the transboundary movement of hazardous wastes, the Basel Convention was initiated since 22 March 1989 and entered into force on 5 May 1992 (UNEP, 1992). A revised version was established in June 2018 (Basel Convention, 2018). The Ban Amendment was introduced in 1995 which prohibits the transboundary movement of hazardous wastes destined for disposal or resource recovery operations from countries stipulated in Annex VII (e.g. OECD, EU, Liechtenstein) to other countries around the world. However, The Ban Amendment entered into full force on 5 December 2019 and new amendments regarding the plastic waste will become effective as of 1 January 2021 (Secretariat of Basel Convention, 2020). International cooperation is crucial to reach common ground among parties of this Convention, but also it can be a lengthy and bureaucratic process (Takayoshi and Huong, 2009; UNTC, 2019). Basel Convention organization compiles the most comprehensive database regarding the transboundary movement of hazardous waste across the globe despite the many gaps in the data. However, a global assessment of hazardous waste flow is made by combining such data from the Basel Convention with UN Statistics Division, OECD statistics, and Eurostat focusing on import and exporting issues around the world. The most recent edition was released in 2018 based on data provided by member parties available between 2007 and 2015. This report reveals that the volume of transboundary movement of wastes increased from 9.3 to 14.4 million tons (MT), but 97 percent of hazardous waste generated in a country stays in that country with some exceptions (Secretariat of The Basel Convention, 2018). These efforts must be continuously supported to meet the aims of the Basel Convention in the medium and long term. Also, African countries initiated the Bamako Convention as a more suitable regional treaty that prohibits any import of hazardous waste (including radioactive wastes) in African countries. However, the lack of a clear definition of "e-waste" in the Basel Convention besides loophole, ambiguous articles favored the informal trades of electronics around the world (Wang et al., 2020). Basel Convention classifies wastes as hazardous or non-hazardous depending on the waste's chemical properties and does not cover all forms of used or end-life electronics (Salehabadi, 2013). Therefore, the Convention principles and the law enforcement may vary from one country/region to another, which makes it more difficult to monitor the formal or illegal e-waste shipments around the world. Baldé et al. (2016) summarize the key issues related to the Basel Convention reporting systems on transboundary movement of waste flows under the Article 13 of the Convention: (i) incomplete reporting – several countries do not submit their annual report, (ii) ambiguous definitions – regarding hazardous and non-hazardous waste fraction among countries which makes it difficult to compile a comprehensive database at global scale, (iii) incorrect categorization – despite the Annex I, Annex VIII, and Annex IX of the Basel Convention which lists the hazardous and non-hazardous waste categories to be controlled, (iv) discrepancies in reporting – related to the amounts of waste in national reports compared to those covered by movement documents, (v) data inaccuracies – as an example the same transboundary shipment may have different amounts registered by the exporting country and the importing country. On the other hand, even data submitted by the same state can vary over the years, therefore, time-series data and cross-country analysis must be made with caution. In fact, only 50 percent of Convention members fulfill their obligations to transmit the annual national reports, of which 20–25 percent provide quantitative data related to the hazardous waste flows (Secretariat of Basel Convention, 2018). Also, the trade-in second-hand EEE and illegal shipments are not captured by the reporting of the Basel Convention (Baldé et al., 2016), therefore, the transboundary movement of e-waste flow is only partly known (Palmeira et al., 2018). The implementation guidelines of the Basel Convention and Waste Shipment Regulation (EU) depend on national and subnational levels by financial resources and training of officials (STEP, 2013). Better cooperation and participation among countries must be supported to improve such basic data, otherwise, the magnitude of transboundary movement of e-waste and other waste categories will remain unknown. Also, the intelligence sector must collaborate on transboundary movements of e-waste flow between the EU and targeted destinations (Africa, Asia, Latin America, or Eastern Europe) via existing platforms such as IMPEL, ENFORCE, and INTERPOL's Environmental Crime programs (BAN, 2019). The lack of uniform global standards for the manufacturing of EEE and recycling of e-waste feed the transboundary movements and illegal trades of electronics (Patil and Ramakrishna, 2020). Reaching a consensus between world countries is difficult and time-consuming, but proper funding in combating illegal e-waste trade and increasing penalties for environmental crimes could enhance the effort of each country in the right direction (Forti et al., 2020).

3.4 TRANSBOUNDARY MOVEMENTS OF E-WASTE IN MAJOR GEOGRAPHICAL AREAS

3.4.1 EXPORT AND IMPORT ISSUES OF E-WASTE FLOW IN EUROPE

According to Epo (2019), based on the Basel Action Network (BAN), European e-waste is still illegally exported to Asian and African countries despite strict regulations and international bans. Employees of the organizations equipped 314 old LCD and tube monitors, PCs, and printers with GPS tracking devices in ten EU countries. Therefore, 19 (6 percent) of the 314 devices were exported, 11 of them to African and Asian countries (BAN, 2019). BAN documented the journeys of illegal electronic waste from delivery to mostly municipal collection points to disposal. People in developing countries who dismantle electronic waste are often unprotected and therefore exposed to harmful substances. For the study, BAN provided old devices in Belgium (29 devices), Denmark (20), Germany (54), Great Britain (39), Ireland (24), Italy (48), Poland (20), Austria (18), Spain (45) and Hungary (17) with GPS trackers (BAN, 2019). Most of the signals came from devices in the UK (5 devices), followed by Denmark (3) and Ireland (3). Five devices arrived in Nigeria, one in Ghana and one in Tanzania. Discarded electronics also went to Thailand, Hong Kong, and Pakistan as well as Romania and Ukraine. Of the ten countries involved, only Hungary did not have any exports. The old devices were equipped with GPS trackers between April 15 and September 2, 2017, and handed over to the collection points. According to the present study, German electrical recycling scores well in a European comparison. Fifty-four old devices were handed in by BAN to collection points and recycling centers in Cologne, Hamburg, Frankfurt, Berlin, Leipzig, Dresden, and Munich. An LCD screen from Dresden traveled almost 11,000 kilometers to an illegal landfill about 134 km southeast of Bangkok. According to the German National Environmental Agency, there are still gaps in the enforcement of international bans (Umweltbundesamt, 2010). The CWIT project estimated that 1.3 MT of e-waste departed the EU through undocumented reports, and 1,6 MT e-waste was exported (Huisman et al., 2015).

3.4.1.1   EU Countries

The waste hierarchy supports the prevention of waste, followed by reuse, recycling, recovery, with landfilling and incineration being the last resort. The needs in the waste sector are many and include the elaboration of national waste strategies, policies, and action plans adapted to the current realities of limited resources and thus the need for resource efficiency, high financial investments, reinforcement of the role of the local and regional authorities and involvement of multiple actors in the implementation of the policy. To address these problems two approaches of legislation were enforced: The Directive on waste electrical and electronic equipment (WEEE Directive 2002/96/EC) and the Directive on the restriction of the use of certain hazardous substances in electrical and electronic equipment (RoHS Directive 2011/65/EU). WEEE does contain a large number of substances and materials. If old electrical and electronic equipment is not properly disposed of, environmental risks can arise due to the pollutants that are still present. In addition to pollutants such as heavy metals and CFCs, WEEE also contains several valuable substances that have to be recovered and thus recycled. If waste electrical and electronic equipment is disposed of properly, primary raw materials (and thus their costly extraction) can be replaced and a significant contribution can be made to conserving natural resources. New EU countries facing the transition from a traditional waste management system based on mixed collection schemes and landfilling towards regional integrated e-waste management systems which promote the circular economy principles (Mihai, 2019). The movement of waste across borders is regulated at both EU (*via the Waste Shipment Regulation* and) global levels *through the Basel Convention.* For the shipment of used electrical and electronic equipment, which is presumably used electrical and electronic equipment, minimum requirements are set which serve the purpose of preventing the export of non-functioning electrical and electronic equipment to developing countries (Nordbrand, 2009). Annex VI of Directive 2012/19/EC specifies minimum requirements for shipments in cases in which the owner of an object claims to want to or to move used electrical and electronic equipment and not waste electrical and electronic equipment. To be able to distinguish used devices from old devices, the owner must submit the following documents to prove this claim if required:

- A copy of the invoice and the contract for the sale of electrical and electronic equipment,
- Proof of functionality for each package within the shipment together with a test report,

- Declaration by the owner that the equipment is not waste,
- Adequate protection against damage during transport and during loading and unloading,
- A relevant transport document, e.g. CMR consignment note and
- A statement by the liable party on his liability.

If the relevant documentation is missing and if there is no adequate protection against damage during transport and during loading and unloading, the device is to be regarded as waste electrical and electronic equipment, i.e. it can be assumed that it is an illegal shipment. Under these circumstances, the cargo is treated by Articles 24 and 25 of Regulation (EC) No 1013/2006, which usually means the monitored return of the waste to the country of dispatch. Exports of all types of waste to the following countries and imports of all types of waste from the following countries are prohibited: Angola, Fiji, Grenada, Haiti, San Marino, Sierra Leone, Solomon Islands, Tajikistan, Timor, Tuvalu (UN members that are not Basel states), USA (only applies to disposal, as the U.S. is an OECD member country), Vanuatu. Kosovo and South Sudan are neither UN members nor contracting parties to the Basel Convention. GPS tracking systems combined with port investigations and intelligence collaborations should be the norm across the EU to ensure that e-waste movements fulfill the legal requirements (BAN, 2018). Also, the CWIT project recommends National Environmental Security Task Force (NEST), formed by different authorities and partners, to enable rapid and collaborative law enforcement response coordinated at international, national, and subnational levels (Huisman et al., 2015).

3.4.1.2   Show Case Lower Bavaria, Germany

In Lower Bavaria, an important transit region for cross-border waste shipments, controls on cross-border waste shipments have been intensified since 2010. Above all on the A 3 federal motorway with an average daily traffic volume of more than 15,000 trucks, numerous waste transports take place in the neighboring Eastern European countries, but also from there via the ports in Hamburg and Antwerp to Africa and Asia (Aiblinger-Madersbacher, 2016). According to Section 14 (2) AbfVerbrG (waste disposal legislation), the government of Lower Bavaria is the competent authority for the control and monitoring of cross-border waste shipments in its own administrative district. According to Aiblinger-Madersbacher (2016), contact with foreign authorities is extremely important in connection with the illegal waste transports. The responsible authorities at the place of dispatch and the destination of the waste transport ultimately decide on the further course of action about the suspected cases reported by the control area authority (usually return or recovery in Germany). The government of Lower Bavaria therefore also participates in EU-wide joint controls and exchange programs run by the IMPEL (Aiblinger-Madersbacher, 2016). IMPEL (Implementation and Enforcement of Environmental Law) is a European network in which all EU member states, as well as Norway and candidate countries of the EU, are represented. Its members are environmental authorities responsible for the implementation of European environmental law. The IMPEL / TFS working group (Transfrontier Shipment of Waste) for the area of cross-border waste shipments exists within the framework of this network. IMPEL has set itself the task of improving the implementation of European environmental law. The network serves the exchange of experience and information between prison practitioners in topic-related events as well as in specific projects (Aiblinger-Madersbacher, 2016). By participating in national and international working groups, the attempt is made to harmonize the classification of used goods as waste/non-waste, at least at the EU level, and to bring in their own experience from enforcement practice. Only through uniform enforcement is it possible to effectively combat illegal waste shipments. In 2014, the annual IMPEL TFS EA Best Practice Meeting was held at the government of Lower Bavaria. Representatives from 22 EU Member States took part. Important items on the agenda were the latest developments in the control of cross-border waste shipments (the type of illegal shipments, transport routes) and the harmonization of waste classifications (Aiblinger-Madersbacher, 2016).

3.4.1.3 Non-EU Countries

Non-EU countries can be divided into countries without intention for EU membership (Iceland, Liechtenstein, Norway, Switzerland, Andorra, San Marino, Belarus), candidate countries on the road to EU membership (Albania, Montenegro, North Macedonia, Serbia, Turkey), and potential candidates to become EU member states (Bosnia and Herzegovina, Kosovo, Ukraine). The United Kingdom left the European Union in 2021. However, the UK implemented all European e-waste management regulations and is supposed to keep them on that level. The European non-EU countries are all contracting parties to the Basel Convention, except Kosovo that is neither a UN member nor a contracting party to the Basel Convention. However, the European Commission, in collaboration with the Kosovo government, works on the integration of Kosovo, being part of the Western Balkan countries into the collaboration with the other Western Balkan countries that are contracting parties to the Basel Convention and involved in the European waste management acquis. Also, such countries have adopted special legislation regarding the e-waste special waste stream. Despite the fact Norway has the highest per-capita e-waste generation rates across the world, 85 percent of e-waste is recovered and used to develop new products instead of using new raw materials (www.environment.no). In Eastern Europe, Ukraine is implementing the EU Directives which will improve the waste management practices including WEEE, Batteries and Accumulators, Waste Packaging, and Municipal Waste. The Western Balkans face two main problems related to *WEEE the disposal which is mostly to landfill and the suboptimal recycling and recovery of WEEE resulting* in a loss of significant valuable recyclable resources and damage to health and the environment. According to Eunomia projections in the report of the European Commission (2017), based on previous work by the United Nations University, an estimated 177 000 tonnes of WEEE is forecasted to rise across the West Balkan Countries in 2030, up from an estimated 130 000 tonnes in 2014. The greatest proportion of WEEE comes from small and large equipment, followed by temperature exchange equipment. As an example, Economic development in Bosnia and Herzegovina brought an increase of hazardous waste materials which is, except for radioactive waste, landfilled together with the rest of municipal solid waste. Bosnia is now trying to solve the problem by shipping hazardous wastes abroad for incineration, but the process is running slow due to low landfilling fees and the absence of hazardous waste law. Kemis BH in

Bosnia is the largest exporter of hazardous waste from Bosnia and Herzegovina to the European Union. They have so far exported 2000 tons of hazardous waste, while their storage capacities include a storage area of 10 000 square meters. They collect all types of hazardous waste, but they focus mostly on WEEE, which has risen significantly in recent years. According to some estimations, Bosnia now produces 27 000 tons of WEEE or 7.8 kg per capita (Forti et al., 2020), for which there are no special landfills or legal disposal and treatment solutions.

3.4.2   EXPORT AND IMPORT ISSUES OF E-WASTE FLOW IN USA AND CANADA

Estimating e-waste flows between developed and developing countries is hard work for several reasons (Ilankoon et al., 2018): from the different national legislations to organizational and technical procedures adopted to move the e-waste stream. Starting from a legislative point of view, Canada and the USA are characterized by similar e-waste management systems, which are mainly based on State/Province legislation and a coordination and control institution, i.e. Canada Environment and US EPA, respectively. Canada's e-waste flow management system is mainly based on the Basel Convention. Canada participated in the definition development of the Basel Convention about the transboundary movement and disposal of hazardous wastes that occurred in 1989; the USA has signed in 1990. Under this legislative framework, several agreements have been defined. As an example, Canada and the USA have defined a comprehensive bilateral agreement on the transboundary movement of hazardous waste also following the guidelines proposed by the OECD: pilot projects have been developed to monitor the effectiveness of this agreement (Government of Canada, 2014). In addition, e-waste flow management is regulated by the NFTA CEC, an agreement between Canada, Mexico and the USA: common procedures and waste quantities exchanged between the three nations are monitored (CEC, 2011). The USA has also defined specific pilot projects with other nations to monitor e-waste flows: one example is the collaboration defined by the US EPA and the Taiwan Environmental Protection Administration (i.e. the Taiwan EPA) aiming to support initiatives for more sustainable recycling of e-waste (Osibanjo and Nnoom, 2019). Although some data is available from these sources, an estimation of the total amount of e-waste exported from the USA and Canada to other nations is still lacking due to several factors: from the multiplicity of organizational systems adopted at local levels (Kahhat et al., 2008) to the complexity of the trade networks (Lepawsky and McNabb, 2010). Also, illegal trades are still affecting e-waste flows (thus contributing to making uncertain the waste flow assessment (Efthymiou et al., 2016). A recent attempt to monitor e-waste trade has been proposed in Petridis et al. (2020) where network analysis was applied on data extracted from the UM COMTRADE database for only a specific e-waste stream.

3.4.3   EXPORT AND IMPORT ISSUES OF E-WASTE FLOW IN AFRICA

E-waste is an increasingly growing global problem because a significant part of e-waste produced worldwide is reported to be recycled by the unregulated activities of the informal sector, mostly in developing countries, resulting in significant nega-tive impacts in terms of human health and the environment (Ezeah et al., 2013). International exports of e-waste occur as part of a wider global waste trade that has been described as indicative of global inequalities and the most glaring evidence of overconsumption in the twenty-first century (Schmidt, 2006).

3.4.3.1   Sources of E-Waste Imported into Africa

Recepient African countries are without doubt negatively impacted by the transboundary movement of e-waste. The region is emerging as a destination of choice for the dumping of such waste. The trend is mostly driven by the high level of con-sumption of used and discarded electrical and electronic equipment (EEE), imported from mostly the US and the EU. In certain cases, e-waste imports have been quite key in bridging the difference in digital access between Africa and Western countries (the digital divide). E-wastes imported into Africa mostly passes through port cities such as Lagos, Mombasa, Dar es Salaam, and Cairo. For instance, it is estimated that 500 shipping containers loaded with used electronic equipment pass through Lagos each month (Terada, 2012). Each container could hold an average of 800 com-puter monitors or central processing units (CPUs), or 350 large TV sets. It has been reported that anywhere between 25 percent to 75 percent of these figures could be useless. Based on these, it is estimated that as much as 100 000 used computers or CPUs, or 44 000 TV sets, enter Africa each month through Lagos alone (Amechi and Oni, 2019; Terada, 2012). Of the 20 million to 50 million tons of e-waste generated globally yearly, it is estimated that 75 percent to 80 percent are shipped to countries in Asia and Africa for "recycling" and disposal (Perkins et al., 2014). It is important to mention also that in absolute numbers, Nigeria dominates sub-Saharan African countries in the total amount of used and new EEE imports, total number of EEE in use, and by extension, total amount of e-waste generated. Collection rates of e-waste vary significantly amongst African countries. Almost all the collected materials are transported to warehouses or waste dumps operated by the informal recycling sector. Some of the most notorious of these 'digital dumping grounds' are the Computer Village in Lagos, Nigeria, and Agbogbloshie, on the outskirts of Accra, Ghana, where young, uneducated, and jobless people are involved in informal recycling activities (Bimir, 2020). According to the Basel Action Network, Nigeria and Ghana should prohibit all imports of non-functional electronics (BAN, 2018).

3.4.3.2   Recycling and Reuse of E-Waste in Africa

Transboundary shipment of e-waste from Developed to African countries, particu-larly Nigeria, Ghana, and Kenya, has become more prevalent over the years. On the other hand, economic situations in some destination countries sometimes encourage the import of e-waste for recycling, to provide some sort of short-term economic benefits. However, most African countries have no access to sound recycling facil-ities and disposing site of e-waste (Perkins et al., 2014). As a result, recycling processes in these countries frequently rely on rudimentary techniques of informal sector recyclers, to extract valuable materials from e-waste (Terekhova, 2012; Baldé et al., 2017). This process, mostly carried out by young men or women, involves the physical dismantling of unserviceable electrical and electronic equipment, by using tools such as hammers, chisels, and screwdrivers (Vaccari et al., 2019). Printed circuit boards are sometimes heated to remove valuable components such as gold and other precious metals. Another popular method involves the stripping of metals in open-pit acid baths. Apart from metals, other recovered items include plastics and non-ferrous materials. These items are either chipped or melted without appropriate personal protective equipment (PPE). Burning electrical cables, and vehicle tires often in open pits to retrieve copper and steel is a very common crude recycling practice which threaten the workers' health and feed environmental pollution (Perkins et al., 2014). Ultimately, these rudimentary methods encourage the recovery of materials that are only worth a fraction of the total potential and rob African countries of significant economic and environmental benefits. When large organizations based in developed countries, export e-waste to African countries without capacities for recycling or other forms of sustainable management, they miss the opportunity to establish safer, cleaner, and sustainable global waste management systems. African countries triggered a regional shift from the Basel Convention, arguing that their particular circumstances were neither considered nor addressed by the Convention. This was the basis for the founding and adoption of the Bamako Convention, which prohibits the import of all hazardous and radioactive wastes into the African continent as well as the ocean or inland waste disposal practices (UNEP Bamako Convention). South Africa and Egypt are developing some formal e-waste recycling activities but the informal sector prevails in Africa (Bimir, 2020).

The management of waste electrical and electronic equipment is a rapidly growing global challenge, mainly because e-waste from major producer regions such as the US and the EU are oftentimes transported to developing countries for reuse, recycling, or disposal. Most receiver developing countries however have little capacity for managing e-waste stream sustainably. Ultimately, this vital function is left for the unregulated informal sector, resulting in significant negative impacts in terms of human health and the environment. Given the critical role of the informal sector in most African countries, there is a need to effectively connect informal collectors to a formal recycling structure, which must provide necessary skills and proper facilities. Finally, there is an urgent need for major e-waste producing nations such as the US as well as major receiving countries such as Nigeria to ratify key e-waste management instruments such as the Basel and Bamako Conventions respectively.

3.4.4 Export and Import Issues of E-Waste Flow in Asia

In countries where regulation lacks and the management is poor, the burden is much more than in countries with better management. For example, China and India as the main producers and exporters of the world's electronic devices have a higher risk of incidence of environmental pollution and water-borne disease caused by heavy metal contaminated soil or water. This is worsened by the fact that China and India import scrap appliances, such as projectors, mobile phones, typewriters, CPUs, etc from different countries worldwide (Borthakur & Govind, 2017). A large number of consumers and weak laws are the reasons for continuing WEEE import through the availability of related regulations (Lu et al., 2015). According to joint study result launched in September 2017, China, Japan, and India are the top three WEEE generators in Asia amounted to 6.0 Mt, 2.2. Mt and 1.7 Mt respectively (ASSOCHAM, 2018). However, China and India also belong to the Asian biggest recipients of WEEE from developed countries beside other Asian countries such as Pakistan and African countries such as Ghana and Nigeria (Mmereki, et al., 2016). In this chapter, WEEE management and movement in four country groups will be discussed.

3.4.4.1   China

It is estimated that about 160 million electronic devices are discarded yearly in China. China had a reputation as the largest e-waste dumping site and the largest importer of e-waste in the world. Data showed that almost 70 percent of global e-waste ends up in China where most of the world's largest dumpsites are homed (Zeng, X, et.al., 2016). Therefore, it is a burden for China because of a twofold problem, namely the national growing WEEE generation and the continuous flow of WEEE import. It is common in China that WEEE can be either discarded when mixed with domestic waste or be sold to the middlemen as a used item or be left abandoned at home. WEEE producer is not compelled to deliver a waste report officially. WEEE generation in China is increasing because of prompt urbanization, technology development, and innovation as well as economic development. In 2030, waste generation in China will be approximately 28.4 million tons putting China as the largest global e-waste producer and it replaces USA (Zeng, X, et.al., 2016). The composition of the WEEE is mainly 26% air conditioners, 24% televisions, 14% computers, 12% refrigerators, 7% washing machines, 9% printers, and 7% fluorescent lamps (Lu et al., 2015). Annually, WEEE production in China increased by 25.7%. This percentage was much beyond the average global increase which is approximately 4−5% (Zeng et al., 2016). Since China is the largest population globally, it becomes the main generator of WEEE. Although the Chinese central government has ratified the Basel Convention in the 1990s and banned the e-waste import by enacting strong regulations in 2002, China is still the leading country in the world for WEEE dumping. Illegal smuggling of WEEE continuously took place through regions with legal lack including Hong Kong (Lu et al., 2015). Estimation of WEEE import is between 1.5 and 3.3 million tons per year which arrive in Chine through illegal channels and Part of it is exported again; Wang et al. 2016). The informal sector is the main recycler in China and most of them are recycled the WEEE in a minimum standard often neglecting their own safety and inefficient (Chi et al, 2011; Orlins and Guan, 2016). Therefore, the government has enacted a WEEE regulation in 2011. In 2012, the Central Government launched the WEEE treatment fund to promote efficient recycling activities through legalized channels. The scheme supports authorized enterprises to improve their competitiveness to get better-standardized treatment and addressed only to five categories of WEEE, i.e., the television, personal computer, Ac and washing machine. In 2015, the government launched a new Catalogue of WEEE Recycling covering more WEEE categories from 5 to 14 mostly are household equipment such as mobile phone, monitor, printer, copy machine, etc. China has its reputation as the world's dumping site until 2018 when regulation prohibits imports of solid waste brought in (Wang et al., 2018). The regulation was tightened by prohibiting the import and informal recycling of WEEE. Import of WEEE is showing a reduced trend since then. It is expected that the WEEE production decreased to 0.32 million tons in 2018 and becomes zero in 2023 (Zeng, X. et.al 2016). Illegal importation of e-waste still enters China due to regulatory loopholes and imperceptible smuggling networks.

3.4.4.2 ASEAN Countries

In all ASEAN countries, except Brunei Darussalam, industrial and hazardous waste has been regulated both at the national and local levels. However, emerging waste streams like WEEE are not specifically addressed. This is one of many reasons why WEEE import and export still occurs in ASEAN. The amount of waste import entering ASEAN has been increasing since China banned the import of 24 types of plastics. In 2018, a further 32 types of solid waste were banned. Consequently, the global waste flows shifted from China to Southeast Asia-including Thailand, Malaysia, Indonesia, and Vietnam. Not all ASEAN countries have adequate facilities to recycle and to treat the WEEE. Myanmar, Laos, and Cambodia have no recycling infrastructure for WEEE recycling and treatment. However, it does not confirm that the other countries have recovery facilities that meet the quantity and quality standards. Indonesia, for example, though the existence of the Extended Producer Responsibility (EPR) mechanism, the implementation is very weak. Only a small amount of the WEEE flows into this mechanism. The rest cannot be traced. A basic Act on Environment and its related regulations such as waste management law, regulation on air and water exist in Indonesia, but no specific regulation regarding WEEE. WEEE plan and projects are at the development stage so far and being covered under a 10-year roadmap for EPR implementation. The government of Indonesia (GoI) does not regulate WEEE import specifically and GoI permits used EEE import but only for computer and monitor after complying the following requirements: usable (proven by certificate), maximum 5-year lifetime, new technology (LCD and LED monitor), one set delivery and proper packaging. In Thailand, the import of WEEE for final discarding is prohibited, while the import of WEEE for recycling is allowed with restriction. Meanwhile, there is no ban on the import of used EEE for direct reuse, repair, and refurbishment. China has been the top exporter to Thailand of used printed circuits boards and used electronic integrated circuits. Other countries belong to the top countries exporting used EEE to Thailand during the period were Hongkong, Canada, and Singapore. The same trend for WEEE, i.e. used batteries, scraps of electrical machinery, used PCs and data processing machines, old circuit boards, and used electronic circuits, is recorded by the Thai Ministry of Commerce. USA, China, and Japan have exported used batteries and scraps of electrical machinery to Thailand with a volume of 25,560 kg, 1.4 million kg, 41,380 kg respectively in 2014. In 2018, this amount imported from the USA, China, and Japan increased to almost 11.8 million kg, 1.84 million kg, 1.64 million kg respectively. The ban to import a wide range of waste including WEEE and used EEE to China after 2017 is one of some factors for the inclining trend of used EEE and WEEE entering into Thailand. Import and re-export WEEE and UEEE are illegal in Vietnam. However, a loophole in Vietnamese legislation allowing the import of second-hand electronics for re-export let the flow of WEEE between Cambodia,

China, and Vietnam. An absence of common definitions of WEEE and UEEE is a factor also why the flow of WEEE between China and Vietnam still exists (Premalatha et al. 2014). In Vietnam informal sector play important role in WEEE collection and recycling to reduce WEEE generated domestically or imported from other countries through illegal channels (Chi et al. 2011; Hai et al., 2015).

3.4.4.3 India

In 2011, Rules on E-waste covering e-waste management and handling were enacted (Lee, 2018). The rule was renewed in 2016 to reduce e-waste production and increase recycling and introduced the concept of EPR. Under the rule, unauthorized WEEE import is banned and producers were targeted to collect back 30 percent to 70 percent (over seven years) of the WEEE they produce and to guarantee that the waste was channelized to authorized recyclers. Targets were revised through the enactment of the E-waste (Management) Amended Rules in March 2018. The target was reduced from 20 to 10 percent for 2016–2017, while a reduction target of 20 percent remains the same for 2017–2018. This target gradually increases to 70 percent in the seventh year. Seven hundred and twenty-six industries have EPR authorization provided by the Central Pollution Control Board (CPCB) in 2018. The authorization is valid for five years from the date of issue and determines the collection targets for the specified time (five years). Though the available WEEE related regulations, the percentage of WEEE recycled by formal recyclers is still low which is only 1.5 percent of the total WEEE. In India, about 75 percent of WEEE are kept domestically instead of bringing them back to producers due to the unclear procedures of disposing WEEE limiting the recycling programs (Borthakur and Govind, 2017b) and about 8 percent of WEEE is useless and disposed of in landfill (Gark and Adhana, 2019). Though the enacted rules in 2011, unauthorized WEEE still flows into India because of weak regulation for used EEE, smuggling, and an increasing trend of informal recycling sector dominated currently in India (Ghosh et al., 2016; Arya and Kumar, 2020). According to Breivik et al. (2014), India imports about 0.85–4.2 Million tons of WEEE from other countries in 2014. This amount is higher than the total domestic WEEE generation which is 0.36 Million tons (Breviek et.al 2014). WEEE increases significantly in India contributed by both WEEE national generation and smuggling (Borthakur and Sinha, 2013; Garlapati, 2016).

3.4.4.4 Japan

Japan is the major WEEE exporter and the export is approximately 30 percent of the total domestic WEEE generation, while only low percentage of WEEE regulated under the Home Appliances Recycling Law is illegally disposed within the country. Asian developed countries such as South Korea) and Asian developing countries i.e., China, the Philippines, India, Thailand, Vietnam, and Pakistan, are the common destinations for WEEE export from Japan (Fuse et al. 2011). About 50 to 60 percent of WEEE generation is collected and recycled within the country (Hotta et al. 2014). Recycling rates are increasing ranging between 75 and 92 percent in 2014 exceeding the legally defined recycling rates, for the four specified household appliances (Kankyōshō and Keizai sangyō shō 2015). Japan is one of the early adopters of targeted WEEE legislative and regulatory frameworks, promoted by a change towards environmental policies in the country. In Japan, the informal sector does not play a significant role as it is in the immensely regulated and well-managed WEEE flow. The informal sector can collect appliances directly from consumers and export them to other countries of the region, including China, Vietnam, Cambodia, and the Philippines because WEEE export is allowed (Hotta et al. 2014). It is estimated that 0.62 million-ton WEEE is exported in 2005 (Breivik et al., 2014). Japan has the most comprehensive and most effective legal framework particularly the Home Appliance Recycling Law (Hotta et al. 2014).

3.5 PATHWAYS TO REDUCE TRANSBOUNDARY MOVEMENTS OF E-WASTE FLOW

3.5.1 DEVELOPMENT OF URBAN MINING PRACTICES

The concept of Urban Mining (UM) is a recent evolution of the well-known concept of landfilling mining based on facilitating resource extraction for maximizing the resource and economic value of urban waste streams, from used products to end-of-life items, such as buildings, WEEE, etc. (Tunsu et al., 2015; Ragazzi et al., 2017).

One promising waste stream for UM is EEE for different reasons: starting from the high and increasing quantity of this waste stream together with the presence of rare and precious metals, which could contribute to providing scarce raw materials essential for hi-tech devices (Binnemans et al., 2013). Even if the quantity used in technological devices of Rare Earth Materials (REE) – such as lanthanum, cerium, etc. – together with critical metals (indium, lithium, and other precious metals) is very small, their absolute demand is quickly increasing over the last years (Zhang et al. 2017) thus contributing to potentially support UM diffusion all over the world. The valuable materials stocked in urban waste, especially WEEE, may currently represent a significant source of resources, with concentrations of elements often comparable to or substituting natural stocks (Lederer et al., 2014). Effectively recovery from end-of-life products could contribute to both reduce the environmental impact of raw materials and to develop new businesses (Gidarakos and Akcil, 2020). UM could also be an "enabling process" for applying circular economy strategies in urban WEEE streams. Xavier et al. (2019) defined UM as a sort of closed-loop supply chain management approach that could contribute to increasing the overall sustainability – from environmental, economic, and social dimensions- of WEEE management. Moreover, it has to be noted that the economic feasibility has to include both the assessment of the recovery process and revenues evaluation from extracted materials (Cossu and Williams, 2015).

Thus, wider and efficient diffusion of UM will provide a significant contribution to the development of sustainable cities (Brunner, 2011; Arora et al., 2017) by increasing the recovery potential of urban areas, especially by focusing on specific waste streams. Recent studies have also pointed out the potential benefits of UM for emerging countries. Quantitative analysis for the metropolitan region of Rio de Janeiro is proposed in Ottoni et al. (2020) outlying the contribution of UM for a more circular and sustainable WEEE management. Recently, Gunarathne et al. (2020)

discussed the potentiality of UM for WEEE in Sri Lanka by focusing on factors –
such as a large number of stakeholders involved and macro-environmental factors
(e.g. the presence of an informal collection system, the financial un-attractiveness
of the recycling industry) – which are contributing to increase the complexity of
its application in emerging economies. A recent critical analysis of the application
of UM for WEEE in emerging countries is in Kazançoglu et al. (2020): authors,
through expert analysis, outlined as the lack of government support as the main barrier to be overcome for a wider diffusion of these practices. From a strategic point
of view, and underdevelopment national legislation for supporting UM for WEEE is
also a contributing factor. Most criticalities of UM, at the operational level, depends
on two main issues: the efficiency of current recycling treatments and reliable data
on material composition. Both issues contribute to determining the actual rate of
recycling of the UM process. The concurrent critical issue to be considered is the
economic feasibility of the recovery process: recent studies evaluate this issue from
a different point of view, i.e. by evaluating technological as well as organizational
implications. Technological issues regard mainly the recovery process from disassembly to purification of extracted materials. Organizational issues affect collection
services, from transportation to storage. A pioneering study (Krook et al., 2011)
analyzed obsolete stocks of copper "hibernated" in the local power grids of two
Swedish cities: in addition, an economic analysis about the cable recovery is also
discussed. Van Eygen et al. (2016) applied the Material Flow Analysis method
to evaluate the global performance of recycling processes adopted for desktop
and laptop computers. Two performance indicators have been quantitatively
evaluated: the effectively recycled weight ratios of specific materials and the recycling of critical raw materials. A comparison with the landfilling scenario is also
discussed. Sun et al. (2016) developed proposed an index-based model to assess
the recovery complexity of waste streams and to determine their potential for metal
recovery developed by crossing industry and literature data. An interesting study
(Zhang et al., 2019) quantified the potential contribution of urban mining located
at university areas by assessing the produced band stocked WEEE in a case study
in China. By focusing only on the technological side of the recovery process, most
important issues regard its efficiency in terms of material recovered and energy
consumption required by the disassembly process. A critical review of the most
widespread recovery technologies for each WEEE stream is in Xavier et al. (2019).
Recent studies proposed innovative solutions for increasing the technological efficacy of the recovery process. Peiró et al. (2020) proposed a feasibility analysis
about adopting non-disruptive technological processes for separating PCB and
permanent magnets from discharged hard disk drives. An economic comparison
between the recovery option and virgin production is also discussed outlining the
efficacy of the proposed recovery process. This tendency is also confirmed by Zeng
et al. (2018): the authors proposed a quantitative analysis of the economic feasibility of extracting copper and gold from discharged TV sets in China. Results
outline the economic convenience of the UM process compared to pure materials
extraction and production.

The organizational issue of UM mainly refers to optimizing the collection process
of WEEE.

Gutberlet (2015) discussed the contribution of informal and organized collection services for improving the UM process in Brazil. Economic, environmental, and social implications have been deeply analyzed outlining the relevant contribution of these collection services to increase the overall efficacy of UM. Nowakowski (2017) analyzed the contribution of different collection systems for improving the performance of UM. A case study optimizing capacity and routing of stationary as well as mobile collections systems for WEEE in Poland is discussed to outline the quantitative benefits of analytical models proposed.

3.5.2 SUPPORTING TRANSITION FROM LINEAR TO THE CIRCULAR ECONOMY

The transition from linear to Circular Economy (CE) is affecting all human activities especially manufacturing ones (Lieder and Rashid, 2016): a huge effort has been dedicated to the waste management sector (Singh and Ordoñez, 2016). Recently, increasing attention is developing for applying pillars of CE strategy in both e-waste management and electronic and electric product (EEE) design and production (Parajuly and Wenzel, 2017; Awasthi, et al., 2019). The basic logic of CE is to promote an economy restorative and regenerative thus involving all life phases of a product/service in a closed-loop approach (MacArthur, 2013). A recent report focused on EEE (World Economic Forum, 2019) has outlined main strategies to be adopted to fully support a quick transition to CE: from the product as a service – e.g. electronics as a service – to sharing of assets, life extension, and materials recycling and recovery. The report has also proposed several fields of intervention to be supported: ones not yet fully developed – e.g. durability and repair of used products, urban mining – and others that require a new impulse, such as buy-back or return systems, reverse logistics, and advanced recycling and recapturing of such material. Several past and recent studies have focused on these issues: in this chapter, only studies, which correlate CE in these fields of intervention, are discussed. The need for a more circular and green design of EEE based on the CE approach is proposed in O'Connor et al., (2016): authors discussed the contribution of different green design alternatives (e.g. design devices for disassembly and/or materials for substitution) to better support CE in electronics products.

By focusing on the re-use process, a quantitative analysis to design a more effective re-use process involving both public and private organizations is analysed in (Cole et al., 2017): a survey analysis is proposed outlining the benefits of effective and coordinate integration of these two options. An interesting recent paper focused on evaluating the technical and economic feasibility of repurposing a mobile pc in this desktop. The repurposing option for electronics is a recent evolution of the concept of re-use for PCs and notebooks (Coughlan et al. 2018). Repurposing refers to a new use for a product that can no longer be used in its original form (Long et al. 2016). This option aims to extend the product life of outdated hardware devices by using them in a different manner through adopting a cloud service.

Furthermore, issues to be considered in the recycling sector are more similar to past studies aiming to increase recycling rates with an additional focus on evaluating the potentiality of secondary markets of recycled materials. Tong et al. (2018) discussed the contribution towards a quicker diffusion of CE in e-waste management of informal recycling plants in China. Results outlined that the relevance of the waste quantity intercepted by this informal sector and the requirement of a more coordinated approach at the policy level. Wagner et al. (2019) deeply analysed the recovery and regeneration potential of e-waste in terms of plastic materials: innovative recycling technologies as well as secondary materials market analysis from a CE point of view have been pointed out.

Recent papers have analysed the potential contribution of emerging technologies in supporting e-waste recovery for CE adoption. Álvarez-de-los-Mozos et al. (2020) described the potentiality of adopting a human-robot collaboration in e-waste recycling lines: economic and environmental assessments of benefits derived by this new innovative solution are also discussed based on a CE approach. Garrido-Hidalgo et al. (2020) critically analysed the potentiality of IoT (Internet of Things) technologies to support CE adoption in the end-of-life management of electric vehicle battery packs: an informative framework for supporting its adoption is also proposed. By focusing on challenges in the reverse logistics process of e-waste, Nowakowski et al (2018) discussed the integration of different collection systems (i.e. mobile and stationary) to support more circular collection services. These issues have been also confirmed by a recent literature review study (Bressanelli et al., 2020), which has analysed the scientific literature from different perspectives (from the tools adopted to the product life phases analysed) and overlapping these perspectives to CE main strategies, such as 4R (Reduce, Reuse, remanufacture and recycle) scheme.

Finally, all recent studies outline that the application of CE in the EEE sector is, from one side, a big challenge, but, from another side, it could determine a huge positive impact for wider CE adoption in waste management as e-waste represents the waste stream with the highest yearly growth rate (Cesaro et al., 2018).

The main challenges for adopting CE in EEE and e-waste management regard their "typical" critical issues. One "traditional" issue to be evaluated is forecasting e-waste waste flow for better supporting CE: Althaf et al., (2019) developed a forecasting model to predict both future waste flow for mature products (i.e. where historic sales data are available as well as for emerging products (i.e. where lack of data is usual) to support the adoption of more integrated CE strategies starting from identifying rapid opportunities as well as risks in the end-of-life management of products, to extend product life and close the loop on critical materials. Furthermore, a recent study (Cordova-Pizarro et al., 2019) analysed how to support CE in e-waste management at the national level by analysing potential material flows that could be interested in such strategies. Another criticality regards the contribution of legislation for supporting more circular e-waste management. Isernia et al. (2019) have analysed the Italian condition from both a legislative and a quantitative point of view.

3.6   CONCLUSIONS

This chapter provides a global overview of geographical disparities in terms of e-waste flows and e-waste legislation coverage under the Basel Convention umbrella. This study points out the gaps in transboundary movement policies and data of e-waste flows which hide the complex geographies around the world and the magnitude of informal shipments. Despite all international regulations and a widespread e-waste legislation coverage the illegal trade of e-waste items from developed countries to those which facing e-waste management deficiencies (Africa, Asia, Eastern Europe) with associated environmental and public health threats are far to be solved in the context of increasing amounts of e-waste in the following years. The Basel Convention provides a global framework guideline that must be further improved and adjusted to include clear definitions related to e-waste fractions, better waste statistics data, and practical solutions. Regional regulations such as Bamako Convention (Africa), WEEE Directive (EU), or national e-waste legislation must be accompanied by stricter enforcement procedures regarding the movement of e-waste including port and border investigations, GPS tracking of e-waste items, and intelligence collaboration between countries. Partnerships with the informal sector should be developed across low and middle-income countries to boost domestic e-waste recycling activities. This chapter argues that the development of EPR policies, urban mining practices, and integration of e-waste management practices into local/regional circular economy mechanisms would divert the amount of e-waste flow towards vulnerable destinations.

REFERENCES

Aiblinger-Madersbacher, K. 2016. Grenzüberschreitende Verbringung von Elektro- und Elektronikaltgeräten, available online: www.vivis.de/wp-content/uploads/RuR9/2016_RuR_327-344_Aiblinger_Madersbacher, accessed 12.09.2020

Álvarez-de-los-Mozos, E., Rentería-Bilbao, A., & Díaz-Martín, F. 2020. WEEE recycling and circular economy assisted by collaborative robots. *Appl. Sci.* 10(14): 4800.

Althaf, S., Babbitt, C. W., & Chen, R .2019. Forecasting electronic waste flows for effective circular economy planning. *Resour Conserv Recycl*, 151: 104362.

Amechi, P. E. and Oni, A. B. 2019. Import of Electronic Waste into Nigeria: the Imperative of a regulatory policy shift. *Chinese Journal of Environmental Law* 3: 141–166.

Arora, R., Paterok, K., Banerjee, A., & Saluja, M. S. 2017. Potential and relevance of urban mining in the context of sustainable cities. *IIMB management review*, 29(3): 210–224.

Arya S., and Kumar, S. 2020. E-waste in India at a glance: Current trends, regulations, challenges and management strategies. *J. Clean. Prod* 271: 1–20. https://doi.org/10.1016/j.jclepro.2020.122707

Associated Chambers of Commerce of India (ASSOCHAM), India, 2018. Electricals and electronics manufacturing in India. NEC technologies in India, 1–56. Assessedon28-12-2019.https://in.nec.com/en_IN/pdf/ElectricalsandElectronicsManufacturinginIndia2018.pdf.

Awasthi, A. K., Li, J., Koh, L., & Ogunseitan, O. A. 2019. Circular economy and electronic waste. *Nature Electronics*, 2(3): 86–89.

Basel Action Network (BAN) 2019. Holes in the Circular Economy. WEEE leakage from Europe.

Basel Convention on the control of transboundary movements of hazardous wastes and their disposal 2018 Update. www.basel.int/Portals/4/Basel%20Convention/docs/text/BaselConventionText-e.pdf

Baldé, C.P., Wang, F., Kuehr, R., Huisman, J. 2015. *The global e-waste monitor – 2014*, United Nations University, IAS – SCYCLE, Bonn, Germany. Available at https://i.unu.edu/media/unu.edu/news/52624/UNU-1stGlobal-E-Waste-Monitor-2014-small.pdf (accessed 19 September 2020).

Baldé, C. P., Wang, F., Kuehr, R., 2016. Transboundary movements of used and waste elec-tronic and electrical equipment, United Nations University, Vice Rectorate in Europe – Sustainable Cycles Programme (SCYCLE), Bonn, Germany.

Baldé, C. P., Forti V., Gray, V., Kuehr, R., Stegmann,P. 2017. The Global E-waste Monitor – 2017, United Nations University (UNU), International Telecommunication Union (ITU) & International Solid Waste Association (ISWA), Bonn/Geneva/Vienna., available online: www.itu.int/en/ITU-D/Climate-Change/Documents/GEM%202017/Global-E-waste%20Monitor%202017%20.pdf, accessed 12.09.2020.

Binnemans, K., Jones, P. T., Blanpain, B., Van Gerven, T., Yang, Y., Walton, A., & Buchert, M. 2013. Recycling of rare earths: a critical review. *J. Clean. Prod*, 51, 1–22.

Bimir, M. N. 2020. Revisiting E-Waste Management Practices in Africa: Selected Countries' Cases. *J. Air Waste Manage. Assoc.* doi:10.1080/10962247.2020.1769769

Borthakur, A., Govind, M., 2017a. Emerging trends in consumers' E-waste disposal behavior and awareness: a worldwide overview with special focus on India. *Resour. Conserv. Recycl.* 117:102–113.

Borthakur, A., Govind, A. 2017b. How well are we managing E-waste in India: evidences from the city of Bangalore. *Energy Ecol. Environ.* 2: 225–235. https://doi.org/10.1007/ s40974-017-0060-0

Borthakur, A., Sinha, K. 2013. Electronic Waste Management in India: A Stakeholder's Perspective. *Electronic Green Journal,* 1(36) https://doi.org/10.5070/G313618041

Borrirukwisitsak, S., Khwamsawat, K. & Leewattananukul, S. 2021. The use of relative poten-tial risk as a prioritization tool for household WEEE management in Thailand. *J Mater Cycles Waste Manag.* https://doi.org/10.1007/s10163-021-01175-x

Bressanelli, G., Saccani, N., Pigosso, D. C., & Perona, M. 2020. Circular Economy in the WEEE industry: a systematic literature review and a research agenda. *Sustainable Production and Consumption.* 23: 174–188.

Breivik, K., Armitage, J. M., Wania, F., Jones, K. C., 2014. Tracking the global generation and exports of e-waste. Do existing estimates add up? *Environ. Sci. Technol.* 48: 8735–8743. http://dx.doi.org/10.1021/es5021313.

Brunner, P. H. 2011. Urban mining a contribution to reindustrializing the city. *J. Ind. Ecol.* 15(3): 339–341.

Cesaro, A., Marra, A., Kuchta, K., Belgiorno, V., & Van Hullebusch, E. D. 2018. WEEE management in a circular economy perspective: An overview. *Glob. NEST J,* 20: 743–750.

Cole, C., Gnanapragasam, A., & Cooper, T. 2017. Towards a circular economy: exploring routes to reuse for discarded electrical and electronic equipment. *Procedia CIRP*, 61: 155–160.

Cordova-Pizarro, D., Aguilar-Barajas, I., Romero, D., & Rodriguez, C. A. 2019. Circular economy in the electronic products sector: Material flow analysis and economic impact of cellphone e-waste in Mexico. *Sustainability* 11(5): 1361.

Coughlan, D., Fitzpatrick, C., & McMahon, M. 2018. Repurposing end of life notebook computers from consumer WEEE as thin client computers–a hybrid end of life strategy for the Circular Economy in electronics. *J. Clean. Prod*, 192: 809–820.

Chi, X., Streicher-Porte, M., Wang, M. Y. L., Reuter, M. A. (2011) Informal electronic waste recycling: a sector review with special focus on China. *Waste Manage* 31(4):731–742.

Cossu, R., Williams, I. D., 2015. Urban mining: Concepts, terminology, challenges. *Waste Manage,* 45:1–3.

CEC (2011). Crossing the Border. Opportunities to Improve Sound Management of Transboundary Hazardous Waste Shipments in North America. Available at: www3.cec.org/islandora/en/item/10158-crossing-border-opportunities-improve-sound-management-transboundary-hazardous-en.pdf

Efthymiou, L., Mavragani, A., & Tsagarakis, K. P. 2016. Quantifying the effect of macroeconomic and social factors on illegal e-waste trade. *Int. J. Environ. Res. Public Health* 13(8): 789.

Ezeah, C., Fazakerley, J. A. and Roberts C.L. 2013. Emerging trends in informal sector recycling in developing and transition countries, *Waste Manage*, 33 (11): 2509–2519.

Epo 2019. Elektroschrott: Altgeräte illegal nach Afrika und Asien verschifft, available online: www.epo.de/index.php?option=com_content&view=article&id=15164:elektro schrott-altgeraete-illegal-nach-afrika-und-asien-verschifft&catid=58&Itemid=100198, accessed 12.09.2020.

European Commission 2017. A Comprehensive Assessment of the Current Waste Management Situation in South East Europe and Future Perspectives for the Sector Including Options for Regional Co-Operation in Recycling of Electric and Electronic Waste, Final Report. Available online.

Forti V., Balde, C. P., Kuehr, R., Bel, G. The Global E-waste Monitor 2020: Quantities, flows and the circular economy potential. United Nations University (UNU)/United Nations Institute for Training and Research (UNITAR) – co-hosted SCYCLE Programme, International Telecommunication Union (ITU) & International Solid Waste Association (ISWA), Bonn/Geneva/Rotterdam.

Fuse, M., E. Yamasue, B. K. Reck, and T. E. Graedel. 2011 "Regional Development or Resource Preservation? A Perspective from Japanese Appliance Exports." *Ecological Eco-nomics*, 70: 788–797.

Garrido-Hidalgo, C., Ramirez, F. J., Olivares, T., & Roda-Sanchez, L. 2020. The adoption of Internet of Things in a Circular Supply Chain framework for the recovery of WEEE: The case of Lithium-ion electric vehicle battery packs. *Waste Manage*, 103: 32–44.

Gark, N., Adhana, D. K. 2019. E-Waste Management in India: A Study of Current Scenario. *International Journal of Management, Technology And Engineering*. 9 (1): 2791–2803.

Garlapati, V. K. (2016). E-waste in India and developed countries: Management, recycling,business and biotechnological initiatives. *Renewable Sustainable Energy Rev.* 54: 877–881.

Gangwar, C., Choudhari, R., Chauhan, A., Kumar, A., Singh, A., & Tripathi, A. 2019. Assessment of air pollution caused by illegal e-waste burning to evaluate the human health risk. *Environment International*, 125:191–199. https://doi.org/10.1016/ j.envint.2018.11.051

Ghosh, S. K., Debnath, B., Baidya, R., De, D., Li, J., Ghosh, S. K., Tavares, A. N., 2016. Waste electrical and electronic equipment management and Basel Conventioncompliance in Brazil, Russia, India, China and South Africa (BRICS) nations. *Waste Manage. Res.* 34 (8): 693–707. https://doi.org/10.1177/0734242X16652956.

Gollakota, A. R. K., Gautam, S., & Shu, C.-M. 2020. Inconsistencies of e-waste management in developing nations – Facts and plausible solutions. *J. Environ. Manage.* 261, 110234. https://doi.org/10.1016/j.jenvman.2020.110234

Gidarakos, E., & Akcil, A. 2020. WEEE under the prism of urban mining. *Waste Manage*, 102, 950–951.

Gunarathne, N., de Alwis, A., & Alahakoon, Y. 2020. Challenges facing sustainable urban mining in the e-waste recycling industry in Sri Lanka. *J. Clean. Prod*, 251, 119641.

Gutberlet, J. 2015. Cooperative urban mining in Brazil: Collective practices in selective household waste collection and recycling. *Waste Manage*, 45, 22–31.

Government of Canada, 2014. Pilot Project on the Transboundary Movement of Municipal Solid Waste between Canada and the United States. Evaluation Report. Available at: http://publications.gc.ca/collections/collection_2014/ec/En14-131-2006-eng.pdf

Kahhat, R., Kim, J., Xu, M., Allenby, B., Williams, E., & Zhang, P. 2008. Exploring e-waste management systems in the United States. *Resour Conserv Recycl*, 52(7): 955–964.

Huisman, J., Botezatu, I., Herreras, L., Liddane, M., Hintsa, J., et al. "Countering WEEE Illegal Trade (CWIT) Summary Report, Market Assessment, Legal Analysis, Crime Analysis and Recommendations Roadmap". August 30, 2015. Lyon, France.

Hai, H. T., Hung, H. V. & Quang, N. D. An overview of electronic waste recycling in Vietnam. 2015. *J Mater Cycles Waste Manag* 19, 536–544. https://doi.org/10.1007/s10163-0150448-x

Hotta, Y., A. Santo, and T. Tasaki. *EPR-based Electronic Home Appliance Recycling System under Home Appliance Recycling Act of Japan*. Tokyo: IGES, 2014.

Ilankoon, I. M. S. K., Ghorbani, Y., Chong, M. N., Herath, G., Moyo, T., & Petersen, J. 2018. E-waste in the international context–A review of trade flows, regulations, hazards, waste management strategies and technologies for value recovery. *Waste Manage*, 82: 258–275.

Isernia, R., Passaro, R., Quinto, I., & Thomas, A. 2019. The reverse supply chain of the e-waste management processes in a circular economy framework: Evidence from Italy. *Sustainability*, 11(8): 2430.

Kaza, S., Yao, L. C., Bhada-Tata, P., Van Woerden, F. 2018. What a Waste 2.0: A Global Snapshot of Solid Waste Management to 2050. Urban Development. Washington, DC: World Bank. © World Bank. https://openknowledge.worldbank.org/handle/10986/30317License: CC BY 3.0 IGO."

Secretariat of Basel Convention. 2018. Waste without frontiers Second Edition. www.basel.int/Portals/4/Basel%20Convention/docs/pub/WasteWithoutFrontiersII.pdf

Kankyōshō 環境省 [Ministry of the Environment] and Keizai sangyō shō経済産業省 [Ministry of Economy, Trade and Industry]. "Heisei 26 nendo niokeru kaden risaikuru jisseki nitsuite" 平成26年度における家電リサイクル実績について [Report on Progress in Recycling Home Appliances in 2014]. 2015. Available at www.env.go.jp/press/101129.html

Kazançoglu, Y., Ada, E., Ozturkoglu, Y., & Ozbiltekin, M. 2020. Analysis of the barriers to urban mining for resource melioration in emerging economies. *Resources Policy*, 68: 101768.

Krook, J., Carlsson, A., Eklund, M., Frändegård, P., & Svensson, N. 2011. Urban mining: hibernating copper stocks in local power grids. *J. Clean. Prod*, 19(9–10), 1052–1056.

Lepawsky, J., & McNabb, C. 2010. Mapping international flows of electronic waste. The *Canadian Geographer/Le Géographe canadien*, 54(2): 177–195.

Lederer, J., Laner, D., Fellner, J., Recheberger, H., 2014. A framework for the evaluation of anthropogenic resources based on natural resource evaluation concepts. In: Proceedings SUM 2014, 2nd Symposium on Urban Mining, Bergamo, Italy; IWWG – International Waste Working Group, ISBN:9788862650311; 016.

Lee, D., Offenhuber, D., Duarte, F., Bidermana, A., Ratti, C., 2018. Monitour: Tracking global routes of electronic waste. *Waste Manage*, 72: 362–370.

Lieder, M., & Rashid, A. 2016. Towards circular economy implementation: a comprehensive review in context of manufacturing industry. *J. Clean. Prod*, 115, 36–51.

Long, E., Kokke, S., Lundie, D. et al. 2016. Technical solutions to improve global sustainable management of waste electrical and electronic equipment (WEEE) in the EU and China. *Jnl Remanufactur 6*, 1. https://doi.org/10.1186/s13243-015-0023-6

Lu, C., Zhang, L., Zhong, Y., Ren, W., Tobias, M., Mu, Z., Xue, B., 2015. An overview of e-waste management in China. *J. Mater. Cycles Waste Manage*. 17 (1), 1–12.

MacArthur, E. 2013. Towards the circular economy. *J. Ind. Ecol*, 2, 23–44.

Matemilola S., Salami H. A. (2020) Basel Declaration on the Control of Hazardous Wastes (Basel Convention). In: Idowu, S., Schmidpeter, R., Capaldi, N., Zu, L., Del Baldo, M., Abreu, R. (eds) *Encyclopedia of Sustainable Management.* Springer, Cham. https://doi.org/10.1007/978-3-030-02006-4_1079-1

Marinello, S., Gamberini, R. 2021. Multi-Criteria Decision Making Approaches Applied to Waste Electrical and Electronic Equipment (WEEE): A Comprehensive Literature Review. *Toxics*, 9, 13. https://doi.org/10.3390/toxics9010013

Mihai, F. C., Gnoni, M. G., Meidiana, C., Ezeah, C., Elia, V., 2019. Waste electrical and elec-tronic equipment (WEEE): flows, quantities, and management, a global scenario. In: Prasad, M. N. V., Vithanage, M. (Eds.), *Electronic Waste Management and TreatmentTechnology*. Elsevier Science & Technology Books, pp. 1-34. Available from: http://doi.org/10.1016/B978-0-12-816190-6.00001-7.

Mihai, F. C., Gnoni, M. G. 2016. E-waste management as a global challenge (Introductory chapter). In: Mihai, F. C. (Ed.), *E-Waste in Transition: From Pollution to Resource.* Intech, Rijeka, Croatia, pp. 1_8.

Mihai, F.-C. 2019. Electronic waste management in Romania: Pathways for sustainable practices. In: *Handbook of Electronic Waste Management* (pp. 533–551). Elsevier. https://doi.org/10.1016/B978-0-12-817030-4.00024-3

Mmereki, D., Baldwin, A., Li, B. 2016. A comparative analysis of solid waste management in developed, developing and lesser-developed countries. *Environ. Technol. Rev* 5: 120–141. doi: 10.1080/21622515.2016.1259357

Singh, N., O. A. Ogunseitan & Y. Tang 2020. Systematic review of pregnancy and neonatal health outcomes associated with exposure to e-waste disposal, *Crit. Rev. Environ. Sci. Technol.*, DOI: 10.1080/10643389.2020.1788913

Nowakowski, P., & Mrówczyńska, B. 2018. Towards sustainable WEEE collection and transportation methods in circular economy-Comparative study for rural and urban settlements. *Resour Conserv Recycl,* 135: 93-107.

Nowakowski, P. 2017. A proposal to improve e-waste collection efficiency in urban mining: Container loading and vehicle routing problems–A case study of Poland. *Waste Manage,* 60: 494-504.

Nordbrand, S. 2009. Out of Control: E-waste trade flows from the EU to developing countries. SwedWatch publications. Avaialable at www.swedwatch.org.

Orlins, S., Guan, D. 2016. China's toxic informal e-waste recycling: Local approaches to a global environmental problem. *J. Clean. Prod.* 114:71–80.

Osibanjo, O. and Nnoom, I. C. 2019. Importing Used Electronics from Developed Countries to Nigeria: Problems and Solutions. Urbanet. Accessed on 22/8/20. Available at www.urbanet.info/e-waste-imports-nigeria/

Ottoni, M., Dias, P., & Xavier, L. H. 2020. A circular approach to the e-waste valorization through urban mining in Rio de Janeiro, Brazil. *J. Clean. Prod,* 120990.

O'Connor, M. P., Zimmerman, J. B., Anastas, P. T., & Plata, D. L. 2016. A strategy for material supply chain sustainability: enabling a circular economy in the electronics industry through green engineering. *CS Sustainable Chem. Eng.* 5879–5888.

Palmeira, V. N., Guarda, G. F., & Whitaker Kitajima, L. F. 2018. Illegal international trade of e-waste—Europe. *Detritus*, 1(0), 48–56. https://doi.org/10.26403/detritus/2018.13

Patil, R. A., & Ramakrishna, S. 2020. A comprehensive analysis of e-waste legislation worldwide. *Environ. Sci. Poll. Res.*, 27(13): 14412–14431. https://doi.org/10.1007/s11356-020-07992-1

Parajuly, K., & Wenzel, H. 2017. Potential for circular economy in household WEEE management. *J. Clean. Prod*, 151: 272–285.

Premalatha, M., Tabassum-Abbasi, T. Abbasi, and S. Abbasi. 2014. The Generation, Impact, and Management of E-Waste: State of the Art. *Crit. Rev. Environ. Sci. Technol.* 44(14) 2014: 1577–1678.

Perkins, D. N., Drisse, M. B., Nxele, T., and Sly, P. D. 2014. E-Waste: A Global Hazard. Icahn School of Medicine at Mount Sinai. *Annals of Global Health.* 80:286-295.

Petridis, N. E., Petridis, K., & Stiakakis, E. 2020. Global e-waste trade network analysis. *Resour Conserv Recycl*, 158: 104742. doi:10.1016/j.resconrec.2020.104742

Peiró, L. T., Girón, A. C., & i Durany, X. G. 2020. Examining the feasibility of the urban mining of hard disk drives. *J. Clean. Prod*, 248: 119216

Ragazzi, M., Fedrizzi, S., Rada, E. C., Ionescu, G., Ciudin, R., & Cioca, L. I. 2017. Experiencing urban mining in an Italian municipality towards a circular economy vision. *Energy Procedia*, 119, 192-200.

Secretariat of Basel Convention 2020. Basel Convention on the control of transboundary movements of hazardous wastes and their disposal & Basel Protocol on Liability and Compensation for damage resulting from transboundary movements of hazardous wastes and their disposal . Text and Annexes (Revised in 2019).

Salehabadi, D., 2013. Transboundary Movements of Discarded Electrical and Electronic Equipment. Solving the E-Waste Problem (StEP) Initiative Green Paper

Sun, Z., Xiao, Y., Agterhuis, H., Sietsma, J., & Yang, Y. 2016. Recycling of metals from urban mines–a strategic evaluation. *J. Clean. Prod*, 112: 2977-2987.

Schmidt, C. W. 2006. Unfair Trade: E-Waste in Africa. *Environmental Health Perspectives*, 114(4).

STEP. 2013. Transboundary Movements of Discarded Electrical and Electronic Equipment. www.step-initiative.org/files/_documents/green_papers/StEP_GP_TBM_20130325.pdf> Solving the E-Waste Problem (StEP) Initiative Green Paper.

Takayoshi, S. and N. T. M. Huong . 2009. The Flow of E-Waste Material in the Asian Region and a Reconsideration of International Trade Policies on E-waste. *Environ. Impact Assess.t Rev.*, 29: 25–31.

Singh, J., & Ordoñez, I. 2016. Resource recovery from post-consumer waste: important lessons for the upcoming circular economy. *J. Clean. Prod* 134: 342–353.

Terada, C. 2012. Recycling Electronic Wastes in Nigeria: Putting Environmental and Human Rights at Risk, 10 Nw. *J. Int'l Hum. Rts.* 154. Available at http://scholarlycommons.law.northwestern.edu/njihr/vol10/iss3/2

Tong, X., Wang, T., Chen, Y., & Wang, Y. 2018. Towards an inclusive circular economy: Quantifying the spatial flows of e-waste through the informal sector in China. *Resour Conserv Recycl*, 135: 163–171.

Tunsu, C., Petranikova, M., Gergorić, M., Ekberg, C., & Retegan, T. 2015. Reclaiming rare earth elements from end-of-life products: A review of the perspectives for urban mining using hydrometallurgical unit operations. *Hydrometallurgy*, 156: 239–258.

Terekhova, T. 2012. E-waste Africa project. Secretariat of the Basel Convention UNEP/SBC.

UNEP Bamako Convention www.unenvironment.org/explore-topics/environmental-rights-and-governance/what-we-do/meeting-international-environmental

UNTC. 2019. A amendment to the Basel Convention on the Control of Transboundary Movements of Hazardous Wastes and their Disposal. Retrieved July 1, 2020, from https://treaties.un.org/Pages/ViewDetails.aspx?src=TREATY&mtdsg_no=XXVII-3-a&chapter=27&clang=_en

Umweltbundesamt. 2010. Export von Elektroaltgeräten - Fakten und Maßnahmen; available online: www.globaleslernen.de/sites/default/files/files/link-elements/4000.pdf, accessed 12.09.2020.

UNEP. 1992. Basel Convention on the control of transboundary movements of hazardous wastes and their disposal. 1992. Available at: www.basel.int/Portals/4/Basel%20 Convention/docs/text/BaselConventionText-e.pdf. Accessed September 9, 2020.

Vaccari, M., Vinti, G., Cesaro, A., Belgiorno, V., Salhofer, S., Dias, M. I., Jandric, A. 2019. WEEE treatment in developing countries: Environmental pollution and health consequences—An overview. *Int. J. Environ. Res.Public Health*, 16: 1595.

Van Eygen, E., De Meester, S., Tran, H. P., & Dewulf, J. 2016. Resource savings by urban mining: The case of desktop and laptop computers in Belgium. *Resour Conserv Recycl*, 107: 53–64.

Wang, K., Qian, J. and Liu, L. 2020. Understanding Environmental Pollutions of Informal E-Waste Clustering in Global South via Multi-Scalar Regulatory Frameworks: A Case Study of Guiyu Town, China. *Int. J. Environ. Res. Public Health* 17: 2802. doi:10.3390/ ijerph17082802

Wang, C., He, P., Zuo, L., 2018. Global responsibility for waste disposal. Science Advances (e-Letters). Available at https://advances.sciencemag.org/content/4/6/eaat0131/tab-e-letters.

Wang, Z., Zhang, B., Guan, D., 2016. Take responsibility for electronic-waste disposal. *Nature* 536 (7614): 23–25.

Wagner, F., Peeters, J. R., De Keyzer, J., Janssens, K., Duflou, J. R., & Dewulf, W. 2019. Towards a more circular economy for WEEE plastics–Part A: Development of innovative recycling strategies. *Waste Manage*, 100, 269–277.

Wiesmeth, H. (2021). WEEE and ELV in a circular economy. In *Implementing the Circular Economy for Sustainable Development* (pp. 255–266). Elsevier. https://doi.org/10.1016/ b978-0-12-821798-6.00021-1

World Economic Forum. 2019. A New Circular Vision for Electronics. Time for a Global Reboot. Available at: https://pacecircular.org/sites/default/files/2019-03/New+Vision+ for+Electronics-+Final%20(1).pdf

Zeng, X., Mathews, J. A., & Li, J. 2018. Urban mining of e-waste is becoming more cost-effective than virgin mining. *Environ. Sci. Technol*, 52(8), 4835–4841.

Zhang, S., Ding, Y., Liu, B., & Chang, C. C. 2017. Supply and demand of some critical metals and present status of their recycling in WEEE. *Waste Manage*, 65: 113–127.

Zhang, L., Qu, J., Sheng, H., Yang, J., Wu, H., & Yuan, Z. 2019. Urban mining potentials of university: In-use and hibernating stocks of personal electronics and students' disposal behaviors. *Resour Conserv Recycl*, 143: 210–217.

Xavier, L. H., Giese, E. C., Ribeiro-Duthie, A. C., & Lins, F. A. F. 2019. Sustainability and the circular economy: A theoretical approach focused on e-waste urban mining. *Resources Policy*, 101467.

Zeng, X., Gong, R., Chen, W. Q., Li, J. 2016 Uncovering the recycling potential of "New" WEEE in China. *Environ. Sci. Technol.* 50 (3), 1347–1358.

Zhavoronkova, K. (2020) Role of the Basel and Bamako Conventions in the Fight Against Wastes in Africa. In: Popkova, E., Sergi, B., Haabazoka, L., Ragulina, J. (eds) *Supporting Inclusive Growth and Sustainable Development in Africa - Volume I.* Palgrave Macmillan, Cham. https://doi.org/10.1007/978-3-030-41979-0_9

Part 2

Benchmark Practices and
Case Studies

4 Sustainable Electronics Waste Management in India

Sandip Chatterjee

CONTENTS

4.1 BACKGROUND

The present pace of production and consumption of electronics and electrical equipment is not sustainable; the endorsement of circular economy tool is a must for sustainability on future development (Shevchenko, et al. 2019). In order to

DOI: 10.1201/9781003095972-6

meet the demand of ever evolving modern gazettes to the consumer's satisfaction, the raw materials are a fundamental requirement. Resource materials are, however, fast depleting from Mother Earth, whereas, electronics waste (e-waste) is the major source of those critical raw materials, if they are managed effectively and efficiently. Various policy measures globally advocate the circular economy concept so as to promote sustainable design of product, optimum use of resources, and waste minimizing through reduce, reuse and recycling (European Parliament 2013, Murray et al. 2015, European Commission 2015, United Nations 2018).

The products are unfortunately designed for 'forced obsolescence' due to obvious business interest in more being bought. Either some of the components or modules will start malfunctioning during use or software support will be ceased. Consumers remain with fewer options to use the existing products and opt for a new gazette. Adequate legal provision like 'Right to Repair' can ensure the requisite consumers' rights to use the existing products as long as their wish (Wiens 2015, United States Copyright Office 2016, European Parliament 2018, Svensson et al. 2019).

The recycling of any end-of-life product should be the last option; refurbishment should be encouraged to elongate their useable life so that it could serve a vast underserved population with better products. One should ensure the repairing and refurbishment cost of the products are not high enough to discourage the consumers (King et al. 2006).

The recycling should only be considered, once repairing and refurbishment are not possible. E-waste recycling is not a viable option for the developed countries due to high labour and logistic cost and expensive capital infrastructure and their operational energy cost. The majority of these wastes are, therefore, either land filled or exported to developing countries for further use or recycling, though trans-boundary movement is illegal due to presence of hazardous materials (Niu and Li 2007, Kaya and Sözeri 2009).

Recycling of e-waste broadly yields metals, plastics and glass. The printed circuit board (PCB), one of the complex components in products, contains precious metals, ceramics and epoxy. PCB requires high-end technology to process. Physical and chemical processes are widely used for recycling PCBs (Bernardes 1997, Li 2004, Das 2009, Grause 2010). The physical process includes magnetic, eddy current, density separations, which is considered environmentally sound and require lesser operational energy. The chemical recycling processes such as hydrometallurgical, pyrometallurgical, electrometallurgical are also extensively used for their ability to maximize the extraction of precious metals.

4.1.1 GENESIS OF THE ARTICLE

India has an adequate knowledge base and a vast pool of skilled manpower to serve as a potential repairing and recycling hub for end-of-life electronics and electrical equipment. Research already reported that the manual dismantling and segregation can enrich extraction yield of precious metals. Manual process with adequate safety measures is feasible with significant skilled workforce in India. India may require requisite infrastructure, affordable machines and process technologies to handhold the end-users. India needs to adopt best-suited semi-automated indigenously developed recycling technology with optimum capacity to address the issue. An attempt is made in this chapter to study the gap areas in legislative provisions, technology readiness, availability of skill sets so as to provide suitable solutions for India to become sustainable global repairing and recycling hub.

4.2 INTRODUCTION

Electronics have become indispensible for human life. Its consumption is growing exponentially with daily needs. Rapid technological upgradation has further resulted in its faster obsolescence and therefore growth of electronics waste (e-waste), which creates a global challenge due to its hazardous nature leading to environmental degradation. India has generated 3.23 million metric tonnes of e-waste in 2019 (Forti et al. 2020).

The major concern of e-waste management in India is lack of awareness, vibrant second hand grey market, and recycling of e-waste in informal units by unscientific, unhealthy and non-environmentally friendly methods (Chatterjee 2016). Nearly, 90 percent of the collected waste in India is recycled by the informal sector. Awareness among the key stakeholders is the important challenge in India to restrict adopting a healthy environmental practice in the entire value chain. A continuous creation of awareness and capacity building may improve the ground situation.

Second, products are having significant residual value after end-of-life in tire II and tier III cities in India and thereby operators in informal sector are active in retrieving this residual value by repairing them with cannibalized components with attractive prices. Fundamental legal and non-legal barriers prevent repairing business in the formal sector. Intellectual protection right (IPR) infringements, designed products for obsolescence rather than longevity or repair are discouraging formal repairing. Lack of skill sets, non-availability of tools, manuals, spare parts from original equipment manufacturers (OEMs) impede repair (Svensson 2019). This residual value is, thus, availed informal units, with a cost of responsible repair and testing. The informal sector often do not follow responsible repairing with proper spare parts or by using safety protocols and therefore pollutes the environment and creates a hazard for workers.

Third, a primitive method of recycling to recover metals by informal operators is common in developing countries including China, India, Vietnam, Philippines and Ghana. Methods used are incineration in the open air exposing toxic materials to the operators, the hazardous method of amalgamation to recover precious metals from segregated components of e-waste that also recover a small amount of precious metals by heating the amalgam on a hot frying pan by inhaling hazardous mercury vapours (Chatterjee and Kumar 2009, Chatterjee and Kumar 2011 and Chatterjee 2018).

The hazardous operations in informal units should be replaced by state of the art technologies and equipment, which are, however, an expensive idea for developing countries and often not suited to the local needs. On the other hand, a falling commodity price is de-incentivizing the recycling sector in the developed world, which forces many developed countries to transport e-waste illegally for processing in the developing countries, where environmental norms are not that stringent (Abdelbasir et al. 2018).

E-waste is considered in a sustainability agenda and is also an important discussion point in the circular economy. International standards are developed to regulate responsible collection, handling, repair, depopulation, dismantling and recycling. India is having adequate legislative provisions to address majority aspects of e-waste management. Right to Repair (R2R) legislation is, however, still a distant dream in India, which lacks the ability to provide legality to repairing business. A regulator needs to be more vigilant so as to ensure stakeholders' responsibility in mandatorily reporting of e-waste management activities. Just legislative provision may not be enough. Affordable technologies, adequate safety and operation skill sets and awareness on toxicity would be important for the informal sector to transform their activities to acceptable sustainable recycling practices.

Ministry of Electronics and Information Technology (MeitY), Government of India, has undertaken many initiatives towards developing cost-effective technology to recycle e-waste in an environmentally friendly manner as well as an awareness outreach program to educate key stakeholders on a pan-India basis on the hazardous effects associated with recycling in the informal sector. This chapter discusses all the initiatives of MeitY, including the indigenous technologies developed up to the demonstration level in laboratories including National Metallurgical Laboratory, Jamshedpur, India, Centre for Materials for Electronics Technology (Hyderabad), a scientific society of MeitY and Central Institute of Plastics Engineering & Technology (CIPET), under Ministry of Chemicals & Fertilizers. The chapter also discusses the key data on impact analysis of the awareness program and behavioral changes among citizens.

4.3   GENERATION OF E-WASTE

E-waste is the fastest-growing domestic waste stream in the world. The Global E-waste Monitor 2020 has estimated the global generation of e-waste as 53.6 million tonnes (MT) in 2019 with an increase of 21 percent in five years. The report found that China is the highest contributor with 10.1 MT of e-waste, followed by the United States (US) with 6.9 MT and India with 3.23 MT in 2019. Out of the generated e-waste, only 17.4 percent was recycled. The report also predicted that global generation of e-waste will reach 74 MT by 2030 (Forti et al. 2020).

4.4   REGULATORY FRAMEWORK

The e-waste in India is regulated through E-Waste (Management) Rules, 2016, notified by the Ministry of Environment and Forest & Climate Change (MoEF&CC) on 1 October 2016. These rules have superseded the earlier version, E-waste (Management and Handling) Rules, 2011, in effect since 1 May 2012. The present Rule introduced Producer Responsibility Organisation (PRO), deposit-refund system (DRS), e-waste exchange, etc. for convenience of the manufacturers (MoEF 2016).

The target based Extended Producers Responsibility (EPR) is another important measure in the Rules. The Rules had subsequently been amended in 2018 (CPCB 2018) to reduce the EPR target from 30 percent to 10 percent during 2017. The

TABLE 4.1
EPR Collection Target Comparison

Year	2017-18	2018-19	2019-20	2020-21	2021-22	2022-23
India						
Old Producers	10% (*)	20% (*)	30% (*)	40% (*)	50% (*)	60% (*)
New Producers	....	5% (**)	5% (**)	10% (**)	10% (**)	15% (**)
EU(***)						
weight wise	45%	45%	65%	65%	65%	65%
generation wise	45%	45%	85%	85%	85%	85%
China	*Presently, no regulation, EPR framework to finalise by 2020, EPR Rule to finalise by2025*					
Japan	Refrigerators/ Washing m/c: 50%, Air conditioners: 60%, TVs: 55%					
South Korea	3.9 Kg/ capita	6.0 Kg/ capita				

*-quantity of generation of waste, **-sales figure of FY2017-18, ***- to be reviewed in 2019

EPR target initiated merely 10 percent in 2017 and would reach 70 percent by 2023. Table 4.1 below gives a comparison of India's collection target with other countries (Porwal and Chatterjee 2019).

4.5 AWARENESS IN INDIA

Awareness of safe disposal of e-waste is a key challenge in our society. Channelization of e-waste for proper recycling and ensuring a system of accountability in its management need greater awareness among all stakeholders. The Ministry of Electronics and Information Technology (MeitY) had implemented a pan-India awareness program on e-waste management during 2015–2020.

The program had been envisaged to create awareness among the public about the hazards of e-waste recycling by the unorganized sector and to educate them about alternate methods of disposing their e-waste. The stakeholders chosen were government officials, school/ college students, resident welfare associations, manufacturers, producers, dealers, re-furbishers, recyclers and the informal sector. The program had been implemented in a total of 31 identified states/UTs, namely, Madhya Pradesh, Uttar Pradesh, Jharkhand, Orissa, Goa, Bihar, Pondicherry, West Bengal, Assam, Manipur, Andhra Pradesh, Andaman and Nicobar Island, Chhatisgarh, Daman and Diu, Delhi, Gujarat, Haryana, Himachal Pradesh, Karnataka, Kerala, Lakshadweep, Maharashtra, Meghalaya, Punjab, Rajasthan, Sikkim, Tamil Nadu, Telangana, Tripura, Uttarakhand and Mizoram. In addition, awareness activities amongst the government officials were covered in Jammu & Kashmir, Arunachal Pradesh and Nagaland as well.

Gap in knowledge and capacity amongst stakeholders observed during execution of the program would be important to mention. Informal sector having low literacy is the most vulnerable stakeholder. They lack awareness on health hazards, safety aspects associated and environmental impact in engaging unscientific e-waste processing activities. De-soldering, acid baths and open burning are often used by them in a poorly ventilated room, without any safety precaution, to extract precious metals from e-waste, which releases dioxins and aromatic hydrocarbons and thus pollute soil, water and air. A large number of women and children are employed for these activities. It is not known to them that e-waste contains toxic substances like mercury, lead, arsenic, chromium and brominated flame retardant.

The consumers expect a return from end-of-life products and thereby, fail to contribute towards their safe disposal. Awareness amongst consumers is also another challenge. Behavioral change among consumers would be imperative. Responsibility of the producers towards compliances of the rules is another concern. Manufacturers find means to get rid of collection targets under extended producer responsibility (EPR). Renewal of EPR, recycler registration, monitoring compliance and action against violations are some of the key challenges for seamless implementation of the Rules. Adequacy of strength and resources within state pollution control boards (SPCBs) and pollution control committees (PCCs) are another area of concern.

The program had created training tools, content materials, films, printed materials, videos and jingles etc. for every strata of the society which are freely available on the dedicated website (MeitY 2020) (http://greene.gov.in/). Further, social media platforms (Twitter handle and Facebook page) and mobile app had also been created to provide online status of the activities and showcase the activities/ workshops/ carnivals etc. conducted under the program. The program was successfully able to conduct 1918 workshops and activities in various cities, which were attended by 16,52,031 participants from school, colleges, RWA, manufacturer, informal operators, bulk consumer, dealers and refurbishers etc. The program had covered 5789 government officials in various states. Besides, 1247 GreenE Champions/trainers had also been trained. The mass awareness amongst youth of the country, nearly 201.2 million audience, had also been created by covering 2813 cinema halls (MeitY 2020) (http:// greene.gov.in/dashboard/).

India being a multilingual country, it was felt that by keeping the content in the local language and suited to local needs would be the game changer. Quality and standard technical contents in Indian language was therefore created in the program for an effective awareness to the citizen. Social media to connect every stakeholder, especially, youth was stressed for cost-effective awareness creation in a sustainable manner. Classroom based awareness programs was preferred for school and college students, whereas, the activities, street dramas and posters, pamphlets were found suitable for RWAs. The dealers, distributors, manpower in manufacturing sector, informal operators were outreached effectively through WhatsApp groups, Facebook and other popular social media by providing suitable content, messages, videos etc.

Creating awareness in a vast country like India in a sustained manner would be challenging. It was, therefore, thought a dedicated website (http://greene.gov.in/) should be created with all relevant information (see Figure 4.1). This website is now available for downloading stakeholder-specific contents, tutorials, videos, audios

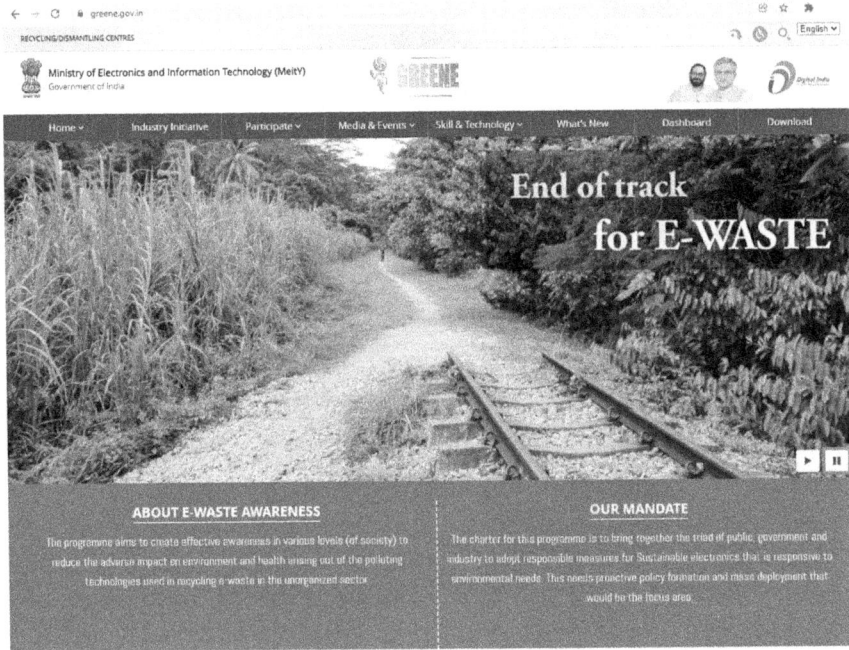

FIGURE 4.1 Green E website.

etc. freely for the citizen. It was also thought that presence of resource persons on a pan-India basis would be important. The program had created 'local champions' as resource persons and their contact details are provided in the website. These champions would assist State Governments, industries and academic institutes in creating future awareness at a local level.

MeitY had also used a massive open online courses (MOOC) platform for uploading relevant courses on e-waste management for school students for the Standard 7 to 9, which are now being used by National Council of Educational Research and Training (NCERT). The MOOC platform is being used for teaching students as it is effective and convenient to reach any part of the country in a seamless manner. The students are those who are in a formal education system as well as school drop outs who would benefit from this initiative.

Third party assessment of the program has been carried out by the Deutsche Gesellschaft für Internationale Zusammenarbeit (GIZ) GmbH and submitted to MeitY (Arora, et al., 2020). The report summarized that the overall impact of the program in society was significant. Evaluation suggested that the objectives of the program were achieved and that the measures taken across all work packages (WPs) were effective at large.

The respondents and interviewees, in particular, expressed their satisfaction with the technical design of the program and indicated positive changes in knowledge and attitudes. The report concluded that the measures taken have led to the fulfillment of the specific objectives and further contributed to the achievement of the overall

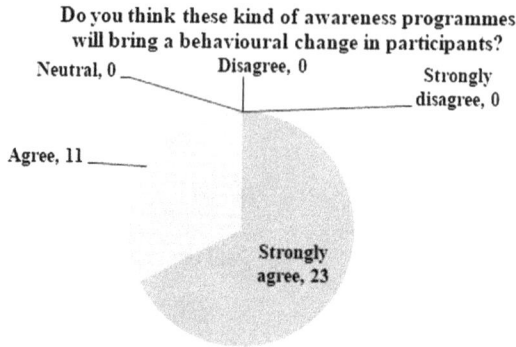

Note: Replies to question 6 of the interviews with school students of Siddaganga Public School on 19.01.2019 (total of 34 interviewees)

FIGURE 4.2   Typical feedback of the e-waste program.

objective – i.e. improved knowledge, attitude and practices (KAP) on e-waste disposal. The stakeholders were satisfied with the quality and relevance of the content, delivery methods and outreaching outcomes. It was, however, observed that in the absence of an effective eco-system of an environmentally sound e-waste management system, it would be challenging to translate awareness in action on a large scale (Arora et al. 2020). A typical feedback assessment undertaken by the project team on 19.01.2019 at the Siddaganga Public School is shown in Figure 4.2.

4.6   REFURBISHED PRODUCTS

Refurbishment is another important strategy for transition towards a circular economy (CE). It is, however, challenging for the companies and customers to overcome existing barriers. Transition to CE demands more efficient use of resources and a relook at the design aspects of the products so as to promote longer lifetimes and be repair friendly. Several factors influence consumers into repairing a product, which include legal and market impediments to factors of cost, convenience and consumer preference. An attempt has been made to examine the current state of Right to Repair (R2R) and perspectives of stakeholders, fundamental challenges to access repair services for consumer electronics.

Fundamental legal and non-legal barriers preventing accessible repair include intellectual protection right (IPR) infringements, products designed for obsolescence rather than longevity or repair. Lack of awareness and knowledge, non-availability of tools, manuals or spare parts also impede repair. Total costs of repair, time and convenience, lack of trust, risk of poor quality and availability of cost-competitive new products make repair a less attractive option. The consumer preferences and attitudes are also another important resisting factor for repair.

The end-user license agreement (EULA), sales contracts are an important legal barrier for consumers forbidding unauthorized repair of software-enabled products or use of parts from original equipment manufacturer (OEM). The violation of terms constitutes breach of contract and goes against IP laws. Lack of awareness on consumer rights, legality of warranty also prevents Right to Repair. The design lifetimes, planned obsolescence also prevents repairing (Svensson et al. 2019).

The EULA / sales contracts also restrict unauthorized repairers to disassemble products and repair using local parts. Disasseembling slim, compact and sleek make products, proprietary screws, non-removable batteries is another greater challenge and thus makes repairing impossible. Legally repairing amounts to 'construction' or 'modification' of the patented article, which constitutes direct infringement under patent law. The protective measures like Technical lock on software including digital rights management (DRM) and technical protection measures (TPM) further de-motivates unauthorized repairing. Software repair activity, copying codes are also an infringement. If these challenges are overcome, a repaired device may still be less functional for lack of updating of embedded software that poses security risks, and probable data loss (Svensson et al. 2019).

The OEM is not obliged to provide spare parts, or software support to other repairers. The manufacturing, selling and importing of unauthorized spare parts may invite patent infringement under patent law and the trademark law. The design measures/ "software doping" also prevents products from functioning with third party parts or equipment compatibility issue (e.g. printers ink cartridges; or electronics battery chargers).

The OEM may refuse to provide proprietary tools to unauthorized repairers or distributors of software for restoration of operation of disks, breaking technical software locks under the copyright law. Creating a replica of a patented tool is an infringement under patent law. The OEM is not obliged to provide manuals and schematics. The unauthorized circulation of copyrighted repair information is against copyright law (Svensson et al. 2019).

It is, therefore, summarized that in order to make repairing job legal in the country, one should arrive a balancing act between OEM's rights and consumers' rights. The IP law protects OEM in exhaustion doctrine on reselling, rental, lending and other third party commercial uses of the products. Anti-circumvention provision protects OEM against infringers from copyrighted works. The copyrighted repair manuals, design and trademarks of complex products/parts are some of the other provisions that protects the OEM's interest. Legal provisions, which should take care of the consumers' right should also be ensured.

The competition and antitrust law is, therefore, provisioned to address the monopolization situation and repair restrictive contractual clauses. The contract law contradicts public policies such as Resource Efficiency (RE). The planned obsolescence of product should be made an offence. Lack of information on warranties, or, misinformation on enforcement about existing consumer rights should not be entertained.

Moreover, the product should be designed for durability, easy repairing, upgrading and remanufacturing. The OEMs should provide a repair service, diagnostics, tools, firmware, service parts, and service documentation at an affordable cost to attract customers to repair their products as well as avoid approaching independent repair technicians.

In order to address the issues in a more holistic manner, developed countries offer the legislative provision, Right to Repair (R2R), which intends to allow consumers the ability to repair and modify their own consumer electronic devices, where otherwise the manufacturers force consumers to only use their services or replace the product with a new device. So far, 25 States in the USA have adopted R2R legal provision by 2019 and similar EU's legislation in 2019 mandates that the manufacturers should supply replacement parts for 10 years from 2021. However, the OEM lobby considers this as a breach of IP, and argues it is a threat to the safety of consumers, hackers could insert vulnerabilities in repaired devices to affect a user's privacy & security etc. Globally it was, however, stressed that this legislative provision would slow down the rate of recycling, reduce energy consumption for recycling as well as fresh production, hence the environment would be protected (Wiens, 2015, United States Copyright Office, 2016, European Parliament, 2018, Svensson et al. 2019).

In India, E-waste Management Rule 2016, encourages refurbishment, however, safety and quality standards are not defined for the refurbished products. The selling of the refurbished products is, therefore, in the formal sector not feasible under Electronics and Information Technology Goods (Requirement for Compulsory Registration) Order, 2012 (MeitY, 2018). Bringing legislative provision like R2R, the refurbished market should be promoted and dependency on the secondary materials should be reduced and circular economy could be further strengthened.

The formal market of refurbished electronics products would definitely encourage the consumers in India's Tier 2 and Tier 3 cities to access the quality brand at an affordable price. It would not only benefit in enhancing the life of the discarded electronics products, but also help to avail Digital India in a more effective way. The significant enhanced requirement of the second hand products would definitely minimize the load of recycling of e-waste and improve the resource efficiency.

4.7 TECHNOLOGY STATUS

The e-waste recycling market in India is still dominated by the informal sector, who are engaged in collecting, dismantling, segregating recoverable metals, plastic and glass from the end-of-life electronics products and finally processing printed circuit boards, Integrated circuits (ICs), connectors etc. through primitive methods and polluting the environment. Few formal recyclers, also, however, are engaged in the business, carrying out the operation till segregation stage and dependent on global metals extraction facilities for PCB recycling. This entails a huge loss of revenue to the country in the form of assaying, sampling, refining and logistics costs besides considerable loss of precious metals resources and loss of employment.

4.7.1 FORMAL SECTOR

India has nearly 312 authorized recyclers, who are having 7.72 Tonnes per annum processing capacity. The formal e-waste recyclers in India are, however, engaged in dismantling, dissembling, segregation and selling the recoverable materials like large and small metal frames, plastic and glass to the local smelters based on the market price for further processing. The recyclers are, however, exporting the most valuable printed circuit boards (PCBs), integrated circuits (ICs) to global smelting facilities with a mutually agreed price, which causes a significant loss of revenue, and also precious metals and other resources from the country and loss of employment. The list of the formal recyclers, their geographical locations, and their processing capacities may be obtained from the CPCB website (CPCB 2019).

Global primary copper and precious metal smelters including Umicore, Belgium; Boliden, Sweden; Aurubis, Germany; Mitsui, Japan; Xstrata, Canada, where generally the formal recyclers are responsible for the end process, are co-processing PCBs and components along with copper ore concentrates. The technologies are highly capital intensive and require substantial power and high input volume of materials, which may not be viable for India. The environmentally sound industrial routes for extraction of metals from e-waste are pyro-metallurgy, hydro-metallurgy and electro-metallurgy (Cui and Zhang 2008, Khaliq et al., 2014, Kaya et al. 2016, Parthsarthy et al. 2018), which are possessed by a handful of companies globally.

4.7.1.1 Case Study: Indian Formal Sector

Out of 312 Indian authorized recyclers, most of them have no basic infrastructure, necessary tools, equipment or knowledge base to process the e-waste in an efficient manner to recover the entire resources available. The following are the top five recyclers, operating more than two decades and having some level of processing capabilities. (CPCB 2019)

(a) TES AMM Private Limited, Tamil Nadu (Capacity: 30000 MT/ annum)
(b) Greenscape Eco Management Pvt.., Rajasthan: (Capacity: 18200 MT/ annum)
(c) E-parisara Pvt. Ltd., Karnataka: (Capacity: 8820 MT/ annum)
(d) Eco Recycling Ltd., Maharashtra: (Capacity: 7200 MT/ annum)
(e) Exigo Recycling Pvt. Ltd., Haryana: (Capacity: 6000 MT/ annum)

In order to understand the activities undertaken by the Indian authorized recyclers, an attempt is made to study the detailed operation of the following two recyclers (Chatterjee 2010).

TES AMM Pvt. Ltd. Tamil Nadu

The company is operational since 2006 in Kancheepuram, Tamil Nadu and also having units in Haryana and Karnataka. The company is also associated with M/s Karosambhav, a recycler and logistics service provider company, for the pan India collection. This unit has obtained all the requisite certificates, licenses and clearance from the MoEF&CC, CPCB and TNPCB (Tamil Nadu Pollution Control Board) to maintain safety standards, environmental hazards etc. The company is certified with ISO 9001, ISO 14001, Responsible Recycling (R2) and OHSAS 18001 to ensure quality, environment, health and safety.

The company has a recycling facility, which is capable of recycling end-of-life electronics, electrical products and Li-ion battery. The processes followed broadly are segregation, dismantling of e-waste till PCBs size reduction with a capacity of 25-30 MT/ day. The company is, however, dependant on foreign smelters for precious metals recovery from PCBs. The plant has a processing capacity of 30000 MT/year.

E-parisara Pvt. Ltd., Bangalore

E-parisara is the first recycling company in India, which started operation in September 2005. The company has requisite certificates from Karnataka State Pollution Control Board (KSPCB) and Control Board and Central Pollution Control Board (CPCB). The company is also certified with ISO 14001:2004 and OHSAS 18001:2007 and Responsible Recycling (R2) to ensure quality, environment, health and safety. The company has collection centres in the major cities including Bangalore, Chennai, Hyderabad, Kolkata, Gurgoan etc. The company has the facility and capability to recycle the majority of e-waste and also Li-ion batteries, dry cells, fluorescent & CFL lamps, industrial, medical, military and space electronics.

The company is innovative and has designed and developed various machineries indigenously to process various categories of e-waste in a cost effective manner. The processes majorly followed are segregation, dismantling of products till PCBs. Earlier, the company was solely dependent on foreign smelters for precious metals recovery from PCBs. However, in recent years, the company has developed indigenous technology jointly with C-MET, Hyderabad for extraction of precious metals from PCBs, Li-ion battery. The company has demonstrated a processing capability of 1000Kg PCB /day to successfully extract precious metals including gold, silver, copper and palladium through pyrolysis, smelting and electrolysis processes. The plant has e-waste processing capacity of recycling 10 MT/ per day.

4.7.2 INFORMAL SECTOR

The major e-waste processing in India is carried out in the informal sector using a primitive method and thereby damaging the environment significantly. The informal recycling sector is vibrant in India and the major materials are being processed in this sector in an illegal manner. The informal sector is located in and around major metropolitan cities like Bangalore, Mumbai, Delhi and Chennai and also in tier-ii cities like Moradabad, Uttar Pradesh (Chatterjee 2012). The biggest informal recycling hub in the country, however, exists in tier-II cities like Moradabad (Centre for Science and Environment 2015) [https://cdn.downtoearth.org.in/pdf/moradabad%20-e-waste.pdf]. Demand for copper by insulated wire, polyvinyl chloride (PVC) wire and cable wire industry, largely clustered in metropolitan cities has promoted these informal units. The recycling processes used by informal operators are unscientific and hazardous. The major activities carried out by the operators are broadly listed under:

A. Manual dismantling with screwdrivers, hammers, chisels, and bare hands, to separate various components and materials

B. De-soldering components of PCBs by heating over coal-fired grills for de-soldering and manual separating components

C. Leaching in open-pit acid baths acid and smelting in small ill-equipped furnaces to extract precious metals including copper, silver and gold

D. Chipping and melting plastics without proper ventilation
E. Open air burning of sleeves of the cables to recover copper and also unwanted materials
F. Disposing of unsalvageable materials in fields and along riverbanks

Though it is illegal, e-waste trade is more profitable in the informal sector as no investment of the sophisticated equipment or cost on regulation compliances. The informal sector has little incentive to convert into formal sectors, as both the bulk consumer and individual consumer prefer auctioning their e-waste to informal dismantlers and get good value for it. The process is, however, highly hazardous and cause air, water and soil pollution besides posing significant health risk (Asante et al. 2012, Annamalai et al. 2015, Amienyo and Azapagic 2016, and Chatterjee 2016, Tran and Salhofer 2018).

4.7.3   SUSTAINABLE OPTION FOR FORMAL RECYCLING

Electronic waste is the great source of valuable resource materials. Figure 4.1 shows the resources materials available in the end-of-life electronics products. A study has shown that assorted e-waste contains nearly 38% of the ferrous metal, 28% non-ferrous metals, 19% of varied plastics, 4% of glass, and 11% of other materials (Khaliq et al. 2014) If we consider India's generation of 3.3 MT the e-waste as in 2019 (Forti et al. 2020), the waste stream has, therefore, already accumulated 1.25MT of ferrous metals, 0.92MT of non-ferrous and precious metals, 0.63MT of varied categories of plastics, 0.13 MT of glass and 0.36 MT of the other materials including hazardous substance (Figure 4.3).

In order to extract the vast pool of resources materials from the waste stream the affordable and environmentally sustainable technologies should be promoted. Indigenous technologies suited to the local environmental condition should be effective for transforming vast informal operation to the formal fold. The Ministry of Electronics and Information Technology (MeitY), being the nodal ministry for Electronics and IT in India, is engaged in nurturing and evolving affordable environmental friendly e-waste recycling technologies. These technologies and processes of e-waste recycling developed through the research and development (R&D) projects must carry out manual dismantling and segregation of the e-waste till the Printed Circuit Boards (PCB) stage. The outcomes of the efforts are discussed below.

4.7.4   INDIGENOUS TECHNOLOGIES

The disassembly and segregation of e-waste recovered mainly small and large structural metal parts, heat sinks, ferrous metal, ferrite and ceramic components, non ferrous metal scrap, mainly Cu and Al as well as cables and wires. Appropriate segregation followed by smelting could be the best possible recycling option in India for these waste materials to transform them into usable raw materials. India has adequate smelter to carry out this part of the recycling. The small and large structural plastic

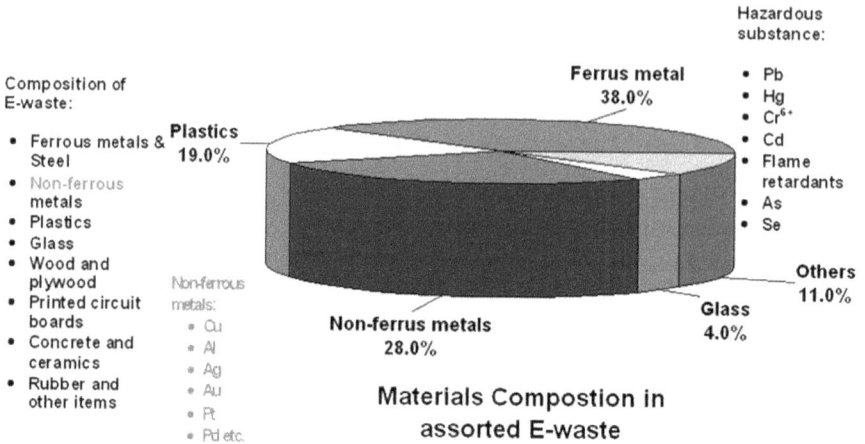

Composition of E-waste:
- Ferrous metals & Steel
- Non-ferrous metals
- Plastics
- Glass
- Wood and plywood
- Printed circuit boards
- Concrete and ceramics
- Rubber and other items

Non-ferrous metals:
- Cu
- Al
- Ag
- Au
- Pt
- Pd etc.

Hazardous substance:
- Pb
- Hg
- Cr^{6+}
- Cd
- Flame retardants
- As
- Se

Ferrus metal 38.0%

Plastics 19.0%

Non-ferrus metals 28.0%

Glass 4.0%

Others 11.0%

Materials Compostion in assorted E-waste

FIGURE 4.3   Materials potential in assorted e-waste.

parts recovered from e-waste could also be recycled in India by chemical recovery (plastics-derived fuels), materials recovery (recycled plastics)-PC, ABS, PC or energy recovery (H2, steam).

Glass smelting is also available in India for transforming waste glass to usable raw materials. Lead extraction from CRT glass would be the key challenge. E-waste also contains hazardous materials including wastes like Chlorofluorocarbons (CFCs), Mercury (Hg) switches, CRT, batteries and capacitor, flame retardants plastic, which require environment-friendly technology and special safety to recover valuable resources presence in these components.

Being the nodal Ministry of Electronics, MeitY is engaged in nurturing and evolving several affordable technologies to recycle e-waste. This ministry felt that consolidation of the recycling technologies available in various research organizations and also promoting indigenous affordable technologies to the recycling sector to recover resource materials and conserve the depleting natural resources would be imperative. MeitY felt that some of the components including PCB recycling, converting plastics to value added products, recycling Li-ion batteries, rare earth materials extraction from hard disc, phosphors, recovering copper from wires, recycling components containing hazardous substance etc. would require special attention.

4.7.4.1   Recycling of Printed Circuit Boards (PCBs)

The populated PCBs constitutes 3 to 5 percent by weight of total e-waste, have rich value of metals such as copper, silver, gold, palladium, platinum, tantalum and other metals in traces level. The recovery of all the metals requires appropriate technology and professional skill, expensive equipment. The various methods have been attempted to develop and establish matured technologies to recover precious metal. The salient features of these technologies are discussed.

A process technology for the recycling of PCB had been developed at National Metallurgical Laboratory (NML), Jamshedpur, where PCB was pulverized and thereafter subjected to a series of physical separation methods to separate metals and epoxy. The PCBs are ground to powder of the desired size (-1.0 mm in size) through various mechanical processes including physical impaction, shredding/fragmentation and granulation, etc. Shredding breaks down the PCBs into pieces via ripping or tearing which may then be sorted into material streams having dissimilar subsequent processing demands. The mechanical process, granulation is used to make PCBs scrap into fine particles. Precious metal particles are further concentrated by means of various separation techniques. Magnetic separation technique is used to separate magnetic materials (iron, nickel and cobalt) from the PCB powder and the aluminium particles are being separated by eddy current separation technique. The metal rich power is then separated from plastic rich particles by electrostatic separation technique. A feed of pulverized PCB containing ~ 23 percent of metal had been successfully enriched to over 78 percent of metal concentrate in this method. Rejectable plastics stream contains only less than 1 percent of metal. Re-circulating intermediate streams during continuous operation could enhance a reasonably high yield of recovery of 90 percent of the metals (Das et al. 2009, Das et al. 2010, Das 2011). The operation cost is cost effective. The harmful effluents are avoided in this technology, as major physical processes are used. The scale of operation tested was 1.0 Metric ton of PCB.

The metal rich feed from the physical process was processed through chemical leaching method for precious metal recovery. After roasting, the feed had been leached by sulphuric acid under atmospheric conditions at a suitable temperature using air and hydrogen peroxide as oxidant. Leaching conditions were optimized to recover ~ 93-98 percent of the constituent base metals including Cu, Ni, Co, Zn, Al and Fe. Maximum recovery of metals in shortest time duration with low acid consumption was achieved under oxygen pressure leaching conditions. More than 98 percent recovery of copper and other base metals was achieved under optimized conditions with about 120 psi oxygen pressure (Das et al. 2009, Das 2011).

Pressure dissolution was carried out in an autoclave and the residue was treated for gold and silver recovery. Nontoxic reagent thiourea leaching methods used to recover gold and silver from the enriched leached residue. Various parameters including thiourea temperature, sulphuric acid and additive concentrations, etc. were optimization to obtain maximum recovery of silver and gold. A maximum of 97 percent silver and 92.5 percent gold recoveries were achieved. The lead from the residue was dissolved in brine solution at 80°C and 96.8 percent recovery was achieved. Subsequently, about 96 percent tin was recovered by hydrochloric acid leaching. The process was tested at 1 kg scale for the recovery of different valuable metals. A tentative process flow chart for recovery of metals/precious metals from the populated PCBs is shown (Chatterjee et al. 2016).

The physical separation methods used in the above project found to be counterproductive for effective recovering other precious metals (like platinum, tantalum, ruthenium etc.), present in the ppm level. The component level segregation at the initial stage was therefore thought to be appropriate before processing PCBs to extract precious metals. The processing of PCBs had therefore been attempted through smelting and electrolysis methods, by the Centre for Materials for Electronics (C-MET), Hyderabad along with formal recycler, M/s E-Parisaraa, Bangalore.

FIGURE 4.4   Smelted black copper.

In this project, the PCB is then subjected to the process of depopulation by vibratory centrifuge to separate the components. Nearly 70 percent of components including small outline transistors (SOTs), ball grid arrays (BGAs), integrated circuits (ICs), crystals, crystal oscillators, chip capacitors, chip resistors and connectors etc. were depopulated and segregated. Thermal incineration combined with pyro-metallurgical treatments was used for metal recovery from PCBs. The bare PCBs were, thereafter, pyrolysed through specially designed system of 500 Kg batch size capacity. The toxic organic components such as brominated flame retardants (BFR) were isolated from the PCB and transferred to the pyrolysis oil with substantial calorific value. Pyrolysis could easily shred and pulverize multilayer circuit boards for liberation of metals to enable chemical leaching process. The pyrolysed metals were further calcinized to obtain fragile materials to facilitate easy leaching and 40-50 percent reduction of mass (Reddy and Parthasarathy 2010). The hydrometallurgical methods in scalable level were used to recover metals from calcinated metal residue. Solid metal so collected was labelled as 'black copper" (Figure 4.4). Electro-refining was then used for separation of black copper from other precious metals and purification (Figure 4.5). Various leaching agents including nitric, hydrochloric, aquaregia and sulphuric acid were used.

Selective recovery of pure metal products directly from residual waste streams was the process for recovery of other precious metals. Leaching of depopulated and pyrolysed components and PCBs satisfactory recover four precious metals such as copper, gold, silver and palladium. The processes have been scaled up and established for demonstration. The process is cost-effective and makes an effective utilization of available lower labour wages. (Reddy and Parthasarathy 2010, Chatterjee, et al., 2016, Chatterjee and Parthasarathy 2018, Parthasarthy et al 2018).

A unique components depopulation followed by pyrolysis and solvent extraction route for the recovery process of precious metals from PCBs has been thus established having 95 percent yield and could recover 150gm gold, 600-700gm silver, 70-80gm of palladium and 200Kg of copper from 1 MT of PCB (ref.). The potential of the technology to recover metals from 1 MT PCB is given at Table 4.2.

FIGURE 4.5 Electrolytic refining of copper.

TABLE 4.2
Economic Potential of MeitY Pilot Plant Technology (1 MT PCB)

Metal	Metal price	Recovery potential
Gold	Rs.27,000/10gm	Rs. 4,05,000/- from 150gm
Silver	Rs.45/gm	Rs.27,000/- from 600-700gm
Palladium	Rs.1500/gm	Rs.3,150/- from 70-80gm
Copper	Rs.400/kg	Rs.80,000/- from 200Kg

4.7.4.2 Recycling of Plastics

Plastics, a significant constituent of electrical and electronics products, are used for insulation, flame retardation, noise reduction, sealing, housing, interior structural parts, electronic components etc. Plastics contain ~ 21 percent by weight in e-waste. Novel recovery and conversion of e-waste plastics to value added product has also been successfully developed at Central Institute of Plastics Engineering and Technology (CIPET), Bhubaneswar, India (Mohanty 2011, Chatterjee et al. 2016).

Among various plastics in e-waste, four homogeneous polymers namely ABS, Polypropylene, Polystyrene and Polyurethane amounted to nearly 70 percent. The epoxy mainly used in printed circuit boards and PVC used in wires could be considered as minority in e-waste. The technology solution discussed here is for the homogenous plastics including PC, ABS, HIPS, PC/ABS blend, ABS/PS blend etc., collected from major structural components like printer housing, monitor stand, monitor cover, computer casing, keyboard bottom plate, TV housing, CPU casing, speaker box, telephone casing etc. Various formulations were attempted with these categories of plastics based on their flow parameters, processability, mechanical performance and cost effectiveness. The process of formulation/ master batch was established. The process broadly includes cleaning, drying and grinding of plastics products, optimization of mixed plastics within virgin General Purpose Polycarbonate (GPPC), improvement of the performance with effect of plasticizer and evaluation of properties (Mohanty 2011, Chatterjee et al. 2016).

The e-waste has various other precious resources that require environment friendly and cost-effective technological solutions suited to the local condition. These components are rechargeable batteries, rare earth extraction from florescent lamps, hard disc etc. Moreover, the design of the products are also changing very fast, conventional materials have been replaced by less costly materials with better functionality. Recycling process should require continuous modification to stay process viable. In order to address this issue, Ministry of Electronics and Information Technology (MeitY), has created a Centre of Excellence (CoE) on E-waste Recycling at C-MET, Hyderabad for working the entire recycling process to provide the viable commercial solutions to society. The following section discusses some of this development.

4.7.4.3 Recycling of Hard Discs

The rare-earth element (REE), Neodymium, is used as permanent magnets in spindles for computer hard disc drives, electric motors and wind turbines. The recovery of neodymium and dysprosium from end-of-life NdFeB magnets would be important to replenish the secondary resource materials. C-MET, Hyderabad has been initiated to develop cost-effective technology to recycle spent magnets at industrial scale. The hard disc drives (HDD) has at least two NdFeB permanent magnets (i.e. NIB magnets). The process involved manually separating the magnets from HDD, using screw drivers and hammer. Analytical study through EDXRF and ICP-OES analysis showed that the magnets contain neodymium, dysprosium and praseodymium. The NIB magnets are thereby subjected to crush to > 200 micron sized powder using jaw crusher and roller disc crusher. The nickel protecting layer was removed during this process. The powder was then subjected to acid leaching using sulphuric acid followed by solvent extraction using DEHPA. So far the purity achieved was 98 percent, further improvement of purity and process up-scaling is in progress.

4.7.4.4 Recycling of Batteries

Lithium ion batteries are the best chosen energy storage devices in most electronic gadgets. The cathode material contains a variety of valuable and scarce metals, such as nickel, cobalt, manganese and lithium. Li-ion Battery Li ion battery contains 15–25% of Al, 0.1–1% of carbon amorphous powder, 5–15% of Cu foil, 1–10% of diethyl carbonate, ethylene carbonate and methyl ethyl carbonate, 1-5% of LiPF6, 10–30% of graphite powder, 25–45% of LiCoO2, 0.5–2% of polyvinyliden fluoride, steel, Ni and inert polymer. Cobalt and Lithium could be targeted for recovery. R&D results are reported to develop environmentally friendly methods for safe dismantling, crushing Li ion batteries and leaching of valuable metals (Petrániková 2011).

C-MET, Hyderabad is engaged in developing cost-effective environmentally friendly technology to recycle spent rechargeable batteries at an industrial scale.

Three methodologies have been attempted to recover cathode materials from spent LIBs, which include i) direct recycling, ii) hydrometallurgical process, and iii) pyrometallurgical process.

The hydrometallurgical process is popular for recycling all types of rechargeable batteries to extract significant metal values. The processes engaged are discharging, dismantling, separating cathode and anode material and separator and then metals are separated. The cathode material is leached with acids and oxidizing agents. The leachate feed is subjected to solvent extraction to extract and separate valuable metals. Final products are precipitated as pure salts and usable for fabrication of new lithium ion batteries. Various solvents such as Cyanex 272, DEPHA are being utilized in this process. The R&D efforts are undergone for identifying more selective solvents for easy scalability.

Alternately, high temperature smelting is also employed for LIBs recycling to extract cobalt, nickel and copper from relatively high grade batteries. The advantage of smelting method is its ability to handle batteries of mixed cathode compositions. The cobalt copper alloy is further separated and purified by sequential electro-refining technique. However, for recovering lithium, hydrometallurgical techniques are being adopted. Major disadvantage of smelting technique is, however, the requirement of substantial capital investment and also loss of lithium in the slag. CMET has already developed processes with both the hydrometallurgical and pyrometallurgical techniques and demonstrated at bench scale. Scaling up of the processes is in progress in CoE.

4.7.4.5 Recycling of Phosphors in Fluorescent Lamps and Color Picture Tube

Phosphors used in fluorescent lamps as well as color picture tube are rich sources of heavy rare earth metals. Recycling of rare earths from phosphors provides an efficient way to recover high value rare earth elements. Separation of REEs is very difficult due to the small difference in ionic radius, the preference for interaction with hard-sphere base donor atoms and the dominance of the trivalent oxidation state across the lanthanide series. Methods such as fractional crystallization or precipitation, ion-exchange, selective oxidation/reduction and solvent extraction were developed and optimized in C-MET, Hyderabad for individual separation of REEs.

The conventional methods such as fractional crystallization and fractional precipitation used in the past were slow and tedious. Presently, the solvent extraction and ion exchange methods are being used. These methods worked on the basis of the lanthanide contraction, i.e. ionic radius is decreasing across the lanthanide series of elements, from lanthanum to lutetium. The heavy members of the series will, therefore, create stronger binding with solute and solvent molecules compared to light members, which allows preferential binding to ion exchange resins, or extraction of the complex into the organic phase. Lamps are initially crushed and phosphor powder is collected. Then the powers are subjected to solvent extraction and followed by the precipitation of the metals and calcination of the precipitate to obtain the pure rare earth oxides. The sulfuric acid is used for acid leaching and DEHPA is used by solvent extraction. The recycling process developed by C-MET, Hyderabad from spent phosphors of CFL and fluorescent lamp is capable extracting > 96 percent purity of yttrium. The batch scale up gradation and further purity improvement is in progress under the CoE.

4.7.4.6 Recycling of LCD Screens

Liquid crystal display (LCD) screen contains indium-tin oxide (ITO). The recycling process involves acid leaching of ITO from the surface of the crushed LCD glass, with various acids (HNO3, HCl, and H2SO4) of different concentrations and with suitable temperature. Leaching is followed by solvent extraction to separate indium from other metals. Indium could be separated into a relatively pure fraction with Di 2 Ethyl Hexyl Phosphoric Acid (DEHPA). Higher purity metal extraction could be achieved by optimization of the solvent extraction and through additional processes like electrolysis (Yang 2012). The process optimization and scale up efforts for indium recovery are being carried out at C-MET, Hyderabad under the CoE.

4.7.5 ECO-PARK

The Eco-park is viable option for a developing country like India in order to promote local e-waste recycling industry, also to mainstream informal recyclers into formal sector framework and finally to ensure environmentally sound e-waste management. Eco-park is a dedicated facility for e-waste management and processing options of end-of-life electronics products for their reuse or recycling of all the parts and components, obtained after segregation for a sustainable solution and to provide impetus to the circular economy. Eco-Park can help to make use of the full potential of the local recycling industry including the informal ones and reduce the dependence on the export of precious recyclable materials from the country.

An ideal eco-park should at least have 9,000 MT material processing capacity per annum in present Indian factory conditions. In order to process the said capacity, the integrated recycling plant should handle 10 tons e-waste per shift and therefore 1 ton circuit boards processing unit per shift. The facility would require 20000 sq m i.e 215278 sq ft (~ 5 acres) of land area. The built up area with non asbestos roof shed for this facility should be 10000 sq m ~ 107639 sq ft.

The eco-park should have a dedicated storage area with shade to provide adequate space for storing e-waste, entering to the park for processing. The shed will avoid toxin seepage into the ground along with rainwater. Area of storage should be at least 40 percent of the total Eco-park. The indoor area of Eco-park should have multiple facilities for dismantling and segregation units, set up by informal operators. The e-waste from the storage area needs to reach these dismantling and segregation units and then output will reach to metal and plastic processing recycling plants in the Eco-park.

The Central Government may initially extend financial support to the promoter company for investment on the capital equipment for state of art facility to ensure processing entire e-waste in the country and retain resource materials and also to ensure zero landfill. It will help in sustaining e-waste processing units in India.

The States can earmark allocation of industrial space or shed for e-waste dismantling and recycling in the existing and upcoming industrial park, estate and industrial areas. The necessary permission and clearance for various processes in the Eco-parks should be provided by State Government agencies. State Pollution Control Board (SPCB) may ensure annual monitoring the compliances on safety and health parameters of the workers engaged in dismantling and recycling. SPCB should also regularly monitor the Eco-park processing facilities for compliances of all environmental guidelines prescribed by the government.

Informal operators will play a pivotal role in the Eco-park. The rag pickers, small scrap dealers, engaged in the informal sector, are primary actors in channelization e-waste from consumers to the recyclers. Informal operators are the driving force for materials collection network in the country. This type of collection mechanism is working effectively due to the presence of suitable financial compensation in the system. Large scrap dealers are then accessing this e-waste from small scrap dealers with appropriate prices. The network is thus significantly reducing the workload of civic agencies. The consumers are also motivated to leave their discarded products instead of storing them or throwing them out as garbage.

The informal operators, are, however, required to be recognized and registered by the State Government before allowing them to set up dismantling and segregating units in the Eco-park. This registration will assist them in forming formal groups so that their activities in the Eco-park are made legal. Industrial skill development and capacity building for informal workers would also be important to ensure in order to improve their understanding of hazardous aspects of the operations.

The registered and formalized informal operators, small and large scrap dealers engaged with the Eco-parks would mainly be responsible for collecting the e-waste from the consumers. The PROs/ formal recyclers can continue to collect e-waste from bulk consumers including government offices, public sector agencies, school, universities, hospitals, corporate houses etc. These robust collection systems attached to a promoter company managing the Eco-park operation will ensure the regular supply of materials to the industrial scale operation for PCBs, rechargeable batteries, plastics and spent magnets and rare earth rich e-waste component.

Well aware and skilled formalized informal operators could set up disassembly and segregation units in the Eco-park and continue their activities with a standard method with environmentally acceptable norms by using adequate safety tools and accessories. The segregated e-waste would provide two types of materials, non-destructive and destructive, shown in Table 4.3, which could be managed by the skilled informal operators. The dissembled items are broadly classified into three main categories viz. small and large structural metal parts and heat sinks; small and large structural plastic parts; and printed circuit boards with IC Chips, electronic components and connectors as shown in Table 4.3 (Chatterjee, 2010).

The non-destructive disassembled parts (Sr. no. 1,2,5 and 6) having a definite market value for reuse, while the destructive disassembly parts (sr. no. 3,4 and 7), requires technology for processing to recover the resource materials. India is having established recycling processing options for the extracted materials, shown in Sr. no. 1,2,5 and 6 at Table 4.3. It would, therefore, be appropriate to sell those materials at market price so that they could reach the smelters for further processing these materials to convert them to virgin materials to boost the circular economy.

TABLE 4.3
Technological Options for Potential Recyclable Materials from E-Waste

S. No.	Potential Recyclable Materials	Technological Options	Uses
1.	Small & large structural metal parts, heat sinks, ferrous metal	Smelting	Secondary raw material
2.	Ferrite & ceramic components Nonferrous metal scrap mainly copper & aluminium	Smelting	
3.	Cables and wires	Striping + Smelting	
4.	Precious metal scrap, PCBs with IC Chips, electronic components and connectors	(i) Smelting + Hydrometallurgy + Electrochemical Process (ii) Mechanical shredding + screening + electrostatic separation + Falcon centrifugal separation + hydrometallurgy	Secondary raw material viz. Copper, silver, gold, palladium etc.
5.	Small and large structural plastic parts	(i) Chemical Recovery- plastics derived fuels (ii) Materials Recovery Recycled plastics viz. PC, ABS, PC- Energy Recovery (H2, steam)	Secondary raw material and energy source
6.	Glass components	Smelting	Secondary raw material
7.	Hazardous wastes like chlorofluorocarbon (CFC), Mercury (Hg) Switches, CRT, batteries and capacitor, flame retardants plastic	Recycling with due environmental care	Secondary raw material

The non-destructive disassembled parts (Sr. no. 1,2,5 and 6) are required to be processed with a proven state of art technology, which requires investment sophisticated equipment. The promoter company in the Eco-park should invest on this technology for processing these parts so that precious metals and valuable resources materials can be extracted in the country and zero landfill can be achieved. The PCBs including small outline transistors (SOTs), ball grid arrays (BGAs), ICs, crystals, crystal oscillators, chip capacitors, chip resistors and connectors, etc., lithium ion and other rechargeable batteries, varied types of the plastics could be processed in Eco-park at industrial scale with established technology. The concentrating e-waste management and processing in single zone would definitely help regulator to monitor the environmental norms in more effective manner.

4.8 DISCUSSION

Effective awareness to the citizen about the hazardous outcome of the unscientific recycling in a sustainable manner would help to streamline materials in the formal channels. The consumers would be more conscious about their environmental responsibility and thereby motivated to encourage formal recycling for their discarded electronics. In order to carry out an awareness program in sustainable mode, Government may provide stakeholder specific standard content in local languages, master trainers, as is already available in (http://greene.gov.in/), through MeitY's initiative. Social media should be utilized effectively to outreach the citizens. Manufacturers under their EPR obligation may use these resources to mobilize the awareness campaign.

India is a growing refurbishment market, though witnessed only in the informal sector. Refurbishment should be encouraged seriously to offer quality products at affordable prices to interested customers and also reduce the recycling load of the products, still having a usable life. Refurbishment is an important step towards achieving circular economy (CE). India should play a proactive role to provide legal provisions to overcome existing barriers. Right to Repair legislation should be enacted in the country to mandate OEMs to design and develop products for durability, easy reparability and re-manufacturability. The customer should have legal options to avail repairing service, for which diagnostics, tools, firmware, service parts, and service documentation should be made available at an affordable cost.

The recycling should be the last option in any effective e-waste management and for which local talents and best practices should be strengthened. India has a significant number of informal operators (~2 Million), who are the asset in the e-waste management chain and having sound knowledge of the materials value chain and the business. Synergies between the informal and formal sector would be immensely important to win each other's trust. Informal operators are innovative in collecting e-waste from consumers at suitable prices. Their dismantling and segregation practices are acceptable at a certain level of environmental standard.

Appropriate training and skill upgrade will convert this willing workforce, who has an adequate understanding of dynamics, complicated materials flow network to engage in the waste management in the country. The presence of a vibrant informal sector is a boon for India. State Government may integrate informal operators into a cooperative entity. Financial support to upgrade the skill sets, and the health and safety practices can be offered to potential industry associations, who may engage their service for ensuring that the EPR obligation of manufactures in effective means.

Due to fast change in product designs and use of advanced materials with lesser recycling value, and also lesser usability of the majority of de-soldered components obsolete, the use of expensive recycling process and sophisticated machineries may lose their relevance in the near future (Chatterjee and Kumar 2009, Chatterjee 2010). The sustainable recycling options for India could, therefore, be proposed in two ways.

One effective way would be to provide low-cost technologies to these informal operators to continue their recycling activities in an environmentally accepted manner. Varied cost-effective technologies to recover precious metals from circuit boards, lithium batteries, WEEE plastics, rare earth materials recovery from hard discs, phosphors etc. have been developed and scaled up at a demonstration level in the country by MeitY through C-MET, CIPET and also similar efforts showcased in few CSIR labs. These low-cost technologies should be offered to the informal operators to transform them to an environmentally acceptable workforce. This may be a game changer idea for converting the huge informal workforce for effective waste management legal business and generate significant employment and also strengthen resource efficiency and circular economy. For example, indigenous technology with a unique method for processing the circuit boards exclusively with a capacity of 100kg/shift capacity developed by MeitY would be ideal for the informal sector with acceptable environmental norms (Chatterjee 2019).

The country needs to prepare itself to manage e-waste in a more responsible and organized manner. The second effective option would be to create Eco-parks in all the Indian states by integrating the formal and informal sector. A private promoter company may set up these parks with initial financial support from the central government, and assistance in obtaining land, subsidized power, water, other utilities and local approvals from the state governments. The formal industries and authorized informal operators may work together to optimize business and revenue earnings. The materials flow could be streamlined from originators to the final destination of recycling centers at a few designated Eco-parks. Higher materials available for operation, improvement of the recovering yields by using appropriate tools, processes and technology, environmental safeguarding are some of the additional benefits of these efforts. Technological solutions to these Eco-parks could be the state-of-art technologies to ensure effective e-waste management in environmental friendly manner. This would aim to achieve cost effective industrial scale recycling technology, while minimizing landfill and zero emission to air, land and water. The recovery of valuable materials such as precious metals and reusable recovered plastics would ensure that the recycling business was an economically profitable venture (Chatterjee 2019).

Benchmarking the processes and technologies, and requisite standards is required in the country. These are important parameters required to assess the efficiency of the technology, return of investment, and profitability. The acquiring cost for improved technology would be significantly high, which would require large industry houses to invest. Micro-medium-small would still depend on domestic technology or low-cost indigenous technology.

The commodities market in India is formalized so that secondary materials can compete with primary ones and their use can be enhanced in products as equivalent to primary materials. This will enhance the envisaged goals of the Make in India and Atmanirbhar Bharat Missions of Government of India. It will also lead to benefits in the Swachh Bharat Mission.

Transition to CE demands for more efficient use of resources and relook in the design aspects of the products so as promote longer lifetimes and repair friendly. Several factors influence consumers for repairing a product, which include legal and market impediments to factors of cost, convenience, and consumer preference.

4.9 CONCLUSION

The present study has provided the most suitable options for an effective e-waste management in the country so that resource efficiency should be ensured. All the important steps including awareness creation, strengthening the existing legislative provisions, creating infrastructure for economical viable recycling options are narrated.

Effective awareness creation would sensitize the citizen to promote channelizing waste materials to the formal sector as well as help operators to know the hazardous effect on informal recycling. Encouraging repairing and refurbishing businesses would ensure the circular economy was strengthened and also engage a vast pool of skilled manpower to new job opportunities.

India is a vast country having various hotspots of e-waste generation near major metro cities. Thus, it requires various recycling units with optimum capacity to effectively manage the waste. Indigenously developed technology discussed here is suitable for the local recycling units, who are otherwise dependent on exporting PCBs to developed countries and losing a substantial amount of revenue due to the lower price of the boards offered by the foreign smelters. The proposed technology could be appropriate and cost-effective for the informal sector towards their transition to the formal sector. Cost effective technology can easily be replicated in other developing countries and export of PCBs to the developed countries can be reduced.

Indigenous technologies discussed to recycle lithium ion batteries, extraction of rare earth elements from spent magnets and phosphors and also various categories of plastics would help local recyclers to extract maximum resources materials, which otherwise might be considered impossible from foreign smelters.

The study also discussed the feasibility of the state of art recycling facilities with integrated efforts at a few dedicated places by creating Eco-parks, which would benefit formal, informal players, help to achieve greater recycling yields and also help regulators to ensure environmental guidelines in a more effective manner.

4.10 DISCLAIMER

The views expressed in this chapter are those of the author and do not represent the official views of the Ministry of Electronics and Information Technology, Government of India. The facts and figures, and data used in this chapter are indicative, and informative and are have been extracted from various research articles and project reports submitted by various research organizations. The validation of these data would be challenging, and not within the scope of the present topic. Use of the data and the recommendations presented are the responsibility of the users of the chapter.

ACKNOWLEDGMENTS

The authors gratefully acknowledge o Dr. S. Rajesh Kumar, Scientist, CMET, Hyderabad and Dr. U. Rambabu, Scientist, CMET, Hyderabad and Dr. P. Parthasarathy, Managing Director, E-Parisaraa Pvt. Ltd. Bangalore for their valuable suggestions on the subject and also thankful for sharing inputs and important data for the electronic waste recycling operations. The author is also grateful to the Ministry of Electronics and Information Technology, New Delhi for financial assistance for the project.

REFERENCES

Abdelbasir S. M., Hassan S. S. M., Kamel A. H. et al. 2018. Status of electronic waste recycling techniques: a review. *Environ. Sci. Pollut Res.*, Springer-Verlag GmbH Germany, part of Springer Nature 2018 https://doi.org/10.1007/s11356-018-2136-6, www.researchgate. net/publication/325019965_Status_of_electronic_waste_recycling_techniques_a_ review (accessed January 4, 2021)

Amienyo, D., Azapagic, A. 2016. Life cycle environmental impacts and costs of beer production and consumption in the UK. *Int. J. Life Cycle Assess.* 492–509. https://doi.org/ 10.1007/s11367-016-1028-6.

Annamalai, J. 2015. Occupational health hazards related to informal recycling of E waste in India: An overview. *Indian J Occup. Environ. Med.*, 19(1): 61–65.

Arora R., Henzler M., Mutz, D. et al., 2020. Feedback/Monitoring and Evaluation of Projects (WP7) for Awareness Program on Environmental Hazards of Electronic Waste by Ministry of Electronics and Information Technology (MeitY), India, March 2020, EU-REI Project, GiZ and adelphi (MeitY, OM No.7(4)/2015–EMCD (Vol.II), dt. 21/06/ 2018- unpublished)

Asante, K. A., Agusa, T., Biney, C. A., el al. 2012. Multi-Trace element levels and arsenic spatiation in urine of e-waste recycling workers from Agbogbloshie, Accra in Ghana. *Sci. Total Environ.* 424, 63–73.

Bernardes, A., Bohlinger, I., Rodriguez, D. et al., 1997. Recycling of printed circuit boards by melting with oxidizing/ reducing top blowing process. *TMS Annual Meeting*, Orlando, pp. 363–375.

Chatterjee, S. and K. Kumar. 2009. Effective Electronic Waste Management and Recycling Process Involving Formal and Non-Formal Sectors, *Int. J. Phys. Sci.*, 4(14): 898–905.

Chatterjee, S. 2010. Electronics Waste Management: An India Perspective, *Lambert Academic Publishing AG&Co,KG*, Saarbrucken, Germany.

Chatterjee, S. and K. Kumar. 2011. Overview of Electronics Waste Management in India, in Zhang L. and G.K. Krumdick (eds.)., *Recycling Electronics Waste II: Proceedings of the Second Symposium*, The Minerals, Metals and Materials Society.

Chatterjee, S. 2012. Sustainable Electronic Waste Management and Recycling Process, *Am. J Environ. Eng.*, 2(1): 23–33.

Chatterjee, S., Kumari, A, and Jha, M. K. 2016. Sustainable Recycling Technology for Electronic Waste, Chapter 11, at p187-201, at the book *Sustainability in the Mineral and Energy Sectors* (K26279), September 8 by CRC Press, Taylor & Francis, USA.

Chatterjee, S. and Parthasarathy, P., 2018, Indigenous technology to recycle scrap circuit boards, *7th International Symposium and Environmental Exhibition, Going Green CARE INNOVATION 2018 Towards a Circular Economy, an event to discuss future strategies, meet your clients and form strategic partnerships*, November 26 – 29, Schoenbrunn Palace Conference Centre, Vienna, Austria.

Chatterjee, S. 2019. E-waste management in India: issues and strategies, Colloquium, *J Decis. Mak.*, 44(3) 127–162.

CPCB. 2019, List of Registered E-Waste Dismantlers/Recyclers in the country (as on 29-12-2016) http://greene.gov.in/wp-content/uploads/2019/09/2019091881.pdf (accessed January 4, 2021).

CPCB. 2018 Ministry of Environment, Forest and Climate Change Notification New Delhi, the 22nd March, 2018 https://cpcb.nic.in/uploads/Projects/E-Waste/e-waste_amendment_ notification_06.04.2018.pdf (accessed January 4, 2021)

Cui, J., Zhang1, L. 2008. Metallurgical recovery of metals from electronic waste: A review. *J. Hazard. Mater.*, 158, 228–256.

Das, A., Vidyadhar, A., Mehrotra, S.P. 2009. A Novel Flowsheet for the Recovery of Metal values from Waste Printed Circuit Boards, *Resour. Conserv. Recycl.*, 53: 464–469.

Das, A. 2011. Development of Processing Technology for Recycling and Reuse of Electronic Waste. (DeitY, OM No.1(7)/2007/M&C, dated 17.09.2007- unpublished).

Das, A., Chatterjee S. and Mehrotra S.P. 2010. Characterisation and Processing of Electronic Waste for the Recovery of Metal Values", Proceedings of the XXV *International Mineral Processing Congress, Brisbane*, Australia.

Electronics and Information Technology Goods (Requirement for Compulsory Registration) Order 2012 www.meity.gov.in/writereaddata/files/Gazette%20Notification%281%29. pdf, (accessed January 4, 2021).

Electronics right to repair, https://en.wikipedia.org/wiki/Electronics_right_to_repair, (accessed January 4, 2021).

European Commission. Closing the Loop—An EU Action Plan for the Circular Economy. Communication from the Commission to the European Parliament, the Council, the European Economic and Social Committee and the Committee of the Regions. Brussels. 2 December 2015. COM (2015) 614 http://eur-lex.europa.eu/legal-content/EN/TXT/ ?uri=CELEX:52015DC0614 (accessed January 4, 2021).

European Parliament. 2013 Decision No 1386/2013/EU of the European Parliament and of the Council of 20 November 2013 on a General Union Environment Action Programme to 2020 Living Well, within the Limits of Our Planet. http://eur-lex.europa.eu/legal-cont ent/EN/TXT/PDF/?uri=CELEX:32013D1386&from=EN (accessed January 4, 2021).

European Parliament. 2018. European Parliament resolution on the implementation of the Ecodesign Directive (2009/125/EC) (2017/2087(INI))'. 31-May-2018 (accessed January 4, 2021)

Forti, V., Baldé, C.P., Kuehr, R., et al., 2020. The Global E-waste, Quantities, flows, and the circular economy potential Monitor 2020, ISBN Digital: 978-92-808-9114-0, United Nations University/United Nations Institute for Training and Research and the International Telecommunication Union. http://ewastemonitor.info/wp-content/uploads/ 2020/07/GEM_2020_def_july1_low.pdf# (accessed January 4, 2021)

Grause, G., Ishibashi, J., Kameda, T. et al. 2010. Kinetic studies of the decomposition of flame retardant containing highimpact polystyrene. *Polym Degrad Stab* 95(6):1129–1137.

Kaya, M. 2016. World Academy of Science, Engineering and Technology *Int. J. Chem. Mol. Eng.* 10(2).

King, A. M., Burgess, S. C., Ijomah, W., et al. 2006. Reducing waste: repair, recondition, remanufacture or recycle? *Sustain. Dev.*, 14 (4): 257–267.

Khaliq, A., Rhamdhani, M. A., Brooks, G. et al. 2014. Metal Extraction Processes for Electronic Waste and Existing Industrial Routes: A Review and Australian Perspective, *Resources* ISSN 2079-9276, 3, 152–179; doi:10.3390/resources3010152 file:///C:/Users/Dell/ Downloads/resources-03-00152.pdf (accessed January 4, 2021).

Li J., Shrivastava P., Z. Gao and H. C. Zhang. 2004. Printed Circuit Board Recycling: A State-of-The Art Survey, *IEEE Trans. Elect. Packag.Manuf.*, 27(1): 33–42.

MeitY. 2018. Electronics and Information Technology Goods (Requirements for Compulsory Registration) Order 2012, March 5, 2018 www.meity.gov.in/writereaddata/files/ Amendment%20Order_2018.pdf (accessed January 4, 2021).

MeitY. 2020 An initiative on E-Waste Awareness. Available at: http://greene.gov.in/. (accessed January 4, 2021)

MoEF. E-Waste (Management) Rules, 2016. Available at: http://moef.gov.in/wp-content/uplo ads/2017/07/notified-ewaste-rule-2015_1_0.pdf. (accessed January 4, 2021)

Mohanty, S. 2011. Novel recovery and conversion of plastics from WEEE to value added products. Central Institute of Plastics Engineering & Technology (CIPET), Bhubaneswar. Deity, OM NO.1(8)/2010/M&C, dated: 01.03.2011- unpublished.

Muammer, Kaya and Ayça, Sözeri. 2009. A review of electronic waste (e-waste) recycling technologies "is e-waste an opportunity or treat?, *Conference: 138th Annual Meeting of TMS, San Francisco,At: San Francisco-California, USA* www.researchgate.net/ publication/260268009_A_REVIEW_OF_ELECTRONIC_WASTE_EWASTE_ RECYCLING_TECHNOLOGIES_IS_EWASTE_AN_OPPORTUNITY_OR_TREAT/ link/0c9605306ea74bd19b000000/download

Murray, A., Skene, K., Haynes, K. 2015. The Circular Economy: An Interdisciplinary Exploration of the Concept and Application in a Global Context. *J. Bus. Ethics*, 140, 369–380.

Niu, X., & Li, Y. 2007. Treatment of waste printed wire boards in electronic waste for safe disposal. *J. Hazard. Mater.*, 145 (3), 410–416.

Porwal, P., Chatterjee, S., 2019, Extended Producer Responsibility on E-waste Management in India: Challenges and Prospects, *Int. J. Sci. Res. (IJSR)*, 8(5), 1026–1034.

Parthasarathy, P., S. Chatterjee, M.R.P. Reddy, et al. 2018. Environmentally Sound Recycling Technology of Scrap Printed Circuit Boards for Developing Countries, *Int. J. Sci. Eng. Res.* 9(3), 1713–1725. www.ijser.org/researchpaper/Environmentally-Sound-Recycling- Technology-of-Scrap-Printed-Circuit-Boards-for-Developing-countries.pdf (accessed January 4, 2021).

Petrániková, M., Miškufová A., Havlík T., Forsén O. et al. 2011. Cobalt recovery from spent portable lithium accumulators after thermal treatment. *Acta Metallurgica Slovaca*, 17: 106–115.

Recommendations to address the issues of informal sector involved in e-waste handling, Centre for Science and Environment, 2015. https://cdn.downtoearth.org.in/pdf/moradabad%20- e-waste.pdf (accessed August 13, 2020).

Reddy, M.R.P., Parthasarathy, P. 2010. Environmentally sound methods for recovery of metals from printed circuit boards. DeitY, OM No.1(2)/2010/M&C dated. 24, May 2010 unpublished.

Shevchenko, T., Laitala, K., and Yuriy Danko. 2019.Understanding Consumer E-Waste Recycling Behavior: Introducing a New Economic Incentive to Increase the Collection Rates. *Sustainability* 11, 2656; doi:10.3390/su11092656. www.mdpi.com/journal/sus- tainability (accessed January 4, 2021).

Svensson, S., Luth, J., Richter, Maitre-Ekern, E., et al. 2018. The emerging 'right to repair' legislation in the EU and the U.S., 7th International Symposium and Environmental Exhibition, Going Green CARE INNOVATION 2018 Towards a Circular Economy, an event to discuss future strategies, meet your clients and form strategic partnerships, November 26 – 29, 2018, Schoenbrunn Palace Conference Centre, Vienna, Austria.

Tran, C.D., Salhofer, S.P. 2018. Processes in informal end-processing of e-waste generated from personal computers in Vietnam. *J. Mater. Cycles. Waste Manag.* 20, 1154. https:// doi.org/10.1007/s10163-017-0678-1.

United Nations. About the Sustainable Development Goals. 2018. Available online: www. un.org/sustainabledevelopment/sustainable-development-goals/ (accessed January 4, 2021).

United States Copyright Office, 'Software-enabled Consumer Products', The Register of Copyrights, Dec. 2016.

Wiens, K. 2015. 'The right to repair [soapbox]', *IEEE Consum. Electron. Mag.*, 4(4): 123–135.

Yang, J. 2012. Recovery of indium from end-of-life liquid crystal displays. Thesis for the Degree of Licentiate of Engineering, Industrial Materials Recycling, Department of Chemical and Biological Engineering, Chalmers University of Technology, Gothenburg, Sweden, publications.lib.chalmers.se/records/fulltext/165702/165702.pdf.

5 E-waste Management in Sri Lanka
Current Status and Challenges

Yasanthi Alahakoon and Nuwan Gunarathne*

CONTENTS

5.1 INTRODUCTION

Municipal solid waste management has become a significant challenge in many developing countries in Asia, creating numerous health, environmental and socio-economic problems (Fernando 2019; Gunarathne et al. 2019; Wilson et al. 2013). Since solid waste management persists as a considerable issue in these countries, the management of special and hazardous waste streams has not yet received the due attention of the policy-makers and regulators, communities, and the public. For instance, many scholars report that e-waste is becoming a significant problem in developing Asian countries such as China (Qu et al. 2019), India (Borthakur and Govind 2018; Sharma et al. 2020), Malaysia (Ismail and Hanafiah 2019), Sri Lanka (Gunarathne et al. 2020a; Mallawarachchi and Karunasena 2012), and Thailand (Sasaki 2020). Management of e-waste is a growing concern as the technological advancements, novel designs and features, and increased consumer purchasing power due to economic growth have significantly reduced the lifespan of electronic goods in these countries (Balde et al. 2017; Gunarathne et al. 2020a; Patil and Ramakrishna 2020). Therefore, e-waste has become one of the fastest-growing waste categories in developed countries and in the developing countries in Asia in terms of its volume and environmental significance (Baxter et al. 2016; Ardi and Leisten 2016; United Nations, UN 2017). However, Balde et al. (2017) reveal that only 20 percent of global e-waste is appropriately collected and recycled, leaving the rest of the e-waste disposed of with other types of waste, dumped, traded, or recycled in improper ways.

Due to these growing trends, the sustainable management of e-waste should receive urgent global attention. As Ismail and Hanafiah 2019 (p. 2) emphasize, "unlike other forms of waste, managing the final disposal of e-waste is complicated because it has a complex structure composed of various materials, including hazardous substances." In particular, e-waste contains precious metals, valuable bulky materials as well as hazardous materials in the form of heavy metals (such as lead, mercury, arsenic, and cadmium) and chemicals (such as CFCs/chlorofluorocarbon or various flame retardants) (Balde et al. 2017; Baxter et al. 2016; Borthakur and Govind 2018; Golev and Corder 2017). From an anthropogenic resource management system perspective, the total value of raw materials in e-waste was estimated at approximately 55 billion Euros in 2016 (Balde et al. 2017). In order to efficiently harvest the resources of this "urban mine" and to overcome the problem of diminishing natural resources, it is necessary to overcome the inefficient "take-make-dispose" economic model (Cossu and Williams 2015). Therefore, the circular economic activities for e-waste should be promoted by encouraging closing the loop of materials as opposed to the conventional unidirectional concept of resources and products in the market economy (George et al. 2015). The e-waste recycling industry could create new business ventures while providing employment opportunities for a large workforce in developing countries in the sectors of collection, primary sorting, dismantling activities, and other recycling operations (Gunarathne et al. 2020a).

Regardless of these opportunities, the management of growing e-waste stocks has impeded the realization of some critical sustainable development goals (SDG), including good health and well-being (SDG No. 3), clean water and sanitation (SDG No. 6), sustainable cities and communities (SDG No. 11) and responsible consumption and production (SDG No. 12) (Ilankoon et al. 2018; UN 2017). Immediate addressal of e-waste is also becoming a pressing need in South Asia, where a comprehensive legal and policy framework is yet to be devised (Gunarathne et al. 2020a).

Countries in South Asia encounter challenges in managing both e-waste generated within the country and those of used electronic and electrical items imported (Sthiannopkao and Wong 2013). In particular, South Asia's rising income levels have resulted in considerably high demands for electronic items. This has led to a noticeable rise in the generation of e-waste. However, the South Asian countries, including Sri Lanka, experience growing environmental challenges, including managing e-waste due to many reasons such as inadequate legal and policy frameworks and enforcement, unavailability of advanced treatment amenities, not having a well-developed formal sector, presence of a large semiformal or informal recyclers, and lack of trustworthy data and infrastructure (Borthakur and Govind 2018; Gunarathne and Lee 2019; Gunarathne et al. 2020a; Gollakota et al. 2020). While much research has been conducted on the current status of e-waste management in countries such as India (for instance, Borthakur and Govind 2018), the context of the other countries in the region is yet to be further explored. Therefore, this chapter aims to present the current status of e-waste management in Sri Lanka and its challenges. It also sheds light on the potential strategies to pursue in addressing these challenges.

The rest of the chapter is organized as follows. The second section presents the legal framework for e-waste management. Methodology followed in developing this chapter is presented in the third section. The fourth section discusses the current status of e-waste management in Sri Lanka, followed by a section on the e-waste management challenges from macro, meso, and micro-level perspectives. The next section explores the different e-waste management strategies for the identified challenges in the Sri Lankan context. The final section provides conclusions.

5.2   SRI LANKAN LEGAL FRAMEWORK FOR E-WASTE MANAGEMENT

Although a surge in the use of electrical and electronic items in Sri Lanka is observed, the regulatory aspects concerning the end-of-life cycle treatment of these items are still at a rudimentary level of development. For proper management of e-waste, there should be regulations governing the e-waste management landscape and effective enforcement. This section aims to outline the existing legal framework for e-waste management in Sri Lanka. The legal framework presented here covers two aspects; first, the initiatives Sri Lanka has taken as a party to global conventions, and second, the country-specific initiatives.

As a part of the global initiatives, Sri Lanka has ratified three global conventions

a)   The Basel Convention,
b)   The Rotterdam Convention, and
c)   The Stockholm Convention

Nearly three decades ago, Sri Lanka ratified the Basel Convention in 1992. The Central Environmental Authority (CEA), which comes under the Ministry of Environment and Natural resources, is the main government institution responsible for e-waste management aspects in Sri Lanka (CEA 2013). Accordingly, the CEA is the competent authority in the implementation of the convention. The aim of the Basel Convention – controlling the transboundary movement of hazardous waste – has the potential to positively contribute to e-waste management of the country (Suraweera 2016). Control of the international movement of hazardous waste is essential for developing countries like Sri Lanka as there is a high tendency of toxic trade inflows to countries in the developing part of the world (Gunarathne et al. 2020a).

Following the Basel Convention, Sri Lanka has taken initiatives to manage transboundary movements (i.e., export and import) of hazardous waste, including e-waste. In terms of imports, Sri Lanka has restricted the inflows of hazardous waste to the country for final disposal and recovery purposes. However, the export of hazardous waste for the final disposal and recovery is not restricted in Sri Lanka, which happens only under the purview of the Basel Convention (Basel Convention Country Fact Sheet 2011).

In addition to the Basel Convention, Sri Lanka ratified the Stockholm Convention in 2005. This Convention aims to manage Persistent Organic Pollutants (POPs), which are directed towards protecting human health and the environment (CEA 2019). To implement this convention, Sri Lanka has drafted a National Implementation Plan (NIP). Further, the Rotterdam Convention was ratified by Sri Lanka in 2006 (CEA 2019). As per this convention, the Ministry of Environment and Natural resources is vested with the responsibility of controlling the harmful impacts of industrial chemicals, including e-waste.

In addition to the pursuance of the above-stated global initiatives, Sri Lanka has also taken country-specific initiatives towards e-waste management as follows

a) Enforcement of the National Environment Act
b) Enforcement of the National Environmental (Protection & Quality) Regulation
c) Adoption of the National Electronic Waste Management Policy

The National Environment Act No. 47 of 1980 is the primary regulation covering all the environmental management aspects, including e-waste management in the country. This Act led to the introduction of two regulatory tools, viz., the Environmental Protection License (EPL) and Environmental Impact Assessment (EIA) scheme. The enactment of these schemes comes under the Central Jurisdiction of the Environment Authority, as the country's regulatory body concerning waste management aspects, including e-waste. The EPL aims to prevent/minimize the negative impacts on the environment from the toxic pollutant emissions from prescribed activities and industrial processes. Accordingly, EPL is issued to those activities/industries that comply with the EPL objectives. Therefore, the e-waste handlers, such as recyclers, must obtain this license (Gunarathne et al. 2020a).

The purpose of EIA is to ensure sustainability through the development projects carried out within the country. EIA is an internationally recognized process that assesses the potential impacts on the natural and social environment by the development activities. Therefore, these impacts will be assessed and measured to prevent/minimize the negative consequences and enhance positive contributions (CEA 2013). The Gazette No. 772/22 of 24.06.1993, which specifies the types of projects that require an EIA, includes the construction of waste treatment plants treating toxic or hazardous waste. As such, the e-waste treatment facilities are required to obtain an EIA.

The National Environmental (Protection & Quality) Regulation prescribes a listed type of e-waste; mercury waste from fluorescent lamps/bulbs, computers removed from use and accessories are treated as hazardous (Gunarathne 2015). Due to the health and environmental threats of handling these wastes, this regulation requires the organizations that generate, collect, store, recycle, recover, or dispose the listed items to get a Scheduled Waste Management (SWM) Licence from the CEA (Gunarathne et al. 2020a; Onuma et al. 2018). Certain private and public institutions have made the Scheduled Waste Management License a primary requirement of the selection criterion in the tender-calling process when they hand over e-waste items. It has encouraged the informal sector to apply for the license and legally operate their activities.

Despite the availability of a regulatory framework, a major challenge for human health and environmental protection in Sri Lanka is the unavailability of an effectively functioning national policy towards e-waste management (Gunarathne et al. 2020a). The CEA has taken initial steps towards a national policy by drafting the National Electronic Waste Management Policy in 2008 (Mallawarachchi and Karunasena 2012). As a country-specific initiative taken by the CEA, it is developed under the Basel Convention provisions. Even though it is intended to be a national strategy towards e-waste management, this has not yet been made public (Gunarathne et al. 2020a). As this policy remains at the draft level, the management of e-waste is conducted without national policy guidance in Sri Lanka.

Apart from the CEA, a few other institutions are responsible for regulating e-waste management in Sri Lanka (Auditor General's Department 2016). Amongst them, the Department of Import and Export Control is a crucial institute to control the importation of electronic equipment and mobile phones that do not comply with the specified criteria as outlined in the Imports and Exports (Control) Act. The Imports and Exports (Control) Act criteria have been established to prevent the importation of electronic and electrical items closer to the end of the life cycle, which would end up in e-waste within a very short-term of usage. Besides, the Consumer Affairs Authority is also expected to monitor the quality and durability of electronic and electrical items imported. A scrutiny of the quality and durability of the importation of electrical and electronic items to the country is expected to minimize premature disposal, leading to e-waste. Apart from these institutions, the Telecommunications Regulatory Commission of Sri Lanka pays specific attention to the importation of mobile phones to the country (Telecommunications Regulatory Commission 2020). However, the Commission is said to have no specific methodology to monitor the discarding of the imported mobile phones, which contributes mainly to the e-waste generation.

5.3 METHODOLOGY

The data for this chapter was collected as a part of an ongoing research project that investigated the waste recycling practices in Sri Lanka. From 2014 onwards, this main project explored various facets of waste recycling and management in Sri Lanka while paying particular attention to the e-waste recycling industry. As such, this project broadly adopted a case study approach by explicitly focusing on the country's recycling industry as the unit of analysis. As highlighted by Yin (2013), the pursuance of a case study method enables to analyze the phenomena in-depth and to provide a well-structured data gathering method from the e-waste management industry. Further, as it allows collecting data from multiple sources and respondents, the case study method facilitates triangulating the data collected from various sources (Yin 2013; Tellis 1997).

This study's data was collected from multiple sources, including semi-structured interviews, site visit observations, and document analysis. The primary data

TABLE 5.1
List of Interviewees

Organization/ Institution/Respondent	Person interviewed
E-waste recycler 1	Managing Director
E-waste recycler 2	CEO
E-waste recycler 3	Factory Manager
E-waste collector	Owner
Waste collector	Owner
Commercial sector organization 1 [From Apparel sector]	Manager- Sustainability and Business Development
Commercial sector organization 2 [From IT sector]	Head of IT and Services
Commercial sector organization 3 [From Plantations sector]	General Manager
End-user of e-waste	Head of Operations
Government institution	Director -Waste Management
Local government 1	Chairman
Local government 2	Revenue Inspector
Environmental activist group	Convener
Households	Family members

collection method, semi-structured interviews, covered several industry partners, including e-waste recyclers, collectors, business organizations, end-users, government associations and local governments, and environmental activists (Table 5.1). Besides, to provide the micro-level view from the e-waste generators' point of view, interviews were conducted with 12 households from three major cities in Sri Lanka. Most of these interviews were conducted face-to-face, and a few were done over the phone. When feasible, these interviews were audio-recorded for later transcription. The research team engaged in the interview process kept during the interviews and after. The site visit observations were made for the recycling centres, municipal waste management facilities, and sorting and processing centres. During these visits, informal discussions were made with the employees and facility managers to obtain additional data and clarifications. These site visits provided the research team with firsthand experience on the e-waste recycling practices, the collectors and recyclers' challenges, and possible solutions. Moreover, the research team extensively used various documents in public domains such as the company and government institutions' web sites, policy documents, government regulations, and newspaper articles. These different data collection methods enabled the research team to synthesize the data collected from multiple sources while triangulating the findings (Golafshani 2003; Yin 2017).

The transcribed interview data were analyzed based on the study's main themes, and the other collected data was utilized to supplement the interview findings, which are presented in the forthcoming sections of the chapter.

5.4 CURRENT STATUS OF E-WASTE MANAGEMENT IN SRI LANKA

This section discusses the current state of e-waste management in the country. First, it provides an overview of the e-waste recycling industry, followed by the e-waste collection aspects currently in existence. Subsequently, the industry value chain is explained, and finally, the e-waste categories and treatment process are briefed.

5.4.1 Sri Lanka's Recycling Industry: An Overview

E-waste recycling is a vital aspect of the whole e-waste management process. It is a global issue on which more contemporary debates have been initiated concerning the negative impacts that e-waste recycling creates, especially in the developing world (Shaikh et al. 2020; Zhang et al. 2012). In developing countries like Sri Lanka, e-waste recycling is becoming important due to environmental and social issues. With the economic growth and increase in consumer expenditure, Sri Lanka has encountered more acute e-waste recycling issues than it did a few decades back (Deyshappriya and Kumari 2019; Gunarathne et al. 2020a; Ranasinghe and Athapattu 2020). The improving economic conditions in the country lead to increased consumption of electrical and electronic products, which ultimately ended up as e-waste, highlighting the recycling industry's importance. Added to this situation is the fast obsolescence of these products due to rapid technological developments (Gunarathne et al. 2020a).

In the global context, e-waste recycling operations have shifted mainly to developing countries due to the ability to recycle them at a lower cost in these countries, primarily due to manual labour. Accordingly, e-waste generated in the developed parts of the world is imported to developing countries such as China, India, Pakistan, and a few other Asian countries (Zhang et al. 2012). In contrast, the industry presents significant opportunities to countries that import e-waste, where they can satisfy the demand for valuable materials through urban mining. However, regardless of these benefits, e-waste recycling comes at the cost of protecting the natural environment and human health. Notwithstanding the global conventions and country-specific initiatives, the e-waste recycling related issues are ongoing as the laws and regulations in these countries are not effectively enforced.

In the Sri Lankan context, the e-waste recycling industry has to cope with both internally generated and imported e-waste (Gunarathne et al. 2020a). However, accurate information about the amount of e-waste generated within the country is not formally known. Being in line with the Basel Convention, the Ministry of Environment and Natural Resources has taken an initial inventory of e-waste in 2008 (Ministry of Environment and Natural Resources 2008). However, there is no published information on any upgrades to this inventory since then, which indicates the inadequate attention paid to the research and developments of e-waste within the country, which is a common observation among developing countries. Ranasinghe and Athapattu (2020) highlight the lack of a database regarding e-waste as a significant constraint towards managing e-waste practically. For instance, designing e-waste management systems and establishing waste management facilities are impeded due to a lack of data. Despite the lack of e-waste quantifications within the country, the United Nations Development Programme (UNDP) has estimated a generation of

70–75 metric tonnes of e-waste per annum within the country, and this is expected to increase over time (UNDP 2015).

5.4.2 REGULATING E-WASTE COLLECTION

As already mentioned, the CEA is the primary regulatory body in Sri Lanka that governs the e-waste management aspects. As of 2020, the CEA has published 14 e-waste collectors in the country with their license status (Table 5.2). Although the CEA labels these organizations as e-waste collectors, detailed information on how they manage e-waste after the collection is not publicly accessible. The CEA has also classified three organizations on the list as licensed e-waste exporters. Further, it is interesting to note that these licensed collectors are concentrated within Western Province except for two e-waste collectors who work outside this province (Gunarathne et al. 2020a). This concentration is justifiable as, at one point, the Western province is recognized as the largest generator of e-waste (Fernando 2019). Nevertheless, it also indicates the lack of a proper nationwide program for e-waste collection. The risk of the unavailability of a national program is e-waste being ended up in landfills in the areas outside the Western province, leading to environmental and health hazards at the regional level.

In addition to regulating the licensed e-waste collectors, the CEA has also adopted a few other measures to manage e-waste in the country. However, most of them represent more fragmented approaches with short-term programs for collecting e-waste. Therefore, these programs' impacts would be more short-lived than the implementation of a full-fledged long-term national policy. For instance, in 2014 and 2020, the CEA conducted two separate short-term programs to collect e-waste island wide. In 2014 a National Corporate E-Waste Management Program under the theme of "Ensuring an e-waste free Sri Lanka" was conducted. It was noteworthy to mention that during this program, CEA had entered into a memorandum of understanding with partner companies to ensure safe collection and disposal of e-waste (CEA 2014). As an extension to this program, an e-waste drop off event was launched island wide.

More recently, in October 2020, another short-term program was launched. To coincide with the postal day, an island wide e-waste collection program was initiated for a period of one week under the theme "a country that breaths, Sri Lanka without e-waste" by the CEA (Economynext 2020). In the initiation of this program, the authorities acknowledged the unavailability of a proper e-waste recycling program. The program invited the public to hand over e-waste to local post offices for collection. However, the CEA has not disclosed a plan for how the collected e-waste is handled or managed.

5.4.3 E-WASTE RECYCLING INDUSTRY'S VALUE CHAIN

The value chain is complicated as various actors are involved, and multiple channels of e-waste flows are available in the industry. A simplified value chain is presented in Figure 5.1. Despite the simplicity, placing the industry players within a value chain provides a better ground on which the recycling industry's issues could be

TABLE 5.2
Licensed E-Waste Collectors in the Country

Company name	Types of e-waste collected	Location (Province)	License to export
Asia Recycling (Pvt) Ltd	CFL bulbs, fluorescent bulbs	Homagama (Western province)	
Ceylon Waste Management (Pvt) Ltd	E-waste excluding CFL bulbs, fluorescent bulbs & CRT monitors	Kelaniya (Western province)	Available
Cleantech (Pvt) Ltd	E-waste excluding CFL bulbs, fluorescent bulbs & CRT monitors	Colombo (Western province)	
Ecogate Lanka Engineering Services	Used batteries & accumulators, waste electric equipment, discarded mobile phones	Wadduwa (Western province)	
E Waste Logistics Solutions (Pvt) Ltd	E-waste excluding CFL bulbs, fluorescent bulbs & CRT monitors	Colombo (Western province)	
Infinity Green International (Pvt) Ltd	E-waste excluding CFL bulbs, fluorescent bulbs & CRT monitors	Kelaniya (Western province)	Available
Inova Environmental Services (Pvt) Ltd	E-waste including computers equipment, mobile phones & batteries	Padukka (Western province)	
INSEE Eco Cycle Lanka (Pvt) Ltd	E-waste excluding CFL bulbs, fluorescent bulbs & CRT monitors	Puttalam (Northwestern province)	
J F Supplier	E-waste excluding CFL bulbs, fluorescent bulbs & CRT monitors	Mawnella (Sabaragamuwa province)	
N.S.Green Links Lanka (Pvt) Ltd	E-waste excluding CFL bulbs, fluorescent bulbs & CRT monitors	Divulapitiya (Western province)	
Recotel Lanka (Pvt) Ltd	E-waste excluding CFL bulbs, fluorescent bulbs & CRT monitors	Colombo (Western province)	Available
SCT Holdings (Pvt) Ltd	E-waste excluding CFL bulbs, fluorescent bulbs & CRT monitors	Kottawa (Western province)	
Think Green (Pvt) Ltd	Not specified	Colombo (Western province)	

Source: Developed based on the CEA's (2020) licensed collectors of e-waste

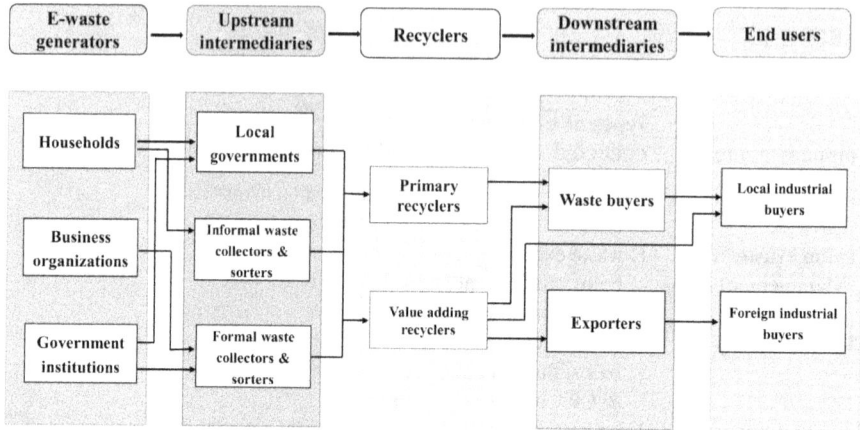

FIGURE 5.1 Recycling industry value chain. (Source: Adapted from Gunarathne et al. 2020a.)

understood. Further, the whole value chain view enables more integrated e-waste management approaches to be adopted rather than initiating isolated and fragmented approaches (Illankoon et al. 2018). The integrated approaches would provide more effective solutions to manage e-waste instead of fragmented systems such as one-off collections, as mentioned above. The network view provided in Figure 5.1 includes the e-waste generators, upstream and downstream intermediaries, recyclers, and end-users. These interrelationships depict the complex network in this value chain (Gunarathne et al. 2020a).

The value chain begins with the e-waste generators, including the electronic equipment consumers at household, corporate, and government levels. During the next stage of the value chain, upstream intermediaries play a crucial role. The upstream intermediaries include the e-waste collectors and sorters, which could be informal or formal. The local governments also play an intermediary role in this value chain by collecting e-waste through their various waste collection mechanisms. Informal collectors play a major role in collecting e-waste from the household sector, while business organizations and government institutions tend to hand over their e-waste to the formal sector (Gunarathne et al. 2019). In managing e-waste, the corporate sector in Sri Lanka uses the formal sector due to the pressure to adhere to socially responsible protocols. Similarly, government institutions also utilize the formal sector to comply with the established waste disposal procedures of the government sector.

Recyclers compose the proceeding section of this value chain. They are two-fold a) primary recyclers who engage in dismantling the components, and b) value-adding recyclers who go a further step by engaging in value-adding activities. The downstream intermediaries are the e-waste buyers and exporters. The e-waste buyers are from both the formal and informal sectors. The final stage of the value chain comprises the end-users who are local and foreign industrial buyers who utilize e-waste as an input.

TABLE 5.3
Major E-Waste Categories in Sri Lanka

Category of E-waste	Size of market (units p.a.)[a]	Growth rate of market
Personal computers	844,000	8%–10%
Printers	293,000	
Televisions	845,000	6%–8%
Mobile phones	26.85 mn[b]	10–12%
Refrigerators	480,000	4 %-6%
Air-conditioners	81,000	4 %–6%
Photocopying machines	12,000	2 %–4%
Washing machines	146,500	6%–8%
Batteries (Auto)	977,500	4%–6%,

Source: Gunarathne et al. (2020a); Ministry of Environment and Natural Resources (2008); Telecommunication Regulatory Commission in Sri Lanka (2020).

[a] These figures were estimated for 2020 based on the data of the Ministry of Environment and Natural Resources (2008) (base year = 2008) by compounding at the average growth rate.

[b] Although this is the number of mobile phone subscriptions as per the latest statistics of the Telecommunication Regulatory Commission in Sri Lanka (2020), the actual number of mobile phones can differ as a user may have dual sim phones.

5.4.4 E-WASTE CATEGORIES AND TREATMENT PROCESS

Various e-waste types are being produced by the household, business, and government institutions – the e-waste generators of the value chain above. The Ministry of Environment and Natural Resources has identified nine e-items used by the above generators, and the components of these items become e-waste ultimately. This assessment was done in 2008 as a part of the initiative towards the "Development of a national implementation plan for electrical and electronic waste management in Sri Lanka." Table 5.3 below lists the e-items identified through this initiative and their market sizes and growth rates.

These e-waste categories include components in large numbers that should be treated appropriately to prevent environmental and health hazards. However, in the developing country context, e-waste treatment faces considerable challenges, including the absence of advanced technologies, proper waste separation mechanisms, and even strongly enforced laws (Bhaskar and Turaga 2018; Ikhlayel 2018). Sri Lanka is no exception to such conditions. More specifically, the use of labour-intensive practices due to the absence of advanced technology and not having treatment solutions to many e-waste types are observed in the Sri Lankan context (Gunarathne et al. 2020a). For instance, information technology and telecommunication appliances such as land phones, mobile telephones, laptop computers, and entertainment devices such as televisions and CD players are dismantled manually or using simple machines. Some components dismantled, such as motherboards and CR ROMs, are exported while less valuable items, such as plastic and wire, are sold in the local market. The lighting equipment, such as CFL bulbs, is first separated into components manually, and machines are used to extract mercury and glass. Apart from these systems, there is no proper mechanism to recycle many other equipment types, such as washing machines, air conditioners, sewing machines, X-ray machines, and CCTV cameras. The e-items mentioned above include many components as well as potentially hazardous materials such as mercury; however, many of these items do not undergo proper recycling mechanisms. Even though dismantling is done for a few items, such dismantling processes largely depend on manual mechanisms. Treatment processes of this nature are ineffective since these processes fail to recover certain valuable materials. Besides, the manual processes expose the workers to health hazards.

5.5   E-WASTE MANAGEMENT CHALLENGES

This section presents the e-waste management challenges in Sri Lanka on three levels; macro-level (challenges arising from the broader regulatory, economic, environmental, technical, and institutional factors), meso-level (challenges at the recycling industry level), and micro-level (challenges associated with the electronic and electrical item consumption level or at an individual level).

5.5.1 MACRO-LEVEL CHALLENGES

Many macro-level factors create challenges for e-waste management in Sri Lanka. A significant issue is the absence of a national policy towards e-waste management. As mentioned in the legal framework, the CEA is the primary institution that oversees the country's e-waste management aspects. However, apart from the fragmented initiatives taken on an ad-hoc basis, the CEA has not yet formulated a proper institutional framework for Sri Lanka's e-waste management. The absence of an institutional framework has led to many of the challenges discussed in this section.

A lack of formal research on e-waste generation, importation, and collection in Sri Lanka is also a common issue in many other developing countries in Asia (Chakraborty 2016; Gunarathne et al. 2020a). Lack of reliable data has been a critical impediment for developing a national policy for e-waste management in Sri Lanka for years. Inadequate information about the e-waste stocks in developing countries has also led to ignorance on the severity of the present and future problems that e-waste creates (Patil and Ramakrishna 2020; Sharma et al. 2020). In Sri Lanka, other than a fragment initiative of HazNet, which is a country-wide e-waste tracking tool developed by the CEA, no further formal mechanisms have been developed for e-waste tracing purposes. Even the Haznet is not operationalized to reach its full potential (Gunarathne et al. 2020a).

Despite the CEA playing the primary role in e-waste management, many other institutions are vested with bits and pieces of e-waste management responsibilities (Gunarathne et al. 2019), including municipal councils, the Ministry of Environment, the Urban Development Authority, and the other provincial-level waste management authorities. The distributed nature of e-waste management initiatives taken by these institutions independently without proper coordination has not improved the e-waste management status of the country. The issues related to the coordination of different institutions would have been minimized had there been a national framework.

In addition to these policy and regulatory issues, many political, technological, and financial factors impose challenges towards e-waste management in Sri Lanka. From a political front, the lack of a visionary political leadership towards environment protection is another issue the country has faced over time (Fernando 2019; Gunarathne and Lee 2019; Gunarathne et al. 2019). As the other more pressing economic and social issues take the priorities in the political agendas, environmental concerns regularly are observed being marginalized by the governments. From a technological perspective, the absence of advanced technology for e-waste management is another major issue. More specifically, the advanced technologies used by the developed countries for e-waste material recoveries are not available in developing countries like Sri Lanka (Golev and Corder 2017; Herat and Agamuthu 2012; Ilankoon et al. 2018; UNEP 2013). Therefore, most of these e-waste mining activities are carried out manually through labour-intensive mechanisms, creating negative impacts, especially on human health and the environment. In financial and economic terms, this industry is not attractive to investors. The sector does not provide lucrative financial returns for the investors, and the existing tax concessions do not promote growth in this industry (Gunarathne et al. 2020a). Accordingly, e-waste management faces many challenges in Sri Lanka owing to the unconducive macro environment. These macro-level challenges have also led to many of the recycling industry's challenges at a meso level.

5.5.2 MESO-LEVEL CHALLENGES

A major issue at the recycling industry level is the absence of mechanisms to collect e-waste separately. Even though the local governments have created mechanisms for collecting general waste, especially in urban areas of the country, no such systems exists to collect e-waste. This problem becomes severe in the rural areas of the country (Onuma et al. 2018). As Table 5.2 illustrates, there are only 14 CEA-approved e-waste collectors, who are primarily concentrated in the Western Province of the country. This signifies the lack of attention paid to e-waste management in rural areas, where e-waste is often mixed up with general waste.

The unavailability of a regulated collection mechanism often leads to the emergence of unregulated informal and semiformal e-waste collectors in the industry (Balasubramanian and Karthikeyan 2016). Sri Lanka's e-waste recycling industry, with the informal and semiformal sector booming, is currently facing the consequence of not having a regulated mechanism. The e-waste collected by this sector may end up in open lands since their collection and subsequent e-waste mining activities are not monitored by any formal authority. This could create health hazards and environmental problems.

The industry players face significant challenges due to a lack of financial-economic support for the waste management industry by the governments (Gunarathne et al. 2019). For example, there is no proper infrastructure developed in the country to facilitate the e-waste recycling industry. Therefore, it imposes additional financial burdens on the individual recyclers to accommodate these facilities, such as noise management and pollution control mechanisms. The unavailability of tax concessions or incentives makes the industry less attractive, as the financial burden of these additional support mechanisms should be borne by the recyclers themselves.

All these factors make the industry financially unattractive. At one point, the lack of investment in advanced technologies minimizes the cost-effectiveness of recycling because the e-waste recycling process relies heavily on labour-intensive mechanisms that reduce material extraction productivity. For instance, the lack of advanced technology prevents recyclers from extracting more valuable materials from e-waste (Sthiannopkao and Wong 2013).

Additionally, the labour-intensive mechanisms adopted by the e-waste recyclers impose threats to the environment and human health (UN 2017; Chi et al. 2011). More specifically, the health and wellbeing of the manual workers exposed to hazardous materials is a significant concern that has not yet received the attention of the industry or government. The problem becomes even more severe as most workers employed in these extracting activities are without proper training for handling hazardous materials. Unfortunately, many recyclers do not dedicate a specific budget to employee training as they are already struggling financially.

5.5.3 MICRO-LEVEL CHALLENGES

The micro-level challenges stem mainly from two reasons; a) lack of awareness and b) absence of regulations and policies (Baxter et al. 2016; UN 2017). More precisely, the consumer or general public is not aware of the life cycle of the electronic and electrical, appropriate disposal methods, and health hazards due to improper disposal. The lack of consumer unawareness of the hazardous nature of e-waste imposes a critical challenge within Sri Lanka, as it leads to irresponsible e-waste disposal behaviour patterns, which are commonly observed in developing country contexts (Borthakur and Govind 2018; UN 2017). On the other hand, low consumer awareness levels do not promote responsible and carbon-conscious consumption behaviours (Gunarathne et al. 2020b).

Lack of awareness, coupled with a lack of e-waste disposal regulations, leads to non-separated e-waste disposal by the e-waste generators (or consumers). Therefore, e-waste is usually mixed up with general waste in Sri Lanka. The governing institutions responsible for e-waste management have not formulated specific regulations or general guidelines on how the e-waste generators should dispose of the e-items at the end of their useful life cycle. Unfortunately, in Sri Lanka, most household e-waste generators have almost no knowledge of this problem.

The absence of knowledge and regulations or guidelines has also led to several other issues. One of the most critical issues is irresponsible e-waste disposal. For instance, in many households, e-waste generally ends up being dumped into open land, mixed with other general waste. These could be either privately owned lands of these households or common waste dumping places, or even roadsides (Chakraborty 2016; Sittampalam 2019). Further, households usually bury lithium batteries with other waste, and broken CFL bulbs are thrown away with other general waste. These e-waste items contain hazardous materials that could contaminate the soil and water sources, causing threats to human health.

Moreover, the households tend to use the e-items much longer than their expected life cycle. Therefore, everyday household items such as televisions, refrigerators, washing machines, rice cookers, and many other items are held in the household beyond their usual lifespan (Gunarathne et al. 2020a). Very often, irrespective of their mal-functionality or complete non-functionality, these items are not disposed of but piled up in households. When these items deteriorate for an extended period, the recycling process becomes more difficult. On the other hand, these excess items piled up in the household are sometimes misused. For instance, these items are used by children as play items leading to many reported accidents such as swallowing batteries. The e-waste stocks piling up within households also prevent the government from accurately calculating the actual e-waste inventories that pose challenges to formulate policy and action plans for e-waste management.

The mixing of e-waste with other waste categories even creates problems for managing the other/general waste as they also get contaminated, creating environmental and health problems for the waste handlers. Conversely, unawareness of e-waste handling at the micro-level leads to other dangers as a result of the actions without knowledge of the consequences (Hettiarachchi and Sarathchandra 2017). Individuals often try to use insecure methods to extract materials from e-waste at the household level. For example, they burn coils or circuit boards at the household level to extract copper and other metals without adopting any proper safety procedures.

5.6   E-WASTE MANAGEMENT STRATEGIES

As discussed in the previous section, e-waste management challenges in Sri Lanka are widespread across the macro, meso, and micro levels. This section discusses a multitude of strategies that could be used to address these challenges and ease out the negative impacts. These strategies spread across the main concerns such as law and policy imposition and enforcement, capacity building initiatives within the industry, and increasing awareness regarding e-waste. The rest of this section briefly discusses how these strategies could be more accomplished explicitly within Sri Lanka.

Institutional policies have been highlighted as an essential concern in overcoming e-waste challenges (Daum et al. 2017). As discussed already, the CEA is playing a pivotal role in e-waste management within the country. However, most of the CEA initiatives have so far been fragmented without the guidance of a national framework. Therefore, developing a national framework for e-waste management is of critical importance to guide and direct the various ad-hoc initiatives adopted by different institutions towards sustainable e-waste management. As shown in Table 5.2, the current e-waste recycling activities are highly concentrated within the Western urban areas of the western province, while in most other areas of the country, e-waste ends up in either the informal sector or in open lands. If a national framework is available, e-waste handling and management across the country could be standardized.

Enhancing the institutional capacity of the relevant national institutions is a recognized strategy for the management of e-waste, especially in countries in the developing world (Gunarathne et al. 2020a). Accordingly, in the Sri Lankan context, the CEA's institutional capacity should be strengthened to minimize the uncoordinated handling of e-waste by several other institutions mentioned earlier, such as municipal councils, the Ministry of Environment, and other provincial-level waste management authorities. Enhancement of the capacity could also result in better-centralized authority on CEA to initiate coordinated efforts at the national level in managing e-waste in Sri Lanka. Furthermore, to address the problem of lack of research data on the e-waste inventories, the existing HazNet information systems could be put into continuous and more effective practice. This information system could facilitate the CEA to increase its waste tracking capabilities towards national e-waste policy formulation and law enforcement.

In addition to these institutional changes, macro-level support is needed for an e-waste recycling industry, specifically in terms of financial and law enforcement aspects (Ardi and Leisten 2016; Anschutz et al. 2004; Chi et al. 2011). Due to the financial non-attractiveness, the e-waste recycling industry in Sri Lanka attracts fewer investments. To improve this condition, government policies should be drafted to provide attractive financial incentives to investors. For example, these incentives could be provided in the form of tax concessions, employee training assistance, or infrastructure facilities. The financial aid for the industry should also be accompanied by the effective enforcement of relevant laws and regulations, especially towards waste separation.

Furthermore, the introduction of new laws to be on par with globally recognized e-waste management practices is also essential to enhance the country's current low e-waste recycling status. For example, introducing the extended producer responsibility (EPR) principle could be one of the critical concerns in the new legislation. According to the ERP principle, the producer or the local distributor takes the overall responsibility for a product. Thus, the responsibility for the management of a product's whole life-cycle is vested with the producer, including collecting, dismantling, and reusing when the product reaches the end of its life (Atasu and Subramanian 2012; Cao et al. 2016). As Sri Lanka is facing acute problems of handling e-items at the end-of-life cycle, the ERP principle can be a catalyst to enhance sustainability design aspects of electronic and electrical items while seeking funds to develop the national level e-waste recycling infrastructure.

Additionally, if the country is to overcome the e-waste recycling challenges, the inadequacies in both the advanced technology and technological know-how should be urgently addressed (Ilankoon et al. 2018; Wilson et al. 2006). For this purpose, dedicated e-waste zones with established infrastructure could be established with training facilities for the recyclers on using advanced technologies. Moreover, general vocational training on e-waste handling can also be provided for the industry workers. Technical expertise exchange with other countries is another essential option to enhance knowledge. Collaborative agreements with foreign industry partners for knowledge sharing and utilizing the opportunities provided by international donor agencies as the ABD, UNDP, and World Bank are some of the other options that could be further evaluated (Gunarathne et al. 2020a). Similarly, as a long term strategy towards e-waste management, e-waste-related research and development should be encouraged. This is important as the current information on the e-waste inventories and handling mechanisms are highly inadequate. The research and development activities could be strengthened through the state university system by providing specific funds for this objective.

In addition, to enhance the technical knowledge of the actors of the e-waste management industry, the general public should also be made aware of the e-waste issues (Baxter et al. 2016; Borthakur and Govind 2018; UN 2017). This is important as the households are a vital category of e-waste generators in the e-waste value chain. A nationwide awareness of different aspects of e-waste, including useful life, proper disposal mechanisms, laws and regulations on e-waste, and health hazards of improper disposals could be created. The awareness programs could also be incorporated into school curricula as a long-term strategy towards producing an e-literate future generation. As mentioned earlier, the majority of the CEA licensed e-waste collectors are located within the Western province of the country. However, the public awareness of the availability of such collectors is not satisfactory even within the province. Therefore, institutions such as the CEA should create awareness programs to encourage households to hand over e-waste to licensed collectors. In this way, issues arising due to the booming, unregulated informal sector could be minimized.

Another problem at the e-waste generator level is the non-separation of waste (Fernando 2019; Gunarathne et al. 2019; UN 2017). Even though some regulations towards waste separation have been introduced, they are not adequately implemented at the household level. Therefore, more targeted actions should be taken concerning waste separation at these generation points. More specifically, regulations should be enforced to encourage e-waste separation from the general waste at the point of origin, which would minimize the waste handling complications and even the health and environmental issues faced at the later stages of the e-waste value chain.

Table 5.4 summarizes the e-waste management challenges and the remedial strategies discussed in the chapter under five broad areas: a) institutional, b) financial, c) legal, d) technology and know-how, and e) awareness related challenges.

5.7 CONCLUSION

The recent economic and social developments in Sri Lanka have led to a noticeable rise in the generation of e-waste, which in turn has led to challenges in managing this growing level of e-waste. Therefore, this chapter presents the current status of e-waste management in Sri Lanka and its challenges while shedding light on the potential remedial strategies. The chapter was developed based on the data collected as a part of an ongoing research project investigating the waste recycling practices in Sri Lanka. This project broadly adopted a case study approach by explicitly focusing on the country's recycling industry as the unit of analysis while collecting data from multiple sources, including semi-structured interviews, site visit observations, and document analysis.

The chapter discusses how the lack of a national-level policy and regulatory framework has given rise to many e-waste management challenges visible at macro, meso, and micro-levels in the country. These interconnected challenges in a vicious cycle act as an impediment that keeps the country from moving towards sustainable management of rising stocks of e-waste. In order to overcome the e-waste management challenges, Sri Lanka needs to adopt several strategies simultaneously and urgently. These remedial strategies broadly include the formulation of national policy and enforcement of the law, pursuing producer responsibility extending principles

TABLE 5.4
Strategies for E-Waste Management

The broad area of challenge	Potential strategies
Institutional challenges	• Development of a national framework to avoid fragmented practices • Enhancement of institutional capacities • Improvement of the existing HazNet information system
Financial challenges	• Enhancement of financial attractiveness of the industry (E.g., tax concessions for investors)
Legal challenges	• Introduction of new laws and effective enforcement of the existing laws • Introduction of the extended producer responsibility principle
Technology and know-how related challenges	• Establishment of dedicated e-waste processing zones with required infrastructure • Provision of training for the recyclers on using the advanced technologies • Introduction of vocational training programs on e-waste handling • Collaboration with foreign industries/partners for knowledge sharing • Utilization of opportunities provided by international donor agencies (e.g., ABD, UNDP, and World Bank) for the capacity building of the industry • Allocation of funds for research and development activities in collaboration with state universities
Awareness related challenges	• Implementation of nationwide awareness creation programs • Incorporation of e-waste awareness into school curricula • Encouraging the public to handover waste to licensed e-waste collectors • Educating consumers and the introduction of regulations on e-waste separation

for e-waste, formalization of the informal e-waste sector, capacity building, consumer awareness, and education incentivization the recycling industry. While the primary responsibility of the remedial actions lies with the government, the support of business organizations, informal and semiformal e-waste sector actors, consumers, and civil society organizations will be essential to move towards sustainable management of e-waste within the country.

REFERENCES

Anschutz, J., gosse, J., and Scheinberg, A. 2004. *Putting integrated sustainable waste management into practice: Using the ISWM assessment methodology as applied in the UWEP Plus Programme (2001–2003)*. Gouda, the Netherlands: WASTE.

Ardi, R., and Leisten, R. 2016. Assessing the role of informal sector in WEEE management systems: a system dynamics approach. *Waste Management* 57:3–16.

Atasu, A., and Subramanian, R. 2012. Extended producer responsibility for e-waste: individual or collective producer responsibility? *Production and Operations Management* 21(6): 1042–1059.

Auditor General's Department. 2016. *Electronic waste management in Sri Lanka- Report No. PER/2016/EW/01*. Colombo: Performance and Environment Audit Division, Auditor General's Department.

Balasubramanian, R., and Karthikeyan, O. P. 2016. E-waste recycling environmental and health impacts. In *Handbook of advanced industrial and hazardous wastes management*, ed. J.P. Chen, L. K. Wang, M. H. S. Wang, Y. T. Hung., and N. K. Shammas. 339–364. Boca Raton: CRC Press.

Balde, C.P., Forti, V., Gray, V., Kuehr, R., and Stegmann, P. 2017. *The global e-waste monitor – 2017*. Bonn/ Geneva/Vienna: United Nations University (UNU), International Telecommunication Union (ITU) & International Solid Waste Association (ISWA).

Basel Convention Country Fact Sheet (2011). Country fact sheet. www.basel.int/countries/countryfactsheets/tabid/1293/default.aspx.

Baxter, J., Lyng, K. A., Askham, C., and Hanssen, O. J. 2016. High-quality collection and disposal of WEEE: environmental impacts and resultant issues. *Waste Management* 57:17–26.

Bhaskar, K., and Turaga, R. M. R. 2018. India's e-waste rules and their impact on e-waste management practices: A case study. *Journal of Industrial Ecology* 22(4): 930–942.

Borthakur, A. and Govind, M. 2018. Public understandings of E-waste and its disposal in urban India: from a review towards a conceptual framework. *Journal of Cleaner Production* 172:1053–1066.

Central Environmental Authority (CEA). 2013. Granting consent for transboundary movement of hazardous waste (import/export/transit) www.cea.lk/web/granting-consent-for-transboundary-movement-of-hazardous-waste-import-export-transit.

Central Environmental Authority (CEA). 2014. National cooperate e-waste management program-collection Centers www.cea.lk/web/en/component/content/article/2-.

Central Environmental Authority (CEA). (2019). Rotterdam and Stockholm Conventions www.cea.lk/web/en/component/content/article?id=1500:chemicals-hazardous-waste-management-unit

Central Environmental Authority. (CEA). 2020. Licensed collectors of electronic waste management in Sri Lanka. www.cea.lk/web/en/index-php-option-com-content-view-article-layout-edit-id-983.

Chakraborty, P., Selvaraj, S., Nakamura, M., Prithiviraj, B., Ko, S., and Loganathan, B. G. 2016. E-waste and associated environmental contamination in the Asia/Pacific region (Part 1): An overview. In *Persistent organic chemicals in the environment: Status and trends in the Pacific Basin Countries I Contamination Status*. ed. B. G. Loganathan., J. S. Khim., P. R. S. Kodavanti., and S. Masunaga. 127–138. Washington, DC: American Chemical Society.

Chi, X., Streicher-Porte, M., Wang, M. Y. L., and Reuter, M. A. 2011. Informal electronic waste recycling: a sector review with special focus on China. *Waste Management* 31:731–742.

Cao, J., Lu, B., Chen, Y., Zhang, X., Zhai, G., Zhou, G., Jiang, B., and Schnoor, J. L. 2016. Extended producer responsibility system in China improves e-waste recycling: government policies, enterprise, and public awareness. *Renewable and Sustainable Energy Reviews* 62:882–894.

Cossu, R., and Williams, I. D. 2015. Urban mining: concepts, terminology, challenges. Waste Management 45:1–3.

Daum, K., Stoler, J., and Grant, R. J. 2017. Toward a more sustainable trajectory for e-waste policy: a review of a decade of e-waste research in Accra, Ghana. *International Journal of Environmental Research and Public Health* 14(2):135.

Deyshappriya, N. R., and Kumari, M. M. T. D. M. 2019. Determinants of the behavioural intentions of households to recycle E-waste in Sri Lanka. *Asian Journal of Empirical Research* 9(8): 202–216.

Economynext. 2020. Sri Lanka Launches country wide e-waste collection project. https://econ omynext.com/sri-lanka-launches-countrywide-e-waste-collection-project-74318/.

Fernando, R. L. S. 2019. Solid waste management of local governments in the Western Province of Sri Lanka: an implementation analysis. *Waste Management* 84:194–203.

George, D. A., Lin, B. C. A., and Chen, Y. 2015. A circular economy model of economic growth. *Environmental Modelling & Software* 73: 60–63.

Golafshani, N. 2003. Understanding reliability and validity in qualitative research. *Qualitative Report* 8:597–607.

Golev, A., and Corder, G. D. 2017. Quantifying metal values in e-waste in Australia: the value chain perspective. Minerals Engineering 107:81–87.

Gollakota, A. R., Gautam, S., and Shu, C. M. 2020. Inconsistencies of e-waste management in developing nations–Facts and plausible solutions. *Journal of Environmental Management* 261:110234.

Gunarathne, A. D. N. 2015. E-waste Management in Sri Lanka. Department of Accounting, University of Sri Jayewardenepura, Colombo, Sri Lanka.

Gunarathne, N., and Lee, K-H. 2019. Institutional pressures and corporate environmental management maturity. *Management of Environmental Quality* 30(1): 157–175.

Gunarathne, A. D. N., de Alwis, A., and Alahakoon, Y. 2020a. Challenges facing sustainable urban mining in the e-waste recycling industry in Sri Lanka. *Journal of Cleaner Production* 251:119641.

Gunarathne, A.D.N., Kaluarachchilage, P.K.H., and Rajasooriya, S.M. 2020b. Low-carbon consumer behaviour in climate-vulnerable developing countries: A case study of Sri Lanka. *Resources, Conservation and Recycling* 154:104592.

Gunarathne, A. D. N., Tennakoon, T. P. Y. C., and Weragoda, J. R. 2019. Challenges and opportunities for the recycling industry in developing countries: The case of Sri Lanka. *Journal of Material Cycles and Waste Management* 21(1):181–190.

Herat, S., and Agamuthu, P. 2012. E-waste: a problem or an opportunity? Review of issues, challenges and solutions in Asian countries. *Waste Management & Research* 30(11): 1113–1129.

Hettiarachchi, H. A. H., and Sarathchandra, K. S. H. 2017. Factors affecting for e-waste recycling in Sri Lanka. *International Journal of Scientific and Research Publications* 10(01): 101–107.

Ikhlayel, M. (2018). An integrated approach to establish e-waste management systems for developing countries. *Journal of Cleaner Production* 170: 119–130.

Ilankoon, I. M. S. K., Ghorbani, Y., Chong, M. N., Herath, G., Moyo, T., Petersen, J., 2018. E-waste in the international context, A review of trade flows, regulations, hazards, waste management strategies and technologies for value recovery. *Waste Management* 82: 258–275.

Ismail, H. and Hanafiah, M. M. 2019. Discovering opportunities to meet the challenges of an effective waste electrical and electronic equipment recycling system in Malaysia. *Journal of Cleaner Production* 238:117927.

Mallawarachchi, H., and Karunasena, G. 2012. Electronic and electrical waste management in Sri Lanka: suggestions for national policy enhancements. *Resources, Conservation & Recycling* 68: 44–53.

Ministry of Environment and Natural Resources (MENR). 2008. Development of a national implementation plan for electrical and electronic waste management in Sri Lanka. Sri Lanka: MENR.

Onuma, Y., Siriwardana, A. N., and Rajapakshe, R. D. D. 2018. Estimation of e-waste flow in Kurunegala Municipal Council, Sri Lanka. Proceedings of the Annual Conference of Japan Society of Material Cycles and Waste Management, Japan Society of Material Cycles and Waste Management.

Patil, R.A., and Ramakrishna, S. 2020. A comprehensive analysis of e-waste legislation worldwide. *Environmental Science and Pollution Research* 27(13): 14412–14431.

Qu, Y., Wang, W., Liu, Y. and Zhu, Q. 2019. Understanding residents' preferences for e-waste collection in China-A case study of waste mobile phones. *Journal of Cleaner Production* 228: 52–62.

Ranasinghe, W. W., and Athapattu, B. C. 2020. Challenges in E-waste management in Sri Lanka. In *Handbook of electronic waste management: International best practices and case studies.* ed. M. N. V. Prasad., M. Vithanage., and A. Borthakur. 283–322. Oxford: Elsevier.

Sittampalam, J. 2019. Dangerous disposal: why Sri Lanka urgently needs to manage its e-waste https://roar.media/english/life/in-the-know/dangerous-disposal-e-waste-sri-lanka (accessed November 05, 2020).

Sasaki, S. 2020. The effects on Thailand of China's import restrictions on waste: measures and challenges related to the international recycling of waste plastic and e-waste. *Journal of Material Cycles and Waste Management* 23:77–83. https://doi.org/10.1007/s10 163-020-01113-3.

Shaikh, S., Thomas, K., Zuhair, S., and Magalini, F. 2020. A cost-benefit analysis of the downstream impacts of e-waste recycling in Pakistan. *Waste Management* 118:302–312.

Sharma, M., Joshi, S., and Kumar, A. 2020. Assessing enablers of e-waste management in circular economy using DEMATEL method: An Indian perspective. *Environmental Science and Pollution Research* 27:13325–13338.

Sthiannopkao, S. and Wong, M.H. 2013. Handling e-waste in developed and developing countries: initiatives, practices, and consequences. *Science of the Total Environment.* 463–464:1147–1153.

Suraweera, I. 2016. E-waste issues in Sri Lanka and the Basel Convention. *Reviews on Environmental Health* 31(1):141–144.

Telecommunication Regulatory Commission in Sri Lanka. 2020. Statistics. http://trc.gov.lk/images/pdf/StatisticalOverViewReportQ32020.pdf.

Tellis, W.M. 1997. Application of a case study methodology. *Qualitative Report* 3:1-19.

United Nations (UN). 2017. *United Nations system-wide response to tackling e-waste.* Geneva, Switzerland: EMG Secretariat.

United Nations Development Programme (UNDP) Sri Lanka. 2015. How to manage your e-waste. https://undpsrilanka.exposure.co/how-to-manage-your-ewaste.

United Nations Environment Programme (UNEP). 2013. *Metal recycling: Opportunities, limits, infrastructure.* Milan: Sustainable Consumption and Production Branch. UNEP.

Wilson, D. C., Velis, C. A., and Rodic, L. 2013. Integrated sustainable waste management in developing countries. *Waste Resources Management* 166:52–68.

Wilson, D. C., Velis, C., and Cheeseman, C. 2006. Role of informal sector recycling in waste management in developing countries. *Habitat International* 30:797–808.

Yin, R. K. 2013. Case study research-design and methods. Thousand Oaks: Sage.

Zhang, K., Schnoor, J. L., and Zeng, E. Y. 2012. E-waste recycling: where does it go from here? *Environmental Sciecne and Technology* 46:10861–10867.

6 E-waste Management in India – A Case Study of Vizag, Andhra Pradesh

Manya Khanna and Ankita Das

CONTENTS

6.1 INTRODUCTION

E-waste has been well-documented as the world's rapidly growing waste stream with growth rate of 3 – 5percent per year (Afroz et al. 2013) and is potentially the biggest challenge to sustainability (Christian 2012; Tsydenova and Bengtsson 2011; Qu et al. 2013). A huge amount of e-waste is being generated every year (Vats and Singh 2014). Globally, about 30–50 million tons of e-waste is disposed of each year (Menikpura et al. 2014). Special attention has always been given to the waste management of plastics. However, the problem of e-waste, remains untouched (Borthakur et al. 2017). E-waste has become the world's fastest growing waste stream which is following an exponential profile (Morton, 2007). Though e-waste contains harmful metals such as lead, mercury etc., some also contain precious metals like gold and silver (Needhidasan et al. 2014). Therefore, it can be said that e-waste has dual identity. Electric and Electronic Equipment (EEE) that have become obsolete and have reached the end of their life are discarded by their owners as waste without the intention of re-using them, can be defined as Electronic waste or e-waste (Baldé et al., 2017).

Disposing e-waste in an improper manner can prove to be hazardous to the environment and result in air, water, and soil pollution (Needhidasan et al., 2014). Many e-waste processing plants are not ethically run i.e., they are not safe. Some e-waste collectors burn wires to get the copper inside which releases hydrocarbons in the air (Ferronato et al., 2019). Chemical stripping of gold-plated chips emits brominated

DOI: 10.1201/9781003095972-8

129

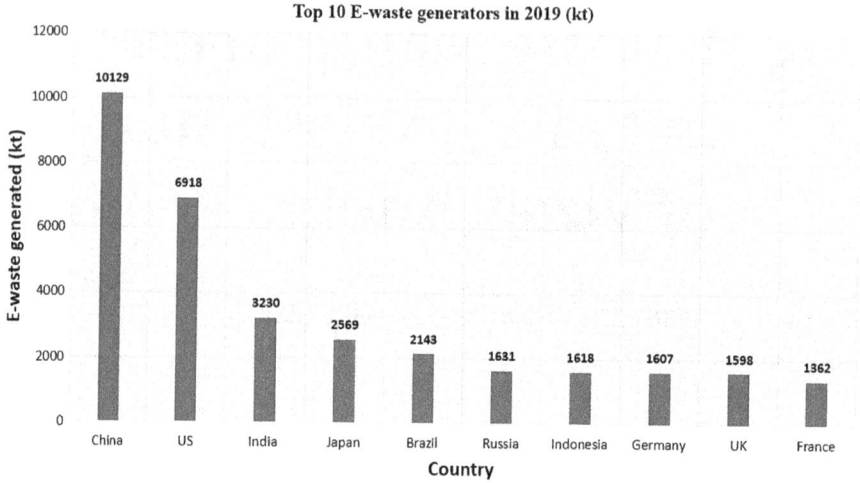

FIGURE 6.1   Top 10 e-waste generating countries in 2019. (Data Source: Forti et al. 2020.)

dioxins and heavy metals (Shanti et al., 2018). All of this results in air pollution. Cathode ray tubes, often found in old televisions, video camera and computer monitors are broken apart and the yoke and shell are separated (Kaushik, 2018). The shell is then landfilled. Components like lead and barium leach through the soil and into ground water that causes soil and water pollution (Zwolak et al., 2019). Consumption of this water has proven to be extremely harmful for both humans and animals. Crops/plants that grow on soil containing heavy metals cause further discrepancies in the food chain and are a potential threat to human beings and animals (Singh et al., 2011).

In addition to its negative impact on the environment, e-waste also has ill-effects on health (Sankhla et al. 2016). Due to unsustainable handling of e-waste several health issues arise, such as – signs of cardiovascular disease like oxidative stress and inflammation: damage of DNA and a possibility of cancer (Dai et al. 2020). Owing to the rudimentary disposal process, various toxins and heavy metals are emitted from e-waste which, through consumption of contaminated water and inhalation of contaminated air get accumulated in the human body (Dai et al. 2020).

As reported by International Solid Waste Association (ISWA), in the year 2019, approximately 53.6 million metric tons of e-waste was generated worldwide (Forti et al. 2020). Nearly 9.3 million metric tonnes of e-waste, which amounts to be 17.4 percent was documented to be collected and recycled properly since 2014. On the other hand, nearly 0.6 million metric tonnes of e-waste ended up in the waste bins of EU (Forti et al. 2020). While Asia was the highest generator of e-waste (24.9 million metric tons), highest amount of e-waste was documented to be collected and properly recycled in Europe (5.1 million metric tons, ~42.5%) (Forti et al. 2020).

The top ten e-waste generating nations are shown in Figure 6.1. From Figure 6.1, it is clear that China is the highest e-waste generator while India is in third position (Technology, International 2020). According to the Rajya Sabha report, India is one of the largest e-waste importers and imports cheap raw materials including

Percentage of E-waste produced (2018)

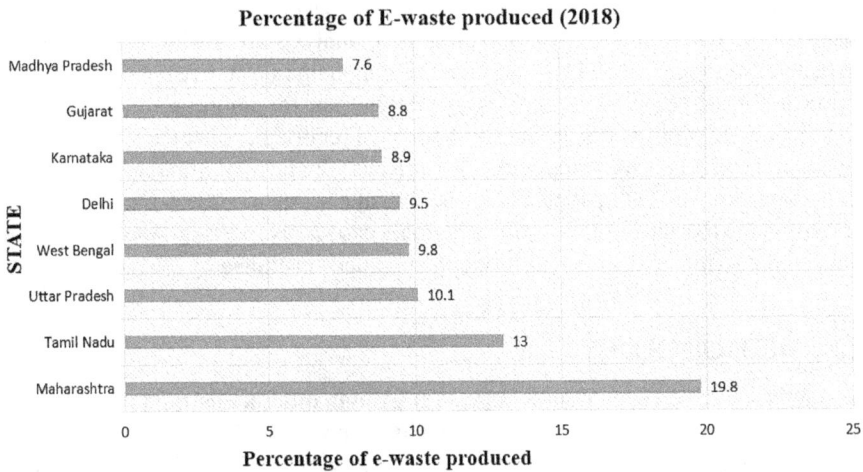

FIGURE 6.2   Leading E-waste generating states in India. (Data source: ASSOCHAM-NEC 2018.)

toxic waste (Rajya Sabha 2011). The ASSOCHAM-NEC report on e-waste reiterates this information. The report mentions that Maharashtra is the highest generator of e-waste, followed by Tamilnadu, Uttar Pradesh and West Bengal (ASSOCHAM-NEC 2018). Figure 6.2 provides information on top e-waste generating states in India. In India, the sectors that are the major contributors of e-waste are the government, public and private (industrial) sectors. They generate approximately 70 percent of total e-waste in the country. A relatively lesser amount of e-waste is generated by individual households (almost 15%) and the remaining is contributed by manufacturers (ASSOCHAM-NEC 2018).

India approximately produced 2 million tons of e-waste in 2018 out of which only 43,80,085 tons were recycled which is just 22% (approx.) (ASSOCHAM-NEC 2018). As a result of the boom in the electronics manufacturing sector in India, engagement in innovating and utilizing new technologies has increased. This is partly the reason why "inbuilt product obsolescence" is observed in India. This has further resulted in excessive quantities of discarded electronic waste and appliances because buying new products is often cheaper than repairing the damaged ones repeatedly (Debnath 2020). A major chunk of the Indian population lacks awareness on the disposal of their old electronic equipment. It is usually observed that people tend to sell scrap to the local scrap man often referred to as the *'kabaddiwalas'* (Sinha 2018; Baidya et al. 2020).

In India, nearly 90 percent of the e-waste is handled by the informal sector (Wath et al. 2010; Tyagi et al. 2015). However, the formal recycling is slowly gaining pace. It is really important to understand and take learning from successful cases in the formal e-waste recycling sector. This will allow us to delve deeper into the issues and challenges which can be helpful in developing the formal recycling scenario in India. Reported literature discusses issues, challenges and even opportunities in the e-waste management sector in India. There are also several instances of policy driven changes. The research questions that arise from analysis are – what is the status of e-waste recycling in India? How can urban mining help in resource circulation? How regional e-waste recyclers are performing? What are the implications towards circular economic proliferation? Some authors have reported case studies, but they are either on informal sector or country comparisons. Studies reporting success stories as regional case studies, specifically in the Indian e-waste context is rare. This study aims to bridge the gap by showcasing a regional case study of Vizag, Andhra Pradesh, India emphasizing on the circular economic practices instituted by the recycling unit.

6.2 METHODOLOGY

First, a thorough literature review was undertaken in order to understand the current status of e-waste in India as well as globally. Second, the literature was screened and further segregated. Thereafter, the knowledge gap was identified which lead to the formulation of the research questions. Thereafter, a semi-structured questionnaire was developed. Information was collected from the representative of the case study organization and the respective recycling facility was visited to understand the operations. Finally, their performance was analyzed based on the circular economy indicators.

6.3 E-WASTE MANAGEMENT IN INDIA

The existing e-waste management system in India is very complex, yet it is important to understand the current system in place. Ideally, a supply chain mapping of the e-waste scenario reveals the existing system in place. Existing literature have captured the existing e-waste management system in India (Wath et al. 2010; Baidya et al. 2019; Debnath et al. 2019a), however very few of them mapped the whole supply chain with technical details (Debnath 2020). We take a relatively unconventional approach to describe the e-waste value chain in India. We employ the running architecture of an algorithm to provide the details of the current e-waste management system in India. It is described below –

Step 1: The electronic equipment which is at the end of its life means the e-waste is collected from the domestic places, schools, colleges, offices and institutions. The collection is carried out either via the formal sectors or the informal sectors. Apart from that, the growing number of Producer Responsibility Organisations (PRO) is also collecting e-waste, primarily from the bulk consumers.

Step 2: After the collection of e-waste by the kabaddiwalas, all are sold to the small scrap dealers.

Step 3: Small scrap dealers dismantle partially and sell those to the big scrap dealers.

Step 4: Big scrap dealers do their partial dismantling and sell them to the third-party informal recyclers. Go to Step 1.

Step 5: In case of formal collection, the approved collector-cum-dismantlers collect e-waste from various sources and perform mechanical dismantling.

Step 6: The dismantled e-waste is sent to the formal recyclers for further recycling. Goto Step 1.

Step 7: PROs collect e-waste from bulk consumers and direct them to the authorised formal recyclers either via own- or third-party logistics.

Step 8: The formal recyclers perform primarily mechanical recycling and recovered fractions are sold to the third-party material recyclers.

The e-waste collection is primarily dominated by the informal sector (Chakraborty and Manis 2019; Baidya et al. 2020; Das et al. 2020). However, the scenario is changing slowly with the involvement of the PROs and increased awareness. In the past few years, initiatives taken by the government as well as the educational institutes have made it possible to slowly move towards a symbiotic model containing both informal and formal sectors (PHD Chamber of Commerce and Industry 2019). This is an indicator that the wheels are slowly turning towards sustainability.

6.4 LEGISLATIVE MEASURES TAKEN IN INDIA

In India, there are many legislations, both published and draft that accommodate e-waste management, directly or indirectly, under its umbrella. Some of these legislations are – i) The Environment (Protection) Act, 1986; ii) The Air (Prevention and Control of Pollution) Act, 1981; iii) The Water (Prevention and Control of Pollution) Act, 1974; iv) E-waste Rules 2016; v) Hazardous and Other Wastes (Management and Transboundary Movement) Rules, 2016; vi) Electronic waste (management) Amendment rules, 2018; vii) Draft National Resource Efficiency Policy (NREP) 2019.

The most vital legislation directly handling e-waste is the Environment (Protection) Act (EPA), 1986. In this act there are three penal provisions i.e., section 15, 16 and 17 respectively. The six waste management rules published in 2016, specifically the e-waste rules of India, are embraced under the umbrella of the said act. Hence, the provisions of penalization mentioned in the EPA are applicable to the electronic waste rules directly. It is clearly mentioned in the section 15 that, infringement of EPA or other rules, orders and directions approved under EPA, is liable to penalization. In the EPA 1986, the three sections which are named numerically 15, 16 and 17 respectively are really important. These sections are responsible for looking at the overall individual liability, company liability and government department liability (Gupta 2017; Chaudhury 2018).

There are several other legislations that are trying to manage e-waste. These may be achieved indirectly as a result of the Air (Prevention and Control of Pollution) Act, 1981 along with the Water (Prevention and Control of Pollution) Act, 1974. The sections of 37, 38, 39, 40 and 41 are of significant importance along with 21, 22 as well as 31A specifically. The sections 37 through 38 are aimed to look at the liabilities of an individual who does not comply with the provisions that have been set up in the 21, 22 and 31A of the same Act. The Section 40 is meant for the company liability along with section 41 aiming to target the government departments (Chaudhury 2018).

The Water Act itself has several sections of which sections 41 through 48 along with 45A particularly and the sections of 20(2), 20 (3), 24, 26, 32(1)(c), 33(2) and 33 A are significant. These sections are important to understand and define the several offences under the Act. The proper application and dismantling of electrical products is a necessity. If that is not done it can lead to contamination and degrades the quality of both air as well as water. Among all the significant sections in both of the Acts, the Section 37 and the Section 41 of the Air Act, 1981 and the Water Act, 1974 specifically is the one that has the penal resolute that has to be taken into consideration in regards to breaking the guidelines set in the section 15 of the EPA, 1986 (Chaudhury 2018).

The way that people tend to live and the way that things are handled in terms of social and city lifestyle has come to the attention of The Department of Parliamentary Standing Committee on Science & Technology, Environment & Forests in its 192nd Report on the Functioning of the Central Pollution Control Board (CPCB) that e-waste will be an important point of concern in the future. Hence, it was stated that it is important to take an active role in the way that these kinds of events happen so as to safeguard the environment or to have some kind of damage control (Singh 2018). On 23 December 2005 a private bill called The Electronic Waste (Handling and Clearance) Bill was actually put forward. This was done by the Hon'ble member of Maharashtra Government, Shri Vijay J. Darda in the Rajya Sabha. It brought to attention the fact that there was a lack of proper law manipulating the e-waste in India and the need of it was brought to light. On 24 September 2008, the "Hazardous Wastes (Management, Handling and Transboundary movement) Rules, 2008" were promulgated and talked about e-waste management. On 12 May 2010, another notice was issued- "E-waste (Management and handling) Rules, 2010" under the Environment (Protection) Act, 1986 (Singh 2018). This Notice addressed issues related to sorting, recycling and, transportation of hazardous waste/e-waste and also to focus on using less hazardous substances while manufacturing electronic equipment. These rules came into effect on 1 May 2012 (Rajput and Nigam 2021).

The E-waste (Management and handling) Rules 2010 was revised in 2015 and was finally published in the Gazette of India in 2016. The rule focuses on Extended Producer Responsibility (EPR) but it lacks clarity in the aspect of informal sector, supply chain management and technology selection. In 2018, amendments were made to this rule which reset the EPR targets for the OEMs (Baidya et al. 2020).

6.5 URBAN MINING OF E-WASTE IN INDIA

Disposal of electronic waste is a major problem faced by many countries and thus scientific disposal of e-waste is necessary or else unscientific processing of e-waste recovery of metals can cause environmental pollution (Chatterjee, 2011). Here's when Urban Mining comes into picture. Urban mining essentially refers to the practice of extracting precious and rare earth metals and energy from urban waste such as – e-waste and putting them back in the economy (Cossu et al., 2012; Bonifazi and Cossu 2013). Urban spaces can be considered as sources of anthropogenic materials that can be used in a cyclic manner, recycled, and reused (Brunner, 2011).

Urban Mining involves extraction of metals such as aluminium, copper, brass and zinc, gold, silver platinum etc., from e-waste products such as computers, motherboards, mobile phones, and server boards (Ramanayaka et al., 2019). Extraction of metals and precious metals from e-waste greatly reduces the amount of waste that is landfilled (Debnath et al. 2018).

A shift from the linear to circular approach with respect to waste management is the need of the hour (Cossu et al., 2013). Urban mining can be a solution for the same. It aims at maximizing the resources and economic value of urban waste, and when applied along with the concept of circular economy can prove to be useful while making policy decisions for sustainable development (Arora et al. 2017; Debnath et al., 2019b).

Mining for rare and precious metals is not only expensive but also harmful for the environment as it causes contamination of land and water, and loss of biodiversity. However, these metals are the vital ingredients of modern technology and as our desire for new technology soars, we tend to throw away our gadgets, which end up being lost in landfills (Williams, 2016). This is like a never-ending vicious cycle. Thus, adopting urban mining to extract secondary resources from urban waste is one step in the right direction to achieve sustainability.

Rapid globalization and urbanization, economic and industrial growth can be considered as some factors for an increase in demand for resources in India. The country is highly dependent on imports of major materials used in the country. Over dependence on imports is associated with economic risks like monopolistic behaviour of exporters, regular price spikes of the materials being imported, and disruptions in supply chain due to conflicts in exporting countries. India is a developing country and therefore shifting to secondary resource extraction can be challenging initially. However, it should be considered as an opportunity towards a more sustainable approach, given that it would save natural resources and in-turn have economic benefits (NITI Ayog, 2017).

Recycling of e-waste is dominated by the informal sector in India that essentially involves collection, segregation and dismantling of the waste. Adding to this, various researchers have shown that informal recycling includes extensive repair and refurbishment activities, which help in extending life of these products. But informal recyclers have little or no control over their activities, which causes severe pollution. Therefore, upliftment of the formal sector that is approved by the government is needed. As a developing country, India should focus on supporting ethical recyclers. This will not only be beneficial for the environment but will also benefit the economy by creating more jobs in this sector. The successive case study is a live example of the formal sector, which is contributing significantly to e-waste management.

6.6   CASE STUDY – RECYCLING COMPANY IN ANDHRA PRADESH

Location: Based in Visakhapatnam, also operates in Hyderabad and Goa

Profitability trends (past three years): Average growth
This multi award-winning company is the first authorized e-waste collection and handling unit of Andhra Pradesh, by the Pollution Control Board, Andhra Pradesh. They provide a comprehensive way of recycling, waste management and sustainable solutions to organizations. India faces many challenges with regards to e-waste disposal, the most common being lack of awareness of the correct procedures of e-waste disposal. Thus, start-ups like these are needed, who conduct awareness programs and workshops in different organizations, schools, and colleges. The company's motto is to *"Reduce, Reuse, Recycle and Recover the assets through continuous implementation*

and innovation in recycling technology". The informal sector is the major contributor of e-waste in the country as it uses primitive techniques of e-waste disposal thus resulting in increased level of pollution. In addition to this, workers working in the informal sector are unaware of the health hazards associated with electronic scrap. Often, partially dismantled electronic appliances are land filled. However, this company has fully trained employees who are aware of the entire e-waste management process including the disposal of unused waste.

The company collects e-waste in an ethical way from Government organizations, SMEs, Corporates, Educational Institutions, Retailers, and Individual households. Their e-waste dismantling process (includes manual, semi-manual and automatic techniques) involves physical segregation of particles like glass, plastic, gauges and circuit boards, wires, steel, non-ferrous materials, and hazardous e-waste like cartridges, tube lights, and sodium vapour lamps (Figure 6.3). Priority is given to the safety of

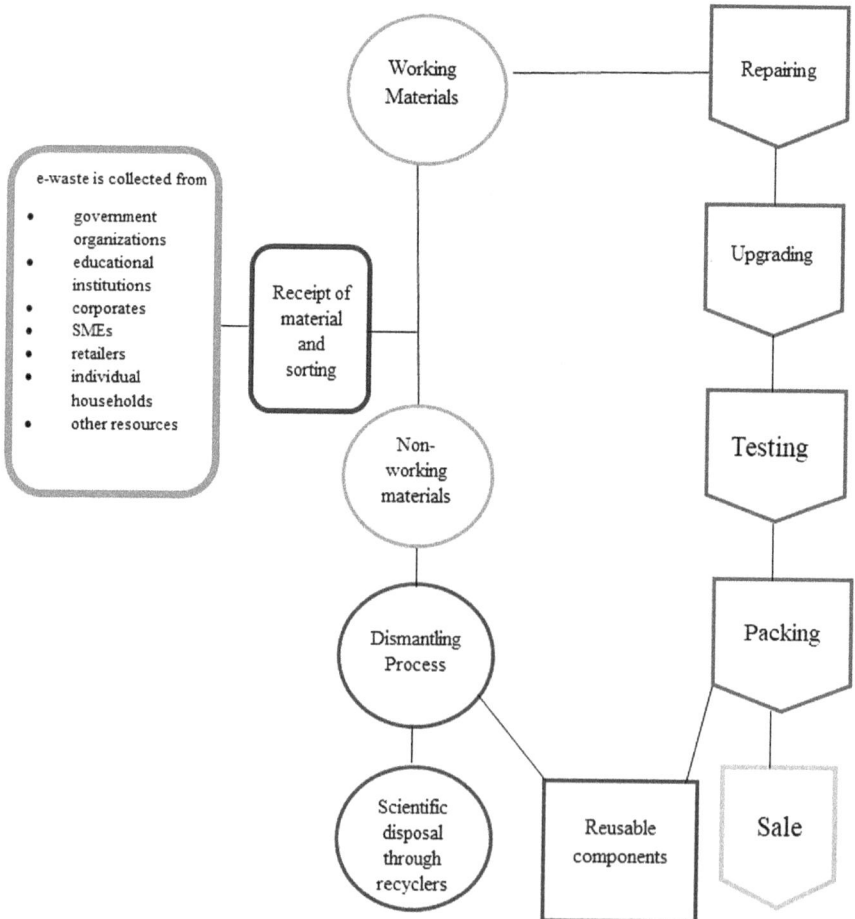

FIGURE 6.3 Process of e-waste collection and management carried out by the company.

the employees and the handling of e-waste is done in a professional manner. They also make unique things by recycling paper like paper pencils, which have seeds attached at the end. Once the pencil is used till the end, it can be inserted upside down in the soil and watered which in turn will grow into a plant. Similarly, they also make paper pens, cloth bags, calendars made of handmade papers with seeds embedded in it etc.

Another initiative taken by the company is installation of 'e-Bins' in selected areas of Visakhapatnam for disposal and recycling of small electronic items. This helps spread awareness about the importance of correct disposal of e-waste and its negative effects towards the environment. Once the bins are filled, the company vehicle collects these items and takes them for recycling.

6.7 CONCLUSION

Seventy-five percent of electronic items are stored in households, offices, industries etc. They lie unattended to and are mixed with domestic waste, which is finally dumped to landfills. As mentioned above, there are two sectors of recycling- formal and informal. As the name suggests the informal sector is less efficient in recycling and harms the environment. The formal sector on the other hand uses professional methods to recycle. Recycling can also help to generate revenue in the economy. It is a known fact that natural resources are limited, and human wants are never satisfied. Thus, it should be our duty to save some resources for future generations. In fact, extracting metals from e-waste (urban mining) consumes lesser energy as compared to extracting from ores. This is one step in the right direction to reduce air, water, and soil pollution. India, being a developing country generates a huge amount of waste and therefore there is a need to set up more formal recycling industries in the country. Workshops regarding e-waste management should be held in schools, colleges, and offices to make people aware of reusing and recycling products. Government organizations and private companies should partner with small-scale industries and start-ups who provide recycling services. This will not only help control pollution and save resources, but also generate employment by funding those small-scale industries. Supporting start-ups in this field would be a big leap towards reducing our carbon footprint and enhancing sustainable development in India. The country has seen some development in this field over the past few years, however, more effort is needed to make people aware of their surroundings and for them to understand consequences of their actions.

ACKNOWLEDGMENT

I would like to thank everyone working in the start-up in Andhra Pradesh for participating in this study as well as the mentors who guided the author while working on this chapter.

REFERENCES

Afroz, R., Muhammad M. M., R. Akhtar, and J. Bt Duasa. "Survey and analysis of public knowledge, awareness and willingness to pay in Kuala Lumpur, Malaysia–a case study on household WEEE management." *Journal of Cleaner Production* 52 (2013): 185–193.

"Assocham India". *Assocham.Org*, 2018. (Access date 5/2/21) www.assocham.org/newsdetail. php?id=6850

Arora, R., K. Paterok, A. Banerjee, and M. S. Saluja. *"Potential And Relevance Of Urban Mining In The Context Of Sustainable Cities"*. *IIMB Management Review*, 2017.

Baidya, R., B. Debnath, S. K. Ghosh, and S-W. Rhee. "Supply chain analysis of e-waste processing plants in developing countries." *Waste Management & Research* 38, no. 2 (2020): 173–183.

Baidya, R., Debnath, B., & Ghosh, S. K. (2019). Analysis of e-waste supply chain framework in India using the analytic hierarchy process. In Waste Management and Resource Efficiency (pp. 867–878). Springer, Singapore.

Baldé, C. P., Forti V., Gray, V., Kuehr, R., Stegmann,P. The Global E-waste Monitor – 2017, United Nations University (UNU), International Telecommunication Union (ITU) & International Solid Waste Association (ISWA), Bonn/Geneva/Vienna.

Bonifazi, G., and Raffaello, Cossu. "The Urban Mining Concept". *Elsevier*, 2013.

Borthakur, A., Govind, M. How well are we managing E-waste in India: evidences from the city of Bangalore. *Energ. Ecol. Environ.* **2**, 225–235 (2017). https://doi.org/10.1007/s40974-017-0060-0

Brunner, P. H. "Urban Mining A Contribution To Reindustrializing The City". *Journal Of Industrial Ecology* 15, no. 3 (2011): 339–341.

Chakraborty, P., and A. Manis. "E-Waste Management In India: Challenges and Opportunities". *Teriin.Org*, 2019.

Chatterjee, S., *Electronic Waste and India* (2011)

Chaudhary, N.. "ELECTRONIC WASTE IN INDIA: A STUDY OF PENAL ISSUES". *ILI Law Review* no. 2018 (2018).

Christian, G. E.. "End-of-life management of waste electrical and electronic equipment for sustainable development." *Centre for International Sustainable Development Law. McGill University* 3644 (2012).

Cossu, R., Salieri, V., Bisinella, V. (Eds.), 2012. Urban Mining – A Global Cycle Approach to Resource Recovery from Solid Waste. CISA Publisher, ISBN 9-788862-650014.

Dai, Q., X. Xu, B. Eskenazi, K. Ansong Asante, A. Chen, J. Fobil, Å. Bergman et al. "Severe dioxin-like compound (DLC) contamination in e-waste recycling areas: An under-recognized threat to local health." *Environment International* 139 (2020): 105731.

Das, A., Biswajit Debnath, Nipu Modak, Abhijit Das, and Debasish De. "E-waste Inventorisation for Sustainable Smart Cities in India: A Cloud-based Framework." In *2020 IEEE International Women in Engineering (WIE) Conference on Electrical and Computer Engineering (WIECON-ECE)*, pp. 332–335. IEEE, 2020.

Debnath, B.. "Towards Sustainable E-Waste Management Through Industrial Symbiosis: A Supply Chain Perspective." In *Industrial Symbiosis for the Circular Economy*, pp. 87–102. Springer, Cham, 2020.

Debnath, B., S. Das, and A. Das. "Study Exploring Security Threats in Waste Phones a Life Cycle Based Approach." (2019a).

Debnath B., Chowdhury R., Ghosh S. K. (2019b) *Urban Mining and the Metal Recovery from E-Waste (MREW) Supply Chain*. In: Ghosh S. (eds) Waste Valorisation and Recycling. Springer, Singapore. https://doi.org/10.1007/978-981-13-2784-1_32

Debnath, B., R. Chowdhury, and S. K. Ghosh. "Sustainability of metal recovery from E-waste." *Frontiers of Environmental Science & Engineering* 12, no. 6 (2018): 1–12.

Rajya Sabha (2011), E-Waste in India, *Research Unit (Larrdis), Rajya Sabha Secretariat, India*.

Ferronato, N., and V. Torretta. "Waste Mismanagement In Developing Countries: A Review Of Global Issues". *International Journal Of Environmental Research And Public Health* 16, no. 6 (2019).

Forti, V., B. C.P., Kuehr, R., Bel, G. The Global E-waste Monitor 2020: Quantities, flows and the circular economy potential. United Nations University (UNU)/United Nations Institute for Training and Research (UNITAR) – co-hosted SCYCLE Programme, International Telecommunication Union (ITU) & International Solid Waste Association (ISWA), Bonn/Geneva/Rotterdam.

Gupta, S. K. 2017. "Legal Control of E Waste Pollution In India". Patna University, Department of Law.

Kaushik, V. "E-Waste And Its Management". *International Journal For Research In Applied Science & Engineering Technology (IJRASET)* 6, no. (2018).

Menikpura, SNMS NM, A. Santo, and Y. Hotta. "Assessing the climate co-benefits from Waste Electrical and Electronic Equipment (WEEE) recycling in Japan." *Journal of Cleaner Production* 74 (2014): 183–190.

Needhidasan, S., M. Samuel, and R. Chidambaram. "Electronic Waste – An Emerging Threat to The Environment of Urban India". *Journal of Environmental Health Science and Engineering* 12 (2014).

NITI Ayog, Government of India, *Strategy on Resource Efficiency* (2017)

PHD Chamber of Commerce and Industry. E-Waste Mass Awareness Programme through Ad Film in Cinema/Theatres (Phase-II). Ministry of Electronics and IT (MeitY), Government of India. 2019. Available at: http://greene.gov.in/wp-content/uploads/2019/03/2019032040.pdf

Qu, Y., Q. Zhu, J. Sarkis, Y. Geng, and Y. Zhong. "A review of developing an e-wastes collection system in Dalian, China." *Journal of Cleaner Production* 52 (2013): 176–184.

Rajput, R., and N. A. Nigam. "An overview of E-waste, its management practices and legislations in present Indian context." *Journal of Applied and Natural Science* 13, no. 1 (2021): 34–41.

Ramanayaka, S., S. Keerthanan, and M. Vithanage. "Urban Mining of E-Waste: Treasure Hunting for Precious Nanometals". In *Handbook Of Electronic Waste Management*, 19–54, 1st ed. Butterworth-Heinemann, 2019.

Shanti, M., and O. Sujana. "Impact Of E-Waste On Human Health And Environment- An Overview". *International Journal For Research In Applied Science & Engineering Technology (IJRASET)* 6, no. (2018).

S. Sankhla, Mahipal, M. K., Manisha, N., S. Mohril, G. Pratap Singh, B. Chaturvedi, and Dr. R. Kumar. 2016. "Effect of Electronic Waste on Environmental & Human Health- A Review". *IOSR Journal of Environmental Science, Toxicology and Food Technology (IOSR-JESTFT)* 10 (Issue 9 Ver. I (Sep. 2016): 98–104.

Singh, J., and A. S. Kalamdhad. "Effects Of Heavy Metals On Soil, Plants, Human Health And Aquatic Life". *International Journal Of Research In Chemistry And Environment* 1, no. 2 (2011).

Singh, V. P. "Law Relating to E-Waste Management in India: A Critical Study." *Available at SSRN 3176909* (2018).

Sinha, A. "E-Waste: A Growing Challenge For Waste Management And Environmental Sustainability In India". *Journal Of Advances And Scholarly Researches In Allied Education* 15, no. 11 (2018). http://ignited.in/I/a/232121 (Access date: 7/2/21)

Technology, International. 2020. "Which Countries Produce the Most E-Waste?". *Envirotech Online* www.envirotech-online.com/news/health-and-safety/10/breaking-news/which-countries-produce-the-most-e-waste/46470 (Access date: 6/2/21)

Tyagi, N., S.K. Baberwal, and N. Passi. 2015. "E-Waste: Challenges and Its Management". *DU Journal of Undergraduate Research And Innovation* Volume 1 (Issue 3): 108–114.

Tsydenova, O., and M. Bengtsson. "Chemical hazards associated with treatment of waste electrical and electronic equipment." *Waste management* 31, no. 1 (2011): 45–58.

Vats, M. and Singh, S., 2014. Status of E-Waste in India – A Review. *International Journal of Innovative Research in Science, Engineering and Technology*, 3(10).

Wath, S. B., A. N. Vaidya, P. S. Dutt, and T. Chakrabarti. "A roadmap for development of sustainable E-waste management system in India." *Science of the Total Environment* 409, no. 1 (2010): 19–32.

Zwolak, A., M. Sarzyńska, E. Szpyrka, and K. Stawarczyk. "Sources of Soil Pollution By Heavy Metals And Their Accumulation In Vegetables: A Review", 2019. (access date: 7/2/21) https://link.springer.com/article/10.1007/s11270-019-4221-y.

Part 3

Technologies for E-waste
Valorization and Management

7 E-Waste Recycling Technologies

An Overview, Challenges and Future Perspectives

Tanvir Alam[1], Rabeeh Golmohammadzadeh[1],
Fariborz Faraji[1], and M. Shahabuddin

[1] These authors contributed equally.

CONTENTS

7.1 INTRODUCTION

Growing population, advanced lifestyle and technological advancement are the key factors leading to increasing the e-waste more than any other waste in the world. However, e-waste recycling is becoming increasingly difficult because of the significant difference in the price of the recoverable materials and technological barriers

(Copani et al., 2019). One of the great challenges of e-waste recycling is the collection of e-waste or waste electrical and electronic equipment (WEEE) itself. Unlike other waste, a significant effort is required to collect and sort out the e-waste due to the variety of materials being used in compact form for electrical and electronic equipment (Sahajwalla and Hossain, 2020a).

E-waste is a highly complex waste containing metals, plastic polymers, ceramics, as well as toxic and hazardous substance (Cayumil et al., 2016). Overall, e-waste contains 40% metal, 30% plastic and the rest 30% is refractory oxides (Sahajwalla and Hossain, 2020b). The most common metal elements in e-waste include Cu of 20%, followed by Fe of 8%, Sn of 4% Ni of 2%, Pb of 2%, Zn of 1%, Au of 0.1%, Ag of 0.02% and Pd of 0.005% (Sum, 1991). In contrast, polyethylene, polycarbonates, polypropylene and polyesters are the commonly used plastic polymers in the e-waste (Sodhi and Reimer, 2001). However, the types and quantity of these materials largely vary with respect to the product lifetime and sources.

Inappropriate dumping of e-waste due to the insufficient standards and cost-effective technologies for recycling, dismantling and recovery, causes a significant danger to human life and the environment. Currently, a small fraction (20% in 2016) of e-waste is recycled appropriately, while others are dumped to landfills or exported to developing countries, which cause ever-increasing health and environmental challenges to the disadvantaged nations (Sahajwalla and Hossain, 2020b). An estimated loss of USD 65 billion has been reported due to the illegal dumping, landfilling and stockpiling of e-waste in 2019 (Sahajwalla and Hossain, 2020b).

Statistics show that globally about 44.7 million tons of e-waste was discarded in 2016, which increased to 52 million tons in 2021 with an increasing rate of 3-4% per year. While considering worldwide generation, Asia is the main contributor of e-waste (18.2%), followed by Europe (12.3%), America (11.3%), Africa (2.2%) and Oceania (0.7%). However, the collection and recycling rate of e-waste in Asia, Europe, America and Oceania are 15%, 35%, 17% and 6%, while the recycling rate is unknown in Africa (Balde et al., 2015).

E-wastes contain a large variety of materials with a big price gap. For example, e-waste from IT and telecommunication products (mobile, telephone, GPS, calculator, computer, PC) contain valuable metals of Ag, Al, Pb, Fe, Zn, Cu, Au, Ni, and Co (Lee et al. 2007; Hageluken 2006; Sahajwalla and Hossain 2020b). On the other hand, e-waste from large household product mainly contains Fe, Ti, Cu, Ni, Al, Si, Mg, Ce, Mg, Ba and Mg (Wu et al., 2014; Xiao et al., 2016).

Since e-waste is complex in nature, it is very difficult to retrieve the materials found in e-waste by recycling. Conventionally two methods are used to recycle the e-waste such as pyrometallurgy and hydrometallurgy (Sahajwalla and Hossain, 2020b). Both of them have particular pros and cons. For example, pyrometallurgy applies to almost all types of e-waste via thermal treatment. However, the challenges include the requirement of extensive energy and investment, difficult to recover Fe and Al and the generation of a huge amount of solid waste and toxic fumes (Mark and Lehner, 2000; Sahajwalla and Hossain, 2020b). Furthermore, the difficulties of pyrometallurgy include, controlling the quality of materials and emission due to the difference in the melting point of several materials embedded in the e-waste. For example, recycling of waste printed circuit board (WPCB) requires the temperature of 1250°C via a pyro-metallurgical method which unavoidably releases harmful Pb into the environment during the process due to a lower melting temperature of 327.5°C for Pb (Veldhuizen and Sippel, 1994).

On the other side, hydrometallurgical techniques are well developed for the recycling of metals from e-waste. These techniques are operated at low temperatures, resulting in high purity products, and are producing little emission (Mukherjee, 2019). Nevertheless, the main constraints of hydrometallurgical techniques a need for a large amount of chemicals, slow recovery rate, several steps to extract the metal and generation of wastewater (Quinet et al., 2005; Sahajwalla and Hossain, 2020b; Sheng and Etsell, 2007). Another promising method of e-waste recycling is the bio-metallurgical method, which is divided into *biosorption* and *bioleaching* (Debnath et al., 2018). Several studies have been conducted on the bio-metallurgical process for e-waste recycling (Bas et al., 2013; Faramarzi et al., 2004; Ilyas and Lee, 2014; Kumar et al., 2018). However, there is no industrial-scale plant using the bio-metallurgical route and there is plenty of room for the development before making it widely popular for the e-waste recycling (Debnath et al., 2018).

Although there are experimental as well as review studies available in the literature (Zhang and Xu, 2016b; Kumar et al., 2017; Copani et al., 2019; Awasthi et al., 2019; Ding et al., 2019) still, there is a need for a comprehensive review for different conventional and emerging new technologies considering pre-treatment. This chapter provides a critical review of different e-waste recycling technologies with a particular focus on novel technologies. Moreover, this chapter provides a comparative analysis of novel technologies with those of conventional ones including recent development and challenges. Furthermore, an outlook on the future advancement of e-waste recycling technologies has been discussed in this chapter.

7.2 E-WASTE CLASSIFICATION

E-waste comprises a number of items and is a piece of equipment operating with a cord or battery (Smith, 2015). It is different from other industrial residues or municipal waste streams in terms of physical and chemical properties as it is valuable and hazardous at the same time. Table 7.1 shows ten various categories of e-waste that have been introduced in Directive 2002/96/EC in January 2010, with defined recovery and recycling goals (Goosey and Goosey, 2020). Generally, computer and mobile phones are in abundance in e-waste streams due to their short life span (Robinson, 2009).

Categories of Table 7.1 have been further expanded by the UK with the addition of four categories of display equipment, cooling appliances containing refrigerants, gas discharge lamps and LEDs and solar panels. Sixty different metals have been identified in the e-waste streams mentioned in Table 7.1. From these 60 metals, 13 are critical raw materials in small e-waste streams such as cell phones, personal computers, laptops, monitors, TVs, batteries and solar panels (Bakas et al., 2014). These elements are beryllium, cobalt, gallium, germanium, gold, indium, lithium, palladium, ruthenium, silver, tantalum, tellurium, tungsten. About 3000 metric tonnes of these

TABLE 7.1

Classification of E-Waste Based on Their Average Percentage

No.	E-waste categories	Examples	Average percentage (wt. %)
1.	Household electronics (large)	Refrigerator, air conditioner, washing machines, dryers, electric radiators, ovens, microwaves, etc.	42.1
2.	Household electronics (small)	Media players, alarm clock, grinder–juicer–mixer, electric kettles, electric vents, vacuum cleaners, iron steam, toothbrushes, toasters, etc.	4.7
3.	Information technology (IT) and telecommunication devices	Laptops, notebooks, network cables, modems, mobile phones, landline phones, fax, printer-scanner, communication satellite, calculators, computers, etc.	33.9
4.	Consumer devices	Radio receivers, television sets, MP3 players, video recorders, digital cameras, camcorders, personal, video-cameras, musical instruments, recorders, etc.	13.7
5.	Health care devices	Medical thermometer, radiotherapy equipment, laboratory equipment, respiratory ventilators, dialysis machine, etc.	1.9
6.	Lightning and illumination equipment items	Ballast, halogen, neon, LED, compact, and different types of lamps (sodium, fluorescent, high-intensity) etc.	1.4
7.	Tools and machinery equipment items	Drills, sewing machines, tools for riveting, nailing or screwing, transistors, diodes, integrated circuits, wires, motors, generators, batteries, switches, relays, transformers and resistors, etc.	1.4
8.	Monitoring and control equipment items	Thermostat, microcontrollers, fire alarm, weighing machines, etc.	0.1
9.	Automatic dispenser machines	Soap dispenser, beverage dispensers, spray dispenser, automatic dispenser etc.	0.7
10.	Entertainment and athletics	Videogames, electric trains, car racing toy set, etc.	0.2

Source: Dave et al., 2016; Garlapati, 2016; Widmer et al., 2005.

TABLE 7.2
Recovered Constituents from Different Sources of E-waste and Their Possible Applications

Recovered constituents from e-waste streams	Source	Possible applications
Base metals	WPCBs	Various industrial and commercial applications such as electrical equipment, batteries, transportation industry, household fixtures
Precious metals	Chip resistors of cell phones and WPCBs	Jewellery, industrial purposes, dental technology, electronics, etc.
Nickel	Ni-Cd batteries	Alloying element for making stainless steel, batteries, coatings and household furniture
Lithium and cobalt	Lithium-ion batteries (LIBs)	Batteries, resistance glass and ceramic, magnets, electroplating
Barium phosphate and heavy metals	CRTs	optical applications, metal alloys, solar panels, particle accelerators, mobile phones, etc.
Chromium	CRTs and spent batteries	Stainless steel manufacturing, controlling blood sugar in people with diabetes
Plastics	Cable and computer casings	Packaging, plastic bags, fuel oils
Mercury	Switches and WPCBs	Alloy making, batteries, fluorescent lights and thermometers
Cadmium	Semiconductors and chip resistors	Batteries, corrosion protection coating, solar cells, pigments, plastic stabilizers
Arsenic	CRTs, LCD and WPCBs	semiconductors, glass, bronzing and hardening
Rare earth elements (REEs)	Magnets	Catalyst systems, refining crude oils. Carbon lightening applications. Manufacturing of magnets, camera, telescope lenses, wind turbines, hybrid cars, aircraft engines, efficient refrigerant systems.

Source: Dave et al., 2016; Garlapati, 2016; Widmer et al., 2005.

critical raw materials can be recovered, of which 83% comprised cobalt from cell phones and laptop batteries (Reeve and Eduljee, 2020). Table 7.2 shows different constituents from different sources of e-waste can be recovered such as (Thakur and Kumar, 2020; Venkatesan et al., 2018).

Aside from that, there are ten materials with high economic value and those can be recovered from cathode ray tube monitors, smartphones, liquid crystal display TVs, LED and LCD monitors, cell phones and LCD notebooks (Figure 7.1).

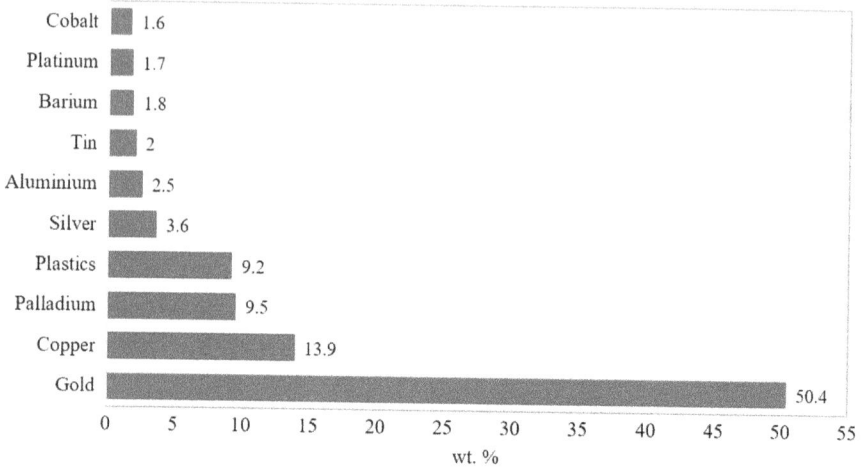

FIGURE 7.1 The commonplace materials that can be recycled from e-waste streams. (Cucchiella et al., 2015.)

7.3 E-WASTE RECYCLING TECHNOLOGIES

7.3.1 Pre-treatment

Some pre-treatment steps usually employed at the beginning of the actual recycling processes to increase efficiency beside minimizing the cost of e-waste recycling (Neto et al., 2016; Li et al., 2009). The choice of steps is based on both characteristics of the e-waste and the subsequent recycling process. There are varieties of pre-treatments such as discharging, disassembling/dismantling, crushing, upgrading, sorting, etc which are explained in three major classifications of thermal, chemical, and physiomechanical pre-treatments (Georgi-Maschlera et al., 2012; Lee and Rhee, 2002; Sun and Qiu, 2012; Chagnes and Pospiech, 2013).

7.3.1.1 Thermal Pre-treatment

Thermal pre-treatment is applicable for a wide range of e-waste, facilitating separation of different electronic components (ECs) and helping the liberation of metals (Wang et al., 2017). Desoldering is an example where heat is applied to melt the solder and disconnect the ECs (like batteries, CPUs, wires, etc.) from the main panel; therefore, each part would be free to undergo a separate recycling process (Wang et al., 2016; Wang et al., 2017). Desoldering is usually performed at temperatures above the melting point of solder, ideally from 183°C to 415°C and is followed by either manual or automatic removal of materials (Kaya, 2016).

Some thermal processes are applied to remove/decrease the carbon-bearing parts of e-waste (casings, the matrix of WPCBs, electrolytes, the anode of spent LIBs, etc). This is usually performed for several reasons such as reducing the size of e-waste, the liberation of metals, better controlling of the emissions over the subsequent processes, decreasing the hardness of the waste in crushing/grinding step, and sometimes recovering the organic parts (Cui and Zhang, 2008; Khaliq et al., 2014; Assefi et al., 2020; Al-Thyabat et al., 2013; Wang et al., 2017). Incineration (combustion), pyrolysis, and molten salt are some common techniques in this field.

Low temperature pyrolysis is a promising thermal pre-treatment method for e-waste recycling. The most commonly used methods are microwave assisted pyrolysis, vacuum pyrolysis, catalytic pyrolysis and co-pyrolysis (Debnath et al., 2018). Among these technologies, the microwave assisted pyrolysis has been reported as an energy efficient process due to its low activation energy (Debnath et al., 2019). On the other hand, vacuum pyrolysis is highly favourable because of its low pressure and temperature requirement (Qiu et al., 2009). Catalytic pyrolysis is advantageous for the polymer fraction of e-waste and can be done using various types of catalysts such as zeolites, FCC catalysts, metal based catalyst, mineral based catalyst and calcium based catalyst (Debnath et al., 2019). Co-pyrolysis is a green and sustainable way of recycling e-waste. It is based on the synergistic effects between different materials and reactions. Co-pyrolysis of e-waste and biomass waste has been reported as a favourable option by Liu et al. (2013). Using incineration technique, organic materials are simply burnt in the presence of oxygen while during pyrolysis, heat is applied in an oxygen-free atmosphere. Both processes can lead to similar results; however, pyrolysis releases less pollution and can recover some of the organic components and ceramics (Wang et al., 2017). In some types of e-waste including WPCBs with bromine in their fire retardant structure, spent LIBs with fluoride in the organic binder, incineration and pyrolysis can result in the poisonous gas emission besides the burning fumes (Wang and Xu, 2014; Shen et al., 2018; Tressaud, 2006). This can be controlled by adding some metal oxides (Na_2O, CaO, ZnO, Fe_2O_3, etc.) as halogen fixing agents and employing a reliable exhausting system (Chen et al., 2018; Terakado, et al., 2013, 2011).

Molten salt is a cleaner thermal treatment technique at which materials are submerged in a mixture of some salts (e.g. NaOH-KOH, Li, Na, K_2CO_3, $AlCl_3$-NaCl, etc) at a relatively lower temperature (usually between 160°C and 700°C). In this procedure, both oxidative and reductive conditions are applicable by sparging air/oxygen or an inert gas (Lin et al., 2017; Flandinet et al., 2012; Wang et al., 2019). As a result, organic compounds are removed, and emissions are controlled. In the case of recycling of spent LIBs, the thermal processes are usually designed to peel off the cathode constituents from the aluminium foil and remove the organic binder, electrolytes, separators and graphite from the structure while in the recycling of WPCBs, the main concern is the decomposition of boards, destruction of the rigid structure, and the liberation of metals (Hassan et al., 2010; Lin et al., 2015; Reddy et al., 2015; Wang et al., 2017).

7.3.1.2 Chemical Pre-treatment

In chemical pre-treatment, acids, bases, or organic compounds are employed to prepare the e-waste for the subsequent steps. For instance, disassembling the ECs from WPCBs by selectively dissolving the solder in acid rather applying heat (Yang et al., 2017). It is also recommended to have a few of the metals selectively removed prior to the hydrometallurgical or pyrometallurgical recycling processes. To prevent some recycling difficulties for used LIBs through acid leaching, aluminium is initially removed by washing it in an alkali (NaOH) solution (Hu et al., 2017). As a result, there would be fewer difficulties over the later processes and aluminium can be recovered.

There are some other chemicals that are utilized in pre-treatment steps for e-waste recycling. Discharging the spent LIBs before starting the process is a crucial phase in most of the recycling procedures, which is normally performed by soaking the batteries in a conductive (NaCl) or cryogenic (liquid nitrogen) solution (Chen et al., 2017; Diekmann et al., 2017). Although this is not a favourable step for the industry, it can decrease the danger of ignition and protect the equipment items and workers (Chen et al., 2017; Contestabile et al., 1999; Diekmann et al., 2017). For industrial causes those spent LIBs are crushed without discharging, different groups of material (metals, ceramics, graphite, etc) are sintered due to some tiny explosions. This makes different groups of materials connected and causes difficulties in the separation steps later (He et al., 2019).

Effective removal of the binder in cathode and anode materials of used LIBs is a critical step in the recycling of batteries. This process has attracted great attention since the binder (usually PVDF) is a relatively stable phase thermally and chemically (Lovinger, 1982; Ross et al., 2001). N-Methyl-2-pyrrolidone (NMP) is an effective chemical that can effectively dissolve the binder materials. Depending on the efficiency of the reagent, this step is usually performed at temperatures between 60°C and 90°C, which may take up to 24 h. It is found that ultrasound can help to make the process faster and more efficient (Hanisch, et al., 2011; He et al., 2015). In general, chemical pre-treatment step is mainly applied before hydrometallurgical recycling approach and mostly focuses on spent LIBs.

7.3.1.3 Physio-mechanical Pre-treatment

The physio-mechanical pre-treatment process includes a group of physical procedures that aim to grind, upgrade, and sort the e-waste. The techniques are very similar to ore dressing steps in mineral processing but are adopted to be utilized for e-waste preparation (Haldar, 2018; Kumar et al., 2015). This pre-treatment involves varieties of practices such as crushing, grinding, milling, floatation, cyclones, screening, magnetic/electric field separation, and airflow separation techniques (Zeng et al., 2015; Tansel, 2017; Khaliq et al., 2014). For various types of e-waste, shredding is a typical step which is mostly performed before hydrometallurgy to improve the leaching efficiency by reducing the size, liberating metals, and increasing their surface area to be in contact with lixiviant. Depending on the type of e-waste, this step could be accomplished by hammering and milling (Verma et al., 2018). Size reduction, in general, could be challenging for WPCBs as they are hard material with a high tensile strength of up to 200 MPa (Yousef et al., 2017; Kaya, 2016). The WPCBs, are potential to be thermally treated first for easier crushing (Cayumil et al., 2018).

Sieving is a complementary step after crushing and is performed to separate a certain group of materials that are concentrated in a similar size or shape (Haldar, 2018). There are different ways to do sieving such as manual shaking, rotary kiln, and vibrating (Kaya, 2019). In e-waste recycling, because metallic parts are usually accumulated in finer sizes, screening can effectively separate metallic and non-metallic proportions (Kaya, 2016). For near to complete separation, a magnetic field is applied to isolate materials with different magnetic properties of paramagnetic, ferromagnetic, and diamagnetic. Besides, an electric field is employed to deviate conductive and non-conductive parts of e-waste (Pryor, 2012). Screening methods based on differences in densities are also helpful in increasing the separation efficiencies, especially for the separation of fine plastics (Kaya et al., 2019).

Flotation is an effective procedure that can be applied to separate different parts of e-waste based on the physical properties of their surfaces (He and Duan, 2017). In flotation, the hydrophilicity of the particles is strengthened by adding collector. Thereafter, by sparging air, choosing the right medium, and adjusting pH, froth is formed. As a result, hydrophobic and hydrophilic fractions of e-waste can be selectively separated (Jeon et al., 2018). Flotation technique is becoming more interesting for selective separation of graphite as a naturally hydrophobic material from metal oxides as hydrophilic substances in spent LIBs. Considering the main interest in LIBs recycling is in the cathode material, which is preferentially separated at the beginning of most of the recycling processes, flotation can increase the possibility of recycling mixed batteries as it can separate and recover graphite effectively (Yu et al., 2018). It is worth mentioning, a combination of different methods can improve the process; however, a good separation is achievable only if they are applied with the correct order and reagents.

7.3.2 Pyrometallurgical Technology

Pyrometallurgy is the most widely used metal recovery technique for e-waste because of its capability to process larger volumes at a faster rate (Kaya, 2016). Common pyrometallurgical routes include pyrolysis, incineration/combustion, smelting and molten salt. Out of those techniques, smelting is considered as the ultimate step that can reach to final metallic products (Kaya, 2018; Kang and Schoenung, 2005; Antrekowitsch et al., 2006). In general, prior to treating e-waste, they are primarily pre-treated and sorted; then the metal-rich fraction (i.e., PCB) is used for the recovery of metals by pyrometallurgical technique, while the non-metallic part is recycled via other techniques such as bioleaching or hydrometallurgy (Zhang and Forssberg, 1998; Xia et al., 2017).

Using smelting, metals are reduced to their elemental by applying carbon along with air/oxygen to the melt at high temperature. In the case of e-waste, the organic and plastic parts can act as reductant thus less external carbon is needed (Ghodrat et al., 2016). Both carbon (as a direct reductant) and carbon monoxide (CO) are involved in the reduction process. During smelting, an alloy containing valuable and precious metals of Cu, Co, Ni, Au, Ag, etc. are formed. Besides, some metallic fumes of Zn, Pb, Hg, Cd, etc. and a slag phase bearing the impurities are released. Furthermore, some interesting metals such as Mn, Al and Li, are left as a residue (Xakalashe et al., 2012).

Copper is the most dominant metal in the structure of most e-waste streams and its reduction is performed at about 1250°C (Khaliq et al., 2014). It is worth mentioning that, spent LIBs are usually treated separately because cathode materials are relatively stable at high temperature and a satisfactory reduction is only achievable by smelting between 1400°C and 1700°C (Assefi et al., 2020; Liu et al., 2019). Moreover, Li content from LIBs is easily oxidized and transferred to the slag. This metal would not be recyclable if it is in a complex slag with many other impurities (Dang et al., 2018).

For most of the smelters that treat e-waste, the main idea is reducing the oxides and keeping them into one major collector (copper or lead). This is to minimize metal loss and save precious metals as the primer interest of e-waste recycling (Ghodrat et al., 2016; Xakalashe et al., 2012). Because of the higher content of ceramics, glass, and oxides in the e-waste (from screens, boards, casings, etc.), a large amount of slag is expected, which can add difficulty to the overall process and results in loss of a large proportion of metals (Dang et al., 2018; Ghodrat et al., 2016; Shuey and Taylor, 2005). To prevent this, e-waste is usually diluted with ore concentrates (or copper scraps) in a way e-waste does not exceed 50 wt.% of the feed. In large-size e-waste recycling plants, this process has been developed under the name of the black copper route at which rich copper-bearing materials (concentrates or scraps) are added to the feed (Ghodrat et al., 2018, 2016).

Flux or slag making materials are the other crucial components for smelting. They are added to decrease the melting point and reduce the viscosity of the slag, control the purity, and to prevent the metal loss (Bodsworth, 1994; Ray, 2006). Due to the complexity of e-waste, slag making for their smelters is still under development; however, because copper (Cu) is the dominant phase in most of the e-waste smelters, FeO_x-CaO-SiO_2 (FCS) as a flux with minimum loss of Cu and valuable metals, is applicable. A typical composition of slag for general e-waste at 1300°C is shown in Figure 7.2.

In the case of smelters for recycling of spent LIBs, flux is a mixture of CaO, SiO_2, Al_2O_3 and occasionally MgO (Assefi et al., 2020; Anindya et al., 2013). Mn-type rechargeable batteries cannot be recycled through smelters as Mn tends to oxidize and migrate to slag (Ren et al., 2017). Moreover, Li from any source of spent LIBs accumulates as a hard-treatable phase of $LiAl(SiO_3)_2$ in the slag (Dang et al., 2020, 2018). To overcome this, researchers have suggested other fluxes (e.g. pyrolusite) and changing the atmosphere (e.g. vacuum) to have MnO and Li_2O/$LiCO_3$ in the

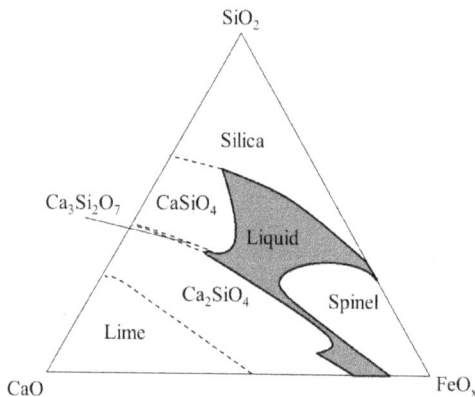

FIGURE 7.2   Suitable region for slag formation in FeO_x-CaO-SiO_2 (FCS) system at 1300°C and PO_2=10^{-6} atm. (Kongoli and Yazawa, 2001.)

slag as some treatable phases for downstream processes (Guoxing et al., 2016; Xiao et al., 2017). Chlorination and sulfation roasting of the slag is also recommended to convert $LiAl(SiO_3)_2$ to a water-soluble compound which later can be refined through hydrometallurgical processes (Dang et al., 2020, 2018).

Although pyrometallurgical methods are fast, easy and capable of treating massive feed, they are energy-intensive. Besides, they generate a substantial amount of solid waste and hazardous gases and fumes. Elements like Fe, Al, Li and Mn are lost as slag which is hard to be recycled (Awasthi et al., 2019; Baniasadi et al., 2019; Mäkinen et al., 2015; Sahajwalla and Hossain, 2020b; Tansel, 2017). Many industrial-scale examples are currently available which we have explained in the later section of this chapter.

7.3.3 Hydrometallurgical Technology

Hydrometallurgy operates at lower temperatures, requires lower capital costs, and is more flexible compared to pyrometallurgy and is capable of efficient recovery of valuable materials using an aqueous system (Tuncuk et al., 2012; Long et al., 2011). Based on the existing elements in the system, various hydrometallurgical techniques for recovering valuable materials can be employed leaching, solvent extraction, chemical precipitation, electrolysis, and electrowinning processes, etc. (Jackson, 1986). Usually, a combination of hydrometallurgical techniques with the use of various reagents is required to recover most of the elements embedded within e-wastes streams as they have a very complex structure.

E-waste streams are composed of both organic and inorganic components; 50% iron steel, 13% non-ferrous metals, 21% plastics, and 16% ceramic, glass and rubbers (Priya and Hait, 2017; Widmer et al., 2005). For metal recovery, they must be in contact with the reagents. This necessitates conduction of pre-treatment step for slight removal of plastics/ceramics encapsulating the metals which are well discussed in a previous section of this chapter (Gramatyka et al., 2007; Tuncuk et al., 2012). Despite being complex, most of the metallic elements are present in either native form or alloy making them suitable for recycling with hydrometallurgical technique.

WPCBs are considered as one of the most significant sources of e-waste as they are composed of precious elements in high concentrations (Nekouei et al., 2019). Chemical leaching of WPCBs can be conducted using cyanide leaching, halide leaching, thiourea leaching, thiosulphate leaching, aqua regia leaching, ionic liquid acid leaching, and other leaching practices using HNO_3, H_2SO_4, and HCl with the assistant of oxidizing reagents (Li et al., 2018; Wu et al., 2017).

Despite being toxic, cyanide as an inexpensive and efficient lixiviant is widely used for recovery of Au, Ag, Pt and Pd from WPCBs. The cyanide is initially oxidized to cyanate and subsequently metal complexes are formed based on the following reactions (Thakur and Kumar 2020; Wu et al., 2017):

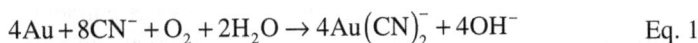

$$4Au + 8CN^- + O_2 + 2H_2O \rightarrow 4Au(CN)_2^- + 4OH^- \qquad \text{Eq. 1}$$

$$4Ag + 4CN^- + O_2 + 2H_2O \rightarrow 4AgCN + 4OH^- \qquad \text{Eq. 2}$$

$$4Pd + 4CN^- + O_2 + 2H_2O \rightarrow 4PdCN + 4OH^- \qquad \text{Eq. 3}$$

$$4Pt + 8CN^- + O_2 + 2H_2O \rightarrow 4Pt(CN)_2^- + 4OH^- \qquad \text{Eq. 4}$$

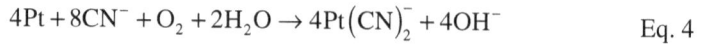

The high concentration of Cu in WPCBs usually increases the cyanide consumption and decreases the leaching efficiency of Au and Ag (Akcil et al., 2015). To overcome this issue and to prevent any damage from the use of toxic cyanide, there have been many attempts to replace cyanide with other reagents. In this context, the halogenation or halide leaching can take place to dissolve the metallic components from WPCBs as follows (Wu et al., 2017):

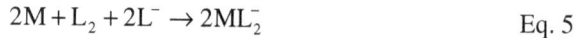

$$2M + L_2 + 2L^- \rightarrow 2ML_2^- \qquad \text{Eq. 5}$$

M can be Au, Ag, Pd, and Pt and L can be chlorine, bromine and iodine. L_2 is an oxidizing reagent and L^- is complexant.

Besides using other reagents, the presence of oxidants such as H_2O_2, Cl_2, Cu^{2+}, Fe^{3+}, air and oxygen along with inorganic acids is necessary for the dissolution of expensive metals because of their high potential of reduction from WPCBs. In this regard, Cu can be completely extracted from WPCBs using H_2SO_4 and H_2O_2 (Birloaga et al., 2014). The following equations represent the mechanism pathway for leaching of Fe, Ni, Cu, Ag, and Pd in a system where Cu^{2+} and Cl^- act as oxidants (Yazici and Deveci, 2013).

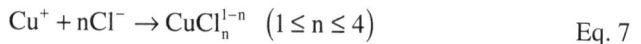

$$Cu^0 + Cu^{2+} \rightarrow 2Cu^+ \qquad \text{Eq. 6}$$

$$Cu^+ + nCl^- \rightarrow CuCl_n^{1-n} \quad (1 \leq n \leq 4) \qquad \text{Eq. 7}$$

$$Cu^0 + Cu^{2+} + 2Cl^- \rightarrow 2CuCl_{(s)} \quad \left(\Delta G^{\circ}_{20^{\circ}C} = -41 \text{ kJ/mol} \right) \qquad \text{Eq. 8}$$

$$Cu^0 + Cu^{2+} + 4Cl^- \rightarrow 2CuCl_2^- \quad \left(\Delta G^{\circ}_{20^{\circ}C} = -25 \text{ kJ/mol} \right) \qquad \text{Eq. 9}$$

Petter et al. (2014) compared conventional reagent for Au and Ag extraction from WPCBs and sodium thiosulfate and ammonium thiosulphate. Conventional cyanide-based reagent and HNO_3 leached 60% and 100% of Au and Ag, respectively. However, sodium thiosulfate and ammonium thiosulphate were able to leach 15% and 3% of the Au and Ag, respectively. In a study performed by Quinet et al. (2005), 95% Au, 93% Ag, and 99% Pd were recovered using cyanide leaching at pH>10, and at a temperature of 25°C while using thiourea as an alternative to cyanide results in low efficiencies.

Additionally, a combination of different reagents was employed to recover Cu and precious metals from WPCBs. H_2SO_4 and H_2O_2 were employed to leach 99% of Cu in a two-step leaching process. Then ca.86% of Au and ca.71% of Ag were leached from the residue using thiourea and ferric ions where they were precipitated from the leach liquor using sodium borohydride. The remaining Pd and Au of the leach liquor were completely dissolved using the $NaClO-HCl-H_2O_2$ system (Behnamfard et al., 2013). H_2SO_4 and H_2O_2 were used to leach more than 95% of the Cu, Fe, Zn, Ni, and Al from WPCBs at 85°C while more than 95% of gold and 100% silver were leached at 40°C using $(NH_4)_2S_2O_3-CuSO_4-NH_4OH$ leaching system. Pb was then extracted using NaCl at room temperature (Oh et al., 2003). It is also reported that Cu can be completely recovered using H_2SO_4 in the presence of $CuCl_2$ and H_2O_2 as oxidants (Yang, et al., 2011).

Using two-step consecutive leaching (3.5 M nitric acid) and solvent extraction (50% LIX 984 N) processes; 99% of Cu along with Fe, Pb, Zn and Ni was dissolved and 99.7% of Cu was extracted (aqueous to the organic ratio of 1:1.5 at pH of 1.5) (Le et al., 2011). It is also reported that 100% of the Cu and Zn, 79% of Fe, 24% of Al, 92% of Ni, and 85% of Sn can be simultaneously leached using 2.7 M HNO_3 in 60 minutes and solid to liquid (S/L) ratio of 1/8 $g.L^{-1}$ (Nekouei et al., 2019).

Hydrometallurgical processes have also been used in the treatment of various e-waste streams such as batteries, magnets, LCDs, cell phones, portable electronic devices, DVD players, calculators, etc. (Hsu et al., 2019; Tuncuk et al., 2012). Organic and inorganic acids with different reducing agents have been widely used for leaching of Co, Li, Ni, Mn, and Cd from used batteries (Porvali et al., 2020; Golmohammadzadeh et al., 2017, 2018). To maximize leaching efficiency, the operation constraints such as solid to liquid (S/L) ratio, acid and reducing agent concentration, type of acid and reducing agent, time of leaching, temperature, etc. have been optimized. It is reported that the addition of reducing agent improves the leaching efficiency of Co by reducing the insoluble Co^{3+} to soluble Co^{2+}. After leaching, metals can be separated using precipitation or solvent extraction method. It is reported that Co can be separated as $CoSO_4$ (Kang et al., 2010), Co_3O_4, CoC_2O_4 (Chen et al., 2011; Nayaka et al., 2016; Hu et al., 2013), $CO(OH)_2$ (Barbieri et al., 2014), $CoCo_3$ (Pagnanelli et al., 2016) and Li can be separated as Li_2CO_3 (Peng et al., 2019; Wang et al., 2019) or Li_3PO_4 (Yang et al., 2017).

Indium tin oxide (ITO) film is a useful portion of LCDs and comprises of various indium and tin oxide forms. It is reported that the SnO_2 and In_2O_3 are the abundant forms of oxides present in ITO film (Li et al., 2011). Indium as a scarce metal is the main metal to be recycled from Indium tin oxide film. Different leaching systems including oxidants have been employed for leaching of ITO film. It has been shown that using acids like concentrated HCl, HNO_3, H_2SO_4, with and without H_2O_2 various elements can be dissolved; In, Sn, Al, Cu, Ba, Zn, Ti, As, Cr, Si, Ca, Fe, K, Sr (Pu et al., 2012). In some studies, it has been reported that 3.2 M HCl can selectively leach In from ITO film while As a toxic substance mostly remained on as residue (Kato et al., 2013). Despite the efficiency of HCl, H_2SO_4 is considered as a better option as it is cost-effective while providing high leaching efficiency for In dissolution (Wang et al., 2013).

Electrochemical processes are considered as highly efficient and clean methods of metals purification and recovery from e-waste materials (Zhang et al., 2020; Xue and Wang, 2020). In this recycling technique, metals are recovered as a result of redox reactions happening in an electrolyte through cathodic and anodic reactions (Petrovic, 2021). Considering electron is the main role player in these systems, consumption of any extra chemical is diminished and consequently, the environmental dangers are minimized (Song et al., 2020). As a typical example, in a study performed by Venkatesan et al. (2018), the magnet waste was partially leached in an HCl media. Then $Fe(OH)_3$ was precipitated using a membrane electrolysis system; two-chamber electrochemical reactor separated using an anion exchange membrane. The partially leached solution and undissolved magnetic particles were added into the anolyte and catholyte (filled with NaCl) chambers, respectively. Through the following reactions in anode $Fe(OH)_3$ was precipitated:

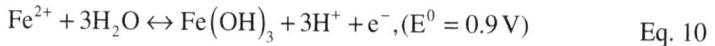

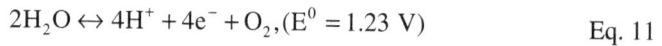

$$Fe^{2+} + 3H_2O \leftrightarrow Fe(OH)_3 + 3H^+ + e^-, (E^0 = 0.9\,V) \qquad \text{Eq. 10}$$

$$2H_2O \leftrightarrow 4H^+ + 4e^- + O_2, (E^0 = 1.23\ V) \qquad \text{Eq. 11}$$

Whereas in cathode:

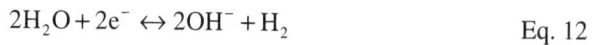

$$2H_2O + 2e^- \leftrightarrow 2OH^- + H_2 \qquad \text{Eq. 12}$$

Subsequently, rare earth oxalate is precipitated from the leachate using oxalic acid and are calcinated to produce rare earth oxides (Venkatesan et al., 2018).

High selectivity is one of the main advantages of this process, allowing effective separation of the metals by adjusting some operational conditions, such as voltage and current density (Ashiq et al., 2020). The selectivity of the process lies in the fact that the conversion of individual metals to their ionic forms happens at a specific redox potential and a certain range of pH (Xue and Wang, 2020). As an example, Barragan et al. (2020) employed an electrochemical process to selectively recover copper and antimony from a multi metal solution that contained over eight different cations and was achieved after leaching of WPCB in 0.5 M HCl and ferric chloride solution (0.074 M). As the first step, 81% of antimony was precipitated by simply adjusting the pH of the medium at 2.4 and later 96% of copper was recovered on the cathode surface through an electrowinning process for 2400 s at electrolysis potential of 0.45 V. As another example, in the recycling of spent LIBs, Wang et al. (2020) managed to re-synthesise a new cathode material in a single electrochemical step by applying electricity current (3.5 A) in ammonium containing medium. In this practice, the spent LIBs were leached in anode while a new $LiCoO_2$ layer was formed on the surface of the cathode.

The electrochemical processes are attracting more attention in the field of e-waste recycling and as a result, more novel processes are being developed. However, it is still early to be able to judge all aspects of this technique.

7.3.4 BIO-METALLURGICAL TECHNOLOGY

Use of microorganisms in the dissolving of the precious components from e-waste is one of the recent recycling methods which is more environmentally benign and economical (Faraji et al., 2018). In this procedure, living species including fungi and bacteria are employed to solubilize metals through certain mechanisms. Because e-waste involves a diverse range of devices and metals, several bioleaching routes are proposed for each group of e-waste streams (Baniasadi et al., 2019). Acidophilic bacteria, fungi, and cyanogenic microorganisms are some of the most used groups of living species that have been reported for e-waste recycling (Srichandan et al., 2019).

7.3.4.1 Acidophilic Bacteria

Sulphur and iron-oxidizing bacteria (e.g. *Acidithiobacillus thiooxidans* and *Acidobacillus ferrooxidans*, respectively) are two famous species that have been widely utilized in mining and vastly used in e-waste recycling (Priya and Hait, 2017). The bacteria are capable of continuous production of sulfuric acid and conversion of Fe^{2+} to Fe^{3+}; therefore, they can provide an excellent dissolution condition for solubilization of many metals available in e-waste (Mostafavi et al., 2018). External provision of sulphur and iron is necessary for these types of bacteria since they are vital for bacteria metabolism. They could be supplied in different forms such as elemental, soluble compounds ($FeSO_4$), or pyrite (FeS_2) (Xin et al., 2009; Mishra et al., 2008; Heydarian et al., 2018).

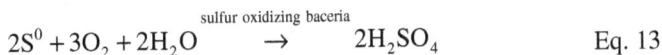

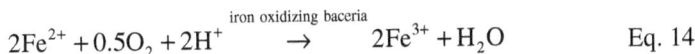

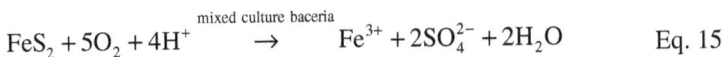

$$2S^0 + 3O_2 + 2H_2O \xrightarrow{\text{sulfur oxidizing baceria}} 2H_2SO_4 \qquad \text{Eq. 13}$$

$$2Fe^{2+} + 0.5O_2 + 2H^+ \xrightarrow{\text{iron oxidizing baceria}} 2Fe^{3+} + H_2O \qquad \text{Eq. 14}$$

$$FeS_2 + 5O_2 + 4H^+ \xrightarrow{\text{mixed culture baceria}} Fe^{3+} + 2SO_4^{2-} + 2H_2O \qquad \text{Eq. 15}$$

The majority of studies about the application of acidophile bacteria is related to recycling of base metals from WPCBs through mechanisms of acid leaching (by sulfuric acid) and oxidation by (Fe^{3+}) (Baniasadi et al., 2019). There are also many reports on the utilization of bacteria in leaching of cathode materials from spent LIBs (Xin et al., 2009; Heydarian et al., 2018; Xin et al., 2016; Liu et al., 2020). Metals in cathode materials are mostly at their higher valence (Ni^{3+}, Co^{3+}, Mn^{4+}) resulting in them being hardly soluble without reductant. However, the presence of iron cations in the biological system (Fe^{2+} and Fe^{3+}) combined with sulfuric acid, lead to reduction solubilization of metals. Desirable recovery is only achievable when a mixture of sulphur-oxidizing and iron-oxidizing bacteria is employed in spent LIBs (Heydarian et al., 2018; Xin et al., 2009). A typical example of bioleaching of cathode materials of spent LIBs by acidophile microorganisms is (Wu et al., 2019):

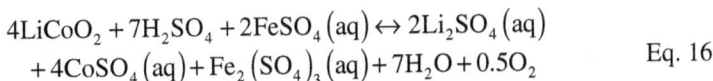

$$4LiCoO_2 + 7H_2SO_4 + 2FeSO_4 (aq) \leftrightarrow 2Li_2SO_4 (aq)$$
$$+ 4CoSO_4 (aq) + Fe_2 (SO_4)_3 (aq) + 7H_2O + 0.5O_2 \qquad \text{Eq. 16}$$

One of the major drawbacks of bacterial bioleaching is the sensitivity of the microorganisms to the amount of heavy metals they are exposed to (Heydarian et al., 2018). This problem could be moderated through a process called adaptation at which the heavy metals are step-wise introduced to a series of bacteria subcultures in little amounts (Baniasadi et al., 2019). Therefore, the successful subculture of bacteria capable of tolerating higher amount of metallic contamination would be selected and employed for actual bioleaching step. It is also important to consider that the condition could be favourable for jarosite formation when sulphur-oxidizing and iron-oxidizing bacteria are employed for recycling of spent LIBs (Zeng et al., 2013). This can precipitate a proportion of the interesting metals; hence a good control over the process in necessary.

7.3.4.2 Fungi

Fungal bioleaching is a more viable technique for e-waste recycling as it can tolerate higher concentration of metallic contamination, metabolize in a wider range of pH with shorter lag phase, and is capable of producing organic acids which are appropriate chelating agents (Bahaloo-Horeh et al., 2019; Faraji et al., 2018; Pant et al., 2012). *Aspergillus Niger* is a well-known fungus that has been used in numerous bioleaching practices as well as e-waste recycling (Islam et al., 2020). This is an acidophile and heterotroph microorganism that relies on the external source of food to survive and metabolize. It converts sugars (e.g. sucrose, glucose, etc) to some organic acids, mainly citric acid, oxalic acid, gluconic acid, and malic acid (Bahaloo-Horeh and Mousavi, 2017). The ratio of each acid depends on the medium condition such as pH, nutrition source, the existence of certain metallic ions, temperature, etc (Baniasadi et al., 2019; Faraji et al., 2018). The organic acids leach metals through three mechanisms of acidolysis, complexolysis, and redoxolysis and considering it is a mixture of acids in fungal bioleaching, the leaching recovery is expected to be higher than the application of individual acids (Seh-Bardan et al., 2012). Fungal bioleaching can cope up with almost every sort of e-waste, including cathode materials from spent LIBs that requires reducing agent. In this process, a part of organic acids act as a reducing agent by decomposing to CO_2 and H_2O; then the reduced form of metals would be mobilized through acidolysis and complexolysis mechanisms of the rest of the organic acids (Das et al. 2012; Bahaloo-Horeh, Mousavi, and Baniasadi 2018).

7.3.4.3 Cyanogenic Microorganisms

Cyanogenic microorganisms (*Chromobacterium violaceum*, *Pseudomonas fluorescens*, *Bacillus megaterium*, etc.) are another important species applicable for metal recycling from e-waste (Liu et al., 2016). Considering there are valuable metals (gold, silver, etc) and platinum group metals (PGMs) in e-waste and knowing the fact cyanidation is the main technique to recover them, bio-cyanidation is recognised as an economic and safer procedure compared to the conventional approaches (Akcil et al., 2015; Garlapati, 2016; Mishra and Rhee, 2010). In bio-cyanidation, the living species can produce cyanide (bio-CN) over their stationary phase of growth (up to about 70 ppm) by being exposed to a precursor (an amino acid-like glycine) (Faraji et al., 2020; Gorji et al., 2020).

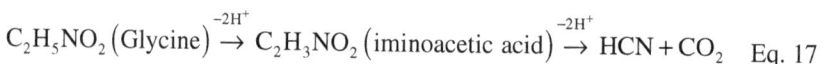

$$C_2H_5NO_2 \left(\text{Glycine}\right) \xrightarrow{-2H^+} C_2H_3NO_2 \left(\text{iminoacetic acid}\right) \xrightarrow{-2H^+} HCN + CO_2 \quad \text{Eq. 17}$$

Because there are many of the other metals in e-waste that can compete with the precious metals and use up the Bio-CN, there is sometimes an acid washing step (or bioleaching aiming for base metals) prior to cyanidation (Pham and Ting, 2009; Das and Ting, 2017). Considering cyanidation is normally performed at about pH 10.5, the cyanogenic microorganisms are developed for an alkaline range of pH. This is originally for the reason that cyanide is volatile (dangerous HCN gas) at pH below 9.3 thus a neutralization pre-step would be essential if acid washing was performed (Marsden and House, 2006).

In general, the microorganism can be helpful in bioleaching of the metals from e-waste. This method is more environmentally friendly, less expensive, safer, and more sustainable than the conventional processing routes. However, it needs longer time, recoveries are not necessarily high, and there is a need to provide some requirements for the microorganisms (food, mineral salts, oxygen, etc) (Banik, 1976; Roshanfar et al., 2019; Trumpy and Millis, 1963).

7.4 COMMERCIAL TECHNOLOGIES

Although there are numerous researches available on different techniques to recycle e-waste, not all of them have come to practice in industries. The viable procedures that end up to the final products are often based on pyrometallurgy (with the core of smelting reduction) or combination of hydrometallurgical and pyrometallurgical techniques (Khaliq et al., 2014). Some of the processes are originally for mineral processing but adjusted to accept e-waste as a part of the feed, some of them are designed for e-waste, and because of the special importance of the lithium-ion batteries, there are some processes specified for their recycling.

Xstrata Copper's Horne Smelter (Noranda) in Quebec Canada (www.glencore.ca/en) is a large size plant that recycles e-waste. A general schematic of this operation is shown in Figure 7.3. The e-waste is mixed with sulfidic copper concentrate and is treated with enriched air at 1250°C to generate the matte phase. A part of the energy required to heat-up the feed is from carbons of the e-waste. At the following step, matte is converted to copper blister, which bears almost entire valuable and precious metals. After a thermal refining step, copper is turned to the anode (99.1% pure and 0.9% precious and PGMs) and is electrowinned. As a result, copper is produced and precious metals are accumulated in the anode slime after the process (Anindya et al., 2009; Khaliq et al., 2014; Xakalashe et al., 2012). Montanwerke Brixlegg in Austria (www.montanwerke-brixlegg.com/en/) is another process that works likewise; in a blast furnace with oxygen lance, a mixture of e-waste, oxide feed, flux, and some coke is added to generate black copper. In the subsequent steps, this product is converted into a copper blister and then to the anode, which is electrowinned at the final step (Anindya et al., 2009).

Umicore in Hoboken Belgium (www.umicore.be/en/) is a well-recognized recycling plant that handles diverse sorts of waste including e-waste. In this route, the part of e-waste with a lower concentration of copper is directed to a blast furnace containing lead concentrate and coke. Because there is already some carbon in the e-waste, less coke than the conventional lead production process is consumed. The copper dross from this step as well as fresh e-waste with high copper content is fed to

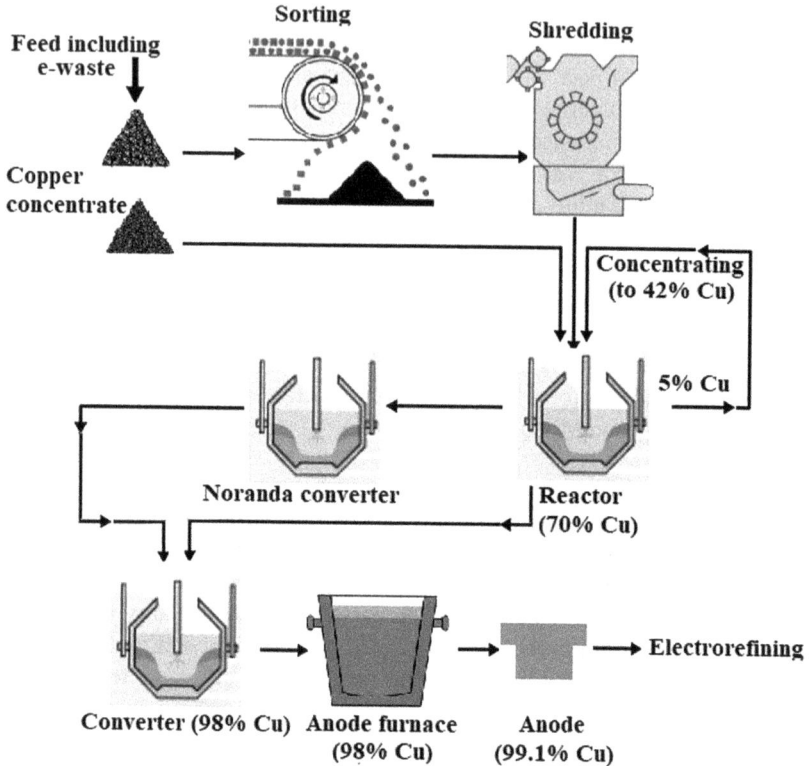

FIGURE 7.3 Schematic view of Xstrata Copper's Horne Smelter (Noranda) in Quebec Canada (www.glencore.ca/en/What-we-do/Metals-and-minerals/Copper).

a copper converter containing copper matte. As a result, metals are reduced mainly by the carbon content of the e-waste and the impurities are transferred to slag. Because multiple sorts of waste are treated through this procedure, the reduction temperature is in a range of 1200–1450°C with a supply of enriched air. The outcome of this process is an alloy of copper-containing valuable metals (Cu, Ni, Co, etc) and precious metals (Au, Ag, PGMs). Over the next steps, this alloy is treated hydro-metallurgically and the metals are separated (Ebin and Isik, 2016; Khaliq et al., 2014; Tesfaye et al., 2017).

In Ronnskar smelter in Sweden (www.boliden.com/) feed (e-waste) with low copper content is mixed with lead oxide concentrate and coke, then it is treated in a tilted furnace called Kaldo. A mixture of air and fuel is also sparged over top of the melt to reduce the toxic fumes. The produced copper from this step and copper-rich e-waste are directed to a convertor, following a route very similar to the Umicore process. In almost all smelters, the volatile metals (Zn, Pb, Sn, etc) are generated, which can be trapped and recycled; in Ronnskar procedure, this step is easier as the combustion is complete and no organic fume is associated with the condensed metals (Kaya, 2019; Shuey and Taylor, 2005; Ebin and Isik, 2016). The schematic diagram

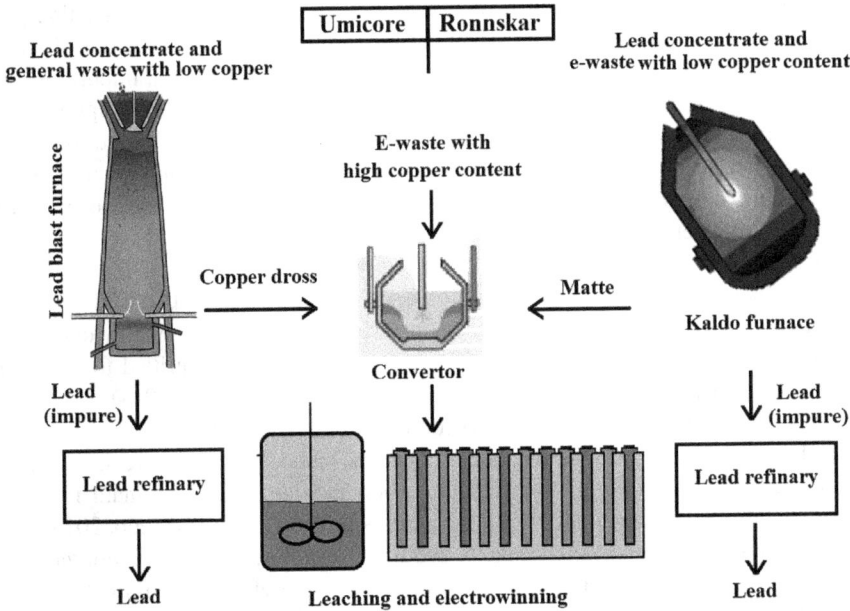

| Umicore | Ronnskar |

FIGURE 7.4 Schematic diagram of the main steps of Umicore and Ronnskar processes for e-waste recycling. (Khaliq et al., 2014.)

of the main steps of both Umicore and Ronnskar processes for e-waste recycling is presented in Figure 7.4.

Atomic Energy Authority (AEA) technology in the UK has a process designed to recycle spent LIBs. Materials are shredded after being dismantled by the assistance of liquid nitrogen. Electrolyte and the electrodes are then treated using acetonitrile and NMP reagents, which lead to the liberation of cathode and anode materials. In the following step, $LiCoO_2$ and graphite are washed by the means of lithium hydroxide solution and solution is electrolyzed. In the end, CoO is formed as the final product of this process (Lain, 2001). Toxco process in Canada is another plant that recycles spent LIBs by using liquid nitrogen to discharge and dismantle. Here, different parts of the battery are firstly separated through some physiochemical treatments and cathode materials are processed hydrometallurgically (McLaughlin and Adams, 1999).

Accure GmbH in Germany (www.accure.net/) uses liquid nitrogen and mechanical pre-treatment to dismantle the spent LIBs. Then by heating up to 250°C in a vacuum furnace, the electrolyte is removed and by the assistance of screening, magnetic field, and air separator, materials are sorted. The particles finer than 0.2 mm, as the fraction bearing valuable cathode materials, are then chosen for the reduction process. The results of smelting are a cobalt-manganese alloy and a slag phase bearing most of the lithium. In the end, the slag is chlorinated, leached, and lithium is selectively precipitated as lithium carbonate (Sojka, 1998).

Recupyl-Valibat technology in France recycles spent LIBs by crushing and physical separation in an oxygen-free atmosphere. After separation, lithium is selectively leached from the valuable fraction of the waste (smaller particles) by adjusting pH over 12. Later, lithium carbonate is selectively precipitated by sparging carbon dioxide into this solution. Rest of the particles are exposed to sulphuric acid, impurities are removed by adding soda to the solution, and pure cobalt hydroxide is precipitated by adding sodium hypochlorite (Tedjar and Foudraz, 2010). Similarly, Batrec Industrie AG in Switzerland (https://batrec.ch/en/) starts the process of recycling spent LIBs by dismantling and crushing in an inert gas environment; however, it continues by acid leaching (Meng et al., 2019).

In Sony-Sumitomo technology at Japan (www.smm.co.jp/E/), the feed is spent LIBs and starts by burning the non-metallic fraction of the e-waste at 1000°C followed by magnetic separation. Next, the cathode materials are reduced to an alloy of cobalt, nickel and iron, which later at hydrometallurgical processes is purified to the final product of cobalt oxide (Gaines and Cuenca, 2000; Meshram et al., 2014).

OnTo Technology in the USA (www.onto-technology.com/) is another plant that recycles spent LIBs. The treatment begins when the cathode materials are dismantled, sorted and separated using supercritical carbon dioxide. Here, the main route is hydrometallurgy and a new lithium-ion battery is the final product (Sloop, 2014).

There are not many details available about the recycling procedures in large-size plants; however, it is always important to improve the efficiency, limit the costs, and control the pollutions. Due to the accelerated rate of e-waste production and increasing demand for raw materials for their reproduction, it is vital to conduct more researches and bringing the novel techniques to the practice. Trials of some promising researches in pilot scale to be adopted for the actual scale seems to be a necessity, for now, to be able to cope up with the future trend.

7.5   NOVEL TECHNOLOGIES

In addition to the conventional and advanced technologies, novel technologies like mechanochemical, ultrasound, molten salts and combination of pyrometallurgy and hydrometallurgy are currently being used for e-waste recycling (Zhang and Xu, 2016a).

The ultrasound technology recovers metals such as Cu and Fe from e-wastes like PCBs by acid leaching aided by ultrasound. Figure 7.5(a) illustrate the schematic diagram of the experimental device. First, the sample is pre-treated and inserted into a beaker with water and 30% H_2O_2 and stirred at 300 rpm at room temperature for 60 minutes. Then, lime is inserted into the specimen and the resultant sludge is filtered, which changes the pH of the solution to below 1.5. The result has shown that this process is efficient in separating Cu from Fe, with the extraction levels of Cu and Fe as high as 93.76% and 2.07% respectively, with increasing of ultrasound energy (Huang et al., 2011).

Efficient recycling of metals from specific types of e-waste (for instance CRT funnel glass) can be achieved using the mechanochemical process. Tan and Li (2015) have done comprehensive research on the application of mechanochemistry in metal recycling from e-waste. Furthermore, they have examined the mechanochemical mechanism and addressed the consequences of this process on reaction and physiochemical properties. An overview of the mechanochemical process can be seen in

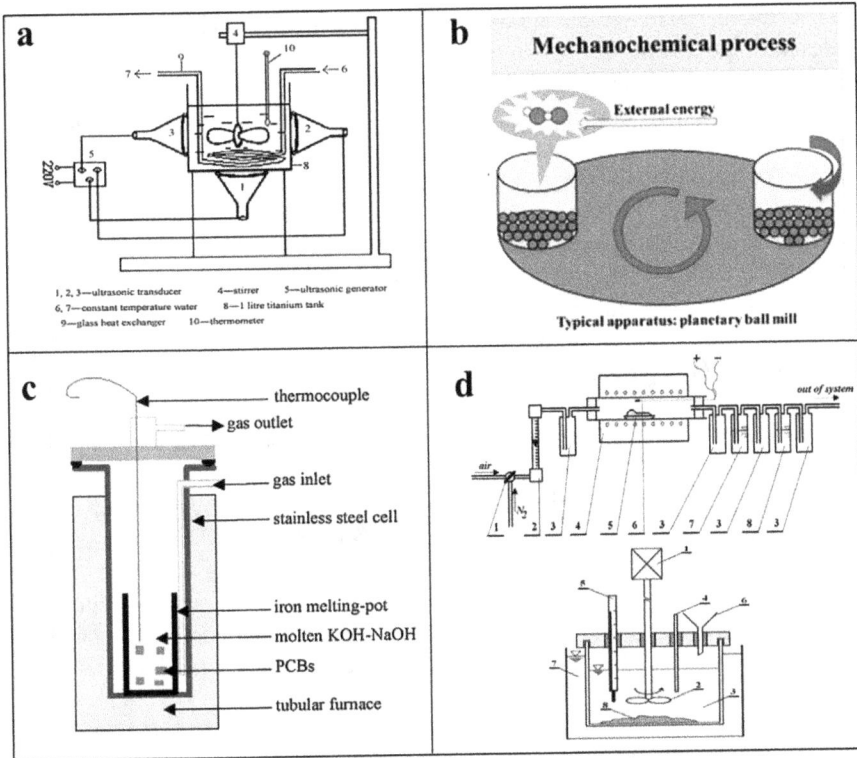

FIGURE 7.5 Schematic diagram of novel e-waste recycling technologies: (a) ultrasound, (b) mechanochemical, (c) molten salts and (d) combination of pyrometallugy and hydrometallurgical. (Adapted from Zhang and Xu 2016a. With permission.)

Figure 7.5(b). However, this technology is most suitable when accompanied by hydrometallurgical technology. As this can greatly improve the metal recovery rate relative to ordinary hydrometallurgy.

Molten salt technology has been suggested by Flandinet et al. (2012) for metal recovery from discarded PCBs. It is an effective way to recover metallic portion from e-wastes. In this method, KOH-NaOH liquid eutectic solutions were used to melt bottles, plastics and oxides without oxidizing the essential metals in PCB waste. Figure 7.5(c) illustrates the schematic of the molten salt process. Specimen of pulverized discarded PCBs were positioned in the melting container consist of salts. Other elements (fibreglass, binder, etc) were liquefied in the KOH-NaOH solutions at 300 °C. Without any decay or melting events, the metal fraction containing copper and other precious metals was extracted.

Havlik et al. (2011) applied a combined pyro-hydrometallurgical tactic to recover Cu and Sn from discarded PCBs. The specimens were leached at 80°C in a solution of 1 mol/l HCl after thermal pre-treatment step. Figure 5(d) portray the schematic of the combined pyro-hydrometallurgical process. The findings revealed that the Cu recovery rate for the burnt specimen had improved dramatically relative to the raw specimen. The reason behind this is the conversion of metallic copper to copper oxides during the burning process. In the hydrochloric acid, copper oxides were also leached, as a result, copper was leached from burnt samples. On the contrary, SnO_2 can be produced after burning and slightly leaching in hydrochloric acid. However, the overall leaching rate does not change much even after applying this procedure.

In order to control and recover metallic fumes generated over different thermal treatments of e-waste, micro recycling technique has been invented by a group of researchers in UNSW Sydney (Hossain et al., 2018). At this practice, the material is recovered step-by-step based on the melting of alloy present in e-waste. It includes the use of 500°C in the first step for the recovery of Sn-based alloy, followed by the application of 1000°C for the recovery of Cu-based alloy from WPCB. This technique does not only provide environmental benefit but also maintains the quality of recovered materials.

Hydrothermal leaching of spent LIBs by the means of organic acids is a novel process in the recycling of e-waste. In this technique, low concentrations of organic acids (like citric acid), and eco-friendly and high-efficiency reagents, are heated up to about 200°C and metals are leached fast and effectively (Zheng et al., 2020). As an example, Zheng et al. (2020) tried to extract metals from mixed cathode substances of used LIBs by using citric acid at 0.4 mol.L^{-1} through a hydrothermal process. Without the addition of any external reductant and with the least amount of secondary pollution, 81% Li, 73% Ni, 100% Co, and 100% Mn were recovered at the temperature of 90°C and over only 20 minutes.

Gasification of e-waste is another novel technology that could be utilized to recover energy from e-waste in form of syngas. However, compared to other available technologies this area is less explored. Only a few reports are currently available on e-waste gasification and most of them are focused on the polymer part of the e-waste. Gurgul et al. (2018) conducted steam gasification on e-waste and found that steam gasification successfully removes all the organics from e-waste. Moreover, the solid product obtained during this process is highly suitable for hydrometallurgical processing and the syngas could be utilized for chemical synthesis. Salbidegoitia et al. (2015) conducted steam gasification on the printed circuit board from e-waste to produce synthetic hydrogen gas. Results showed that the metals within e-waste plays a catalytic role during gasification and enhance the hydrogen yield. Moreover, the presence of nickel powder also accelerates hydrogen production. Zhang et al. (2013) also investigated on the steam gasification of the phenolic circuit board from e-waste to recover clean syngas from the polymer portion and to recover precious metals from the e-waste. Results showed that the production of hydrogen rich gas from the polymer portion of e-waste is viable. Moreover, the presence of molten carbonate enhanced the yield of syngas during gasification. Furthermore, the phenolic board with a particle size less than < 0.15 mm showed more stable syngas output. Yang et al. (2016) conducted gasification experiments on the polyurethane foam obtained from e-waste. Results showed the viability of syngas production from polyurethane foam and ER 0.3 was found the optimum condition. However, the syngas requires additional cleaning due to its higher HCN and NH_3 concentration. The solid residue obtained from these experiments was assessed for a hazardous test by Kang et al.

(2016). Researchers concluded that the solid residues are non-hazardous and can be used for various purposes (metal recovery, brick manufacturing or other recycling products).

7.6 CONCLUSION AND FUTURE PERSPECTIVE

Increase in the demand and shortening the lifetime of electronic and electrical equipment have led to accelerating the production and accumulation of e-waste streams. If e-waste is not correctly handled, it may trigger severe environmental concerns. However, recycling e-waste is a challenging process because of the complex structure of e-waste. Several technologies have been developed to recycle e-waste. Yet, most of them have not been commercialized due to technological barriers and cost-effectiveness.

Pre-treatment of e-waste is an integral part of recycling. Thermal pre-treatment for example is mainly performed to separate electronic components from the boards, helping to decrease the carbon content of some e-waste. Also, pre-treatment facilitates in liberating the metallic parts of e-waste as well as decrease the hardness of plastic components. The chemical pre-treatment however is used to improve the selectivity of the subsequent process and recovering some non-metallic parts of the e-waste. In the physio-mechanical treatment of e-waste, the material is concentrated through varieties of practices such as crushing, grinding, milling, floatation, cyclones, screening, magnetic/electric field separation, and airflow separation techniques.

Among different recycling techniques, the key method of metal recovery from e-waste is pyrometallurgical smelting. This method is capable of treating a huge amount in a short period of time. However, challenges of the pyrometallurgical technique include higher energy consumption and emission compared to others. Hydrometallurgy on the other hand works at lower temperatures with lower capital costs. This technique is more flexible compared to that of pyrometallurgy and is capable of efficient recovering valuable materials using varieties of chemicals in the aqueous system. Bio-metallurgy at which metals are mobilized as a result of microorganisms' metabolism, is an alternative method to mitigate environmental impact beside minimizing the operational cost in recycling. However, prolong processing time, need for extra care, and occasional failing to bring successful recovery are some common drawbacks of this process. There are many large-scale plants around the world, which mainly work based on the combination of pyrometallurgy and hydrometallurgy. Interesting to note that rigorous advancements on recycling technologies are underway. As such, the application of ultrasound, mechanochemistry, molten salt process, micro recycling technique, and hydrothermal leaching are named to be a few.

The increasing number of recently developed e-waste recycling methodologies has demonstrated that scientist has played a significant role in improving the sustainable processing of e-waste worldwide. By integrating recently developed technologies with various currently available technologies offers a significant possibility for the scientist in the direction of determining e-waste recycling technologies by choosing the utmost effective technology depending on their study extent, parameters and data accessibility. Based on the literature, it can be concluded that the future e-waste recycling technologies would be more eco-friendly, cost-effective and efficient by implementing an integrated recycling process accompanied by various recycling technologies.

REFERENCES

Akcil, A., Erust, C., Gahan, C.S., Ozgun, M., Sahin, M., Tuncuk, A., 2015. Precious metal recovery from waste printed circuit boards using cyanide and non-cyanide lixiviants-A review. Waste Manag. 45, 258–71.

Al-Thyabat, S., Nakamura, T., Shibata, E., Iizuka, A., 2013. Adaptation of minerals processing operations for lithium-ion (LiBs) and nickel metal hydride (NiMH) batteries recycling: Critical review. Miner. Eng. 45, 4–17.

Anindya, A., Swinbourne, D.R., Reuter, M.A., Matusewicz, R., 2009. Tin distribution during smelting of WEEE with copper scrap, in: Proc. European Metallurgical Congress. pp. 555–568.

Anindya, A., Swinbourne, D.R., Reuter, M.A., Matusewicz, R.W., 2013. Distribution of elements between copper and FeOx–CaO–SiO2 slags during pyrometallurgical processing of WEEE: Part 1–Tin. Miner. Process. Extr. Metall. 122, 165–173.

Antrekowitsch, H., Potesser, M., Spruzina, W., Prior, F., 2006. Metallurgical recycling of electronic scrap, in: EPD Congress. pp. 899–908.

Ashiq, A., Cooray, A., Srivatsa, S.C., Vithanage, M., 2020. Electrochemical enhanced metal extraction from E-waste, in: Handbook of Electronic Waste Management. Elsevier, pp. 119–139.

Assefi, M., Maroufi, S., Yamauchi, Y., Sahajwalla, V., 2020. Pyrometallurgical recycling of Li-ion, Ni-Cd and Ni-MH batteries: A mini-review. Curr. Opin. Green Sustain. Chem. 24, 26–31.

Awasthi, A.K., Hasan, M., Mishra, Y.K., Pandey, A.K., Tiwary, B.N., Kuhad, R.C., Gupta, V.K., Thakur, V.K., 2019. Environmentally sound system for E-waste: Biotechnological perspectives. Curr. Res. Biotechnol. 1, 58–64.

Bahaloo-Horeh, N., Mousavi, S.M., 2017. Enhanced recovery of valuable metals from spent lithium-ion batteries through optimization of organic acids produced by Aspergillus niger. Waste Manag. 60, 666–679.

Bahaloo-Horeh, N., Mousavi, S.M., Baniasadi, M., 2018. Use of adapted metal tolerant Aspergillus niger to enhance bioleaching efficiency of valuable metals from spent lithium-ion mobile phone batteries. J. Clean. Prod. 197, 1546–1557.

Bahaloo-Horeh, N., Vakilchap, F., Mousavi, S.M., 2019. Bio-hydrometallurgical Methods For Recycling Spent Lithium-Ion Batteries, in: Recycling of Spent Lithium-Ion Batteries. Springer, pp. 161–197.

Bakas, I., Fischer, C., Haselsteiner, S., McKinnon, D., Milios, L., Harding, A., Kosmol, J., Plepys, A., Tojo, N., Wilts, C.H., 2014. Present and potential future recycling of critical metals in WEEE.

Balde, C.P., Wang, F., Kuehr, R., Huisman, J., 2015. The global e-waste monitor 2014: Quantities, flows and resources.

Baniasadi, M., Vakilchap, F., Bahaloo-Horeh, N., Mousavi, S.M., Farnaud, S., 2019. Advances in bioleaching as a sustainable method for metal recovery from e-waste: A review. J. Ind. Eng. Chem. 76, 75–90.

Banik, A.K., 1976. Mineral nutrition of Aspergillus niger for citric acid production. Folia Microbiol. (Praha). 21, 139–143.

Barbieri, E.M.S., Lima, E.P.C., Cantarino, S.J., Lelis, M.F.F., Freitas, M., 2014. Recycling of spent ion-lithium batteries as cobalt hydroxide, and cobalt oxide films formed under a conductive glass substrate, and their electrochemical properties. J. Power Sources 269, 158–163.

Barragan, J.A., Ponce de León, C., Alemán Castro, J.R., Peregrina-Lucano, A., Gómez-Zamudio, F., Larios-Durán, E.R., 2020. Copper and Antimony Recovery from Electronic Waste by Hydrometallurgical and Electrochemical Techniques. ACS Omega.

Bas, A.D., Deveci, H., Yazici, E.Y., 2013. Bioleaching of copper from low grade scrap TV circuit boards using mesophilic bacteria. Hydrometallurgy 138, 65–70.

Behnamfard, A., Salarirad, M.M., Veglio, F., 2013. Process development for recovery of copper and precious metals from waste printed circuit boards with emphasize on palladium and gold leaching and precipitation. Waste Manag. 33, 2354–2363.

Birloaga, I., Coman, V., Kopacek, B., Vegliò, F., 2014. An advanced study on the hydrometallurgical processing of waste computer printed circuit boards to extract their valuable content of metals. Waste Manag. 34, 2581–2586.

Bodsworth, C., 1994. The extraction and refining of metals. CRC Press.

Cayumil, R., Ikram-Ul-Haq, M., Khanna, R., Saini, R., Mukherjee, P.S., Mishra, B.K., Sahajwalla, V., 2018. High temperature investigations on optimising the recovery of copper from waste printed circuit boards. Waste Manag. 73, 556–565.

Cayumil, R., Khanna, R., Rajarao, R., Ikram-ul-Haq, M., Mukherjee, P.S., Sahajwalla, V., 2016. Environmental impact of processing electronic waste–key issues and challenges. E-waste transition—from Pollut. to Resour. InTech 9–35.

Chagnes, A., Pospiech, B., 2013. A brief review on hydrometallurgical technologies for recycling spent lithium-ion batteries. J. Chem. Technol. Biotechnol. 88, 1191–1199.

Chen, L., Tang, X., Zhang, Yang, Li, L., Zeng, Z., Zhang, Yi, 2011. Process for the recovery of cobalt oxalate from spent lithium-ion batteries. Hydrometallurgy 108, 80–86.

Chen, X., Ma, H., Luo, C., Zhou, T., 2017. Recovery of valuable metals from waste cathode materials of spent lithium-ion batteries using mild phosphoric acid. J. Hazard. Mater. 326, 77–86.

Chen, Y., Zhang, Y., Yang, J., Liang, S., Liu, K., Xiao, K., Deng, H., Hu, J., Xiao, B., 2018. Improving bromine fixation in co-pyrolysis of non-metallic fractions of waste printed circuit boards with Bayer red mud. Sci. Total Environ. 639, 1553–1559.

Contestabile, M., Panero, S., Scrosati, B., 1999. A laboratory-scale lithium battery recycling process. J. Power Sources 83, 75–78.

Copani, G., Picone, N., Colledani, M., Pepe, M., Tasora, A., 2019. Highly evolvable e-waste recycling technologies and systems, in: Factories of the Future. Springer, Cham, pp. 109–128.

Cucchiella, F., D'Adamo, I., Koh, S.C.L., Rosa, P., 2015. Recycling of WEEEs: An economic assessment of present and future e-waste streams. Renew. Sustain. Energy Rev. 51, 263–272.

Cui, J., Zhang, L., 2008. Metallurgical recovery of metals from electronic waste: A review. J. Hazard. Mater. 158, 228–256.

Dang, H., Li, N., Chang, Z., Wang, B., Zhan, Y., Wu, X., Liu, W., Ali, S., Li, H., Guo, J., 2020. Lithium leaching via calcium chloride roasting from simulated pyrometallurgical slag of spent lithium ion battery. Sep. Purif. Technol. 233, 116025.

Dang, H., Wang, B., Chang, Z., Wu, X., Feng, J., Zhou, H., Li, W., Sun, C., 2018. Recycled lithium from simulated pyrometallurgical slag by chlorination roasting. ACS Sustain. Chem. Eng. 6, 13160–13167.

Das, A.P., Swain, S., Panda, S., Pradhan, N., Sukla, L.B., 2012. Reductive Acid Leaching of Low Grade Manganese Ores. Geomaterials 02, 70–72.

Das, S., Ting, Y.P., 2017. Improving Gold (Bio) Leaching Efficiency Through Pretreatment Using Hydrogen Peroxide Assisted Sulfuric Acid. Clean – Soil, Air, Water 45, 2–9.

Dave, S.R., Shah, M.B., Tipre, D.R., 2016. E-waste: metal pollution threat or metal resource? J Adv Res Biotech 1, 14.

Debnath, B., Chowdhury, R., Ghosh, S.K., 2019. An analysis of e-waste recycling technologies from the chemical engineering perspective, in: Waste Management and Resource Efficiency. Springer, pp. 879–888.

Debnath, B., Chowdhury, R., Ghosh, S.K., 2018. Sustainability of metal recovery from E-waste. Front. Environ. Sci. Eng. 12, 2.

Diekmann, J., Hanisch, C., Froböse, L., Schälicke, G., Loellhoeffel, T., Fölster, A.-S., Kwade, A., 2017. Ecological recycling of lithium-ion batteries from electric vehicles with focus on mechanical processes. J. Electrochem. Soc. 164, A6184–A6191.

Ding, Y., Zhang, S., Liu, B., Zheng, H., Chang, C., Ekberg, C., 2019. Recovery of precious metals from electronic waste and spent catalysts: A review. Resour. Conserv. Recycl. 141, 284–298.

Ebin, B., Isik, M.I., 2016. Pyrometallurgical processes for the recovery of metals from WEEE, in: WEEE Recycling. Elsevier, pp. 107–137.

Faraji, F., Golmohammadzadeh, R., Rashchi, F., Alimardani, N., 2018. Fungal bioleaching of WPCBs using Aspergillus niger: Observation, optimization and kinetics. J. Environ. Manage. 217, 775–787.

Faraji, F., Wang, J., Mahandra, H., Ghahreman, A., 2020. A Green and Sustainable Process for the Recovery of Gold from Low-Grade Sources Using Biogenic Cyanide Generated by *Bacillus megaterium*: A Comprehensive Study. ACS Sustain. Chem. Eng. acssuschemeng.0c06904.

Faramarzi, M.A., Stagars, M., Pensini, E., Krebs, W., Brandl, H., 2004. Metal solubilization from metal-containing solid materials by cyanogenic Chromobacterium violaceum. J. Biotechnol. 113, 321–326.

Flandinet, L., Tedjar, F., Ghetta, V., Fouletier, J., 2012. Metals recovering from waste printed circuit boards (WPCBs) using molten salts. J. Hazard. Mater. 213, 485–490.

Gaines, L., Cuenca, R., 2000. Costs of lithium-ion batteries for vehicles. Argonne National Lab., IL (US).

Garlapati, V.K., 2016. E-waste in India and developed countries: Management, recycling, business and biotechnological initiatives. Renew. Sustain. Energy Rev. 54, 874–881.

Georgi-Maschlera, T., Friedricha, B., Weyheb, R., Heegnc, H., Rutzc, M., 2012. Development of a recycling process for Li-ion batteries. J Power Sources 207, 173–182.

Ghodrat, M., Rhamdhani, M.A., Brooks, G., Masood, S., Corder, G., 2016. Techno economic analysis of electronic waste processing through black copper smelting route. J. Clean. Prod. 126, 178–190.

Ghodrat, M., Rhamdhani, M.A., Khaliq, A., Brooks, G., Samali, B., 2018. Thermodynamic analysis of metals recycling out of waste printed circuit board through secondary copper smelting. J. Mater. Cycles Waste Manag. 20, 386–401.

Golmohammadzadeh, R., Faraji, F., Rashchi, F., 2018. Recovery of lithium and cobalt from spent lithium ion batteries (LIBs) using organic acids as leaching reagents: A review. Resour. Conserv. Recycl. 136, 418–435.

Golmohammadzadeh, R., Rashchi, F., Vahidi, E., 2017. Recovery of lithium and cobalt from spent lithium-ion batteries using organic acids: Process optimization and kinetic aspects. Waste Manag. 64.

Goosey, E., Goosey, M., 2020. Chapter 1 Introduction and Overview, in: Electronic Waste Management (2). The Royal Society of Chemistry, pp. 1–32.

Gorji, M., Hosseini, M.R., Ahmadi, A., 2020. Comparison and optimization of the bio-cyanidation potentials of B. megaterium and P. aeruginosa for extracting gold from an oxidized copper-gold ore in the presence of residual glycine. Hydrometallurgy 191, 105218.

Gramatyka, P., Nowosielski, R., Sakiewicz, P., 2007. Recycling of waste electrical and electronic equipment. J. Achiev. Mater. Manuf. Eng. 20, 535–538.

Guoxing, R., Songwen, X., Meiqiu, X., Bing, P., Youqi, F., Fenggang, W., Xing, X., 2016. Recovery of valuable metals from spent lithium-ion batteries by smelting reduction process based on $MnO-SiO_2-Al_2O_3$ slag system, in: Advances in Molten Slags, Fluxes, and Salts: Proceedings of the 10th International Conference on Molten Slags, Fluxes and Salts 2016. Springer, pp. 211–218.

Gurgul, A., Szczepaniak, W., Zabłocka-Malicka, M., 2018. Incineration and pyrolysis vs. steam gasification of electronic waste. Sci. Total Environ. 624, 1119–1124.

Hageluken, C., 2006. Improving metal returns and eco-efficiency in electronics recycling-a holistic approach for interface optimisation between pre-processing and integrated metals smelting and refining, in: Proceedings of the 2006 IEEE International Symposium on Electronics and the Environment, 2006. IEEE, pp. 218–223.

Haldar, S.K., 2018. Mineral exploration: principles and applications. Elsevier.

Hanisch, C., Haselrieder, W., Kwade, A., 2011. Recovery of active materials from spent lithium-ion electrodes and electrode production rejects, in: Glocalized Solutions for Sustainability in Manufacturing. Springer, pp. 85–89.

Hassan, M.F., Rahman, M.M., Guo, Z.P., Chen, Z.X., Liu, H.K., 2010. Solvent-assisted molten salt process: A new route to synthesise α-Fe2O3/C nanocomposite and its electrochemical performance in lithium-ion batteries. Electrochim. Acta 55, 5006–5013.

Havlik, T., Orac, D., Petranikova, M., Miskufova, A., 2011. Hydrometallurgical treatment of used printed circuit boards after thermal treatment. Waste Manag. 31, 1542–1546.

He, J., Duan, C., 2017. Recovery of metallic concentrations from waste printed circuit boards via reverse floatation. Waste Manag. 60, 618–628.

He, K., Zhang, Z.Y., Alai, L., Zhang, F.S., 2019. A green process for exfoliating electrode materials and simultaneously extracting electrolyte from spent lithium-ion batteries. J. Hazard. Mater. 375, 43–51.

He, L.P., Sun, S.Y., Song, X.F., Yu, J.G., 2015. Recovery of cathode materials and Al from spent lithium-ion batteries by ultrasonic cleaning. Waste Manag. 46, 523–528.

Heydarian, A., Mousavi, S.M., Vakilchap, F., Baniasadi, M., 2018. Application of a mixed culture of adapted acidophilic bacteria in two-step bioleaching of spent lithium-ion laptop batteries. J. Power Sources 378, 19–30.

Hossain, R., Nekouei, R.K., Mansuri, I., Sahajwalla, V., 2018. Sustainable recovery of Cu and Sn from problematic global waste: exploring value from waste printed circuit boards. ACS Sustain. Chem. Eng. 7, 1006–1017.

Hsu, E., Barmak, K., West, A.C., Park, A.H.A., 2019. Advancements in the treatment and processing of electronic waste with sustainability: A review of metal extraction and recovery technologies. Green Chem. 21, 919–936.

Hu, C., Guo, J., Wen, J., Peng, Y., 2013. Preparation and electrochemical performance of nano-Co3O4 anode materials from spent Li-ion batteries for lithium-ion batteries. J. Mater. Sci. Technol. 29, 215–220.

Hu, J., Zhang, J., Li, H., Chen, Y., Wang, C., 2017. A promising approach for the recovery of high value-added metals from spent lithium-ion batteries. J. Power Sources 351, 192–199.

Huang, Z., Xie, F., Ma, Y., 2011. Ultrasonic recovery of copper and iron through the simultaneous utilization of Printed Circuit Boards (PCB) spent acid etching solution and PCB waste sludge. J. Hazard. Mater. 185, 155–161.

Ilyas, S., Lee, J., 2014. Biometallurgical recovery of metals from waste electrical and electronic equipment: a review. ChemBioEng Rev. 1, 148–169.

Islam, A., Ahmed, T., Awual, M.R., Rahman, A., Sultana, M., Abd Aziz, A., Monir, M.U., Teo, S.H., Hasan, M., 2020. Advances in sustainable approaches to recover metals from e-waste-A review. J. Clean. Prod. 244, 118815.

Jackson, E., 1986. Hydrometallurgical extraction and reclamation. Chichester: Horwood; New York: Wiley.

Jeon, S., Ito, M., Tabelin, C.B., Pongsumrankul, R., Kitajima, N., Park, I., Hiroyoshi, N., 2018. Gold recovery from shredder light fraction of E-waste recycling plant by flotation-ammonium thiosulfate leaching. Waste Manag. 77, 195–202.

Kang, H.-Y., Schoenung, J.M., 2005. Electronic waste recycling: A review of US infrastructure and technology options. Resour. Conserv. Recycl. 45, 368–400.

Kang, J., Senanayake, G., Sohn, J., Shin, S.M., 2010. Recovery of cobalt sulfate from spent lithium ion batteries by reductive leaching and solvent extraction with Cyanex 272. Hydrometallurgy 100, 168–171.

Kang, J.J., Lee, J.S., Yang, W.S., Park, S.W., Alam, M.T., Back, S.K., Choi, H.S., Seo, Y.C., Yun, Y.S., Gu, J.H., 2016. A study on environmental assessment of residue from gasification of polyurethane waste in e-waste recycling process. Procedia Environ. Sci. 35, 639–642.

Kato, T., Igarashi, S., Ishiwatari, Y., Furukawa, M., Yamaguchi, H., 2013. Separation and concentration of indium from a liquid crystal display via homogeneous liquid–liquid extraction. Hydrometallurgy 137, 148–155.

Kaya, Muammer, 2019. Sorting and Separation of WPCBs BT – Electronic Waste and Printed Circuit Board Recycling Technologies, in: Kaya, M. (Ed.), . Springer International Publishing, Cham, pp. 143–176.

Kaya, M, 2019. Electronic Waste and Printed Circuit Board Recycling Technologies, The Minerals, Metals & Materials Series. Springer International Publishing.

Kaya, M., 2018. Current WEEE recycling solutions, in: Waste Electrical and Electronic Equipment Recycling. Elsevier, pp. 33–93.

Kaya, M., 2016. Recovery of metals and nonmetals from electronic waste by physical and chemical recycling processes. Waste Manag. 57, 64–90.

Khaliq, A., Rhamdhani, M.A., Brooks, G., Masood, S., 2014. Metal extraction processes for electronic waste and existing industrial routes: a review and Australian perspective. Resources 3, 152–179.

Kongoli, F., Yazawa, A., 2001. Liquidus surface of FeO-$Fe2O_3$-SiO_2-CaO slag containing Al 2 O 3, MgO, and Cu 2 O at intermediate oxygen partial pressures. Metall. Mater. Trans. B 32, 583–592.

Kumar, A., Holuszko, M., Espinosa, D.C.R., 2017. E-waste: an overview on generation, collection, legislation and recycling practices. Resour. Conserv. Recycl. 122, 32–42.

Kumar, A., Saini, H.S., Kumar, S., 2018. Bioleaching of gold and silver from waste printed circuit boards by Pseudomonas balearica SAE1 isolated from an e-waste recycling facility. Curr. Microbiol. 75, 194–201.

Kumar, V., Lee, J., Jeong, J., Jha, M.K., Kim, B., Singh, R., 2015. Recycling of printed circuit boards (PCBs) to generate enriched rare metal concentrate. J. Ind. Eng. Chem. 21, 805–813.

Lain, M.J., 2001. Recycling of lithium ion cells and batteries. J. Power Sources 97, 736–738.

Lee, C.K., Rhee, K. in, 2002. Preparation of LiCoO2 from spent lithium-ion batteries. J. Power Sources 109, 17–21.

Lee, J., Song, H.T., Yoo, J.-M., 2007. Present status of the recycling of waste electrical and electronic equipment in Korea. Resour. Conserv. Recycl. 50, 380–397.

Li, H., Eksteen, J., Oraby, E., 2018. Hydrometallurgical recovery of metals from waste printed circuit boards (WPCBs): Current status and perspectives–A review. Resour. Conserv. Recycl. 139, 122–139.

Li, J. hui, Li, X. hai, Zhang, Y. he, Hu, Q. yang, Wang, Z. xing, Zhou, Y. yuan, Fu, F. ming, 2009. Study of spent battery material leaching process. Trans. Nonferrous Met. Soc. China (English Ed. 19, 751–755.

Li, Y., Liu, Zhihong, Li, Q., Liu, Zhiyong, Zeng, L., 2011. Recovery of indium from used indium–tin oxide (ITO) targets. Hydrometallurgy 105, 207–212.

Lin, C., Chi, Y., Jin, Y., 2017. Experimental study on treating waste printed circuit boards by molten salt oxidation. Waste and Biomass Valorization 8, 2523–2533.

Lin, N., Han, Y., Wang, L., Zhou, Jianbin, Zhou, Jie, Zhu, Y., Qian, Y., 2015. Preparation of Nanocrystalline Silicon from SiCl4 at 200° C in Molten Salt for High-Performance Anodes for Lithium Ion Batteries. Angew. Chemie Int. Ed. 54, 3822–3825.

Liu, C., Lin, J., Cao, H., Zhang, Y., Sun, Z., 2019. Recycling of spent lithium-ion batteries in view of lithium recovery: A critical review. J. Clean. Prod.

Liu, R., Li, J., Ge, Z., 2016. Review on Chromobacterium violaceum for gold bioleaching from e-waste. Procedia Environ. Sci. 31, 947–953.

Liu, W.-W., Hu, C.W., Yang, Y., Tong, D.M., Zhu, L.F., Zhang, R.N., Zhao, B.H., 2013. Study on the effect of metal types in (Me)-Al-MCM-41 on the mesoporous structure and catalytic behavior during the vapor-catalyzed co-pyrolysis of pubescens and LDPE. Appl. Catal. B Environ. 129, 202–213.

Liu, X., Liu, H., Wu, W., Zhang, X., Gu, T., Zhu, M., Tan, W., 2020. Oxidative Stress Induced by Metal Ions in Bioleaching of LiCoO2 by an Acidophilic Microbial Consortium. Front. Microbiol. 10, 3058.

Long Le, H., Jeong, J., Lee, J.C., Pandey, B.D., Yoo, J.-M., Huyunh, T.H., 2011. Hydrometallurgical process for copper recovery from waste printed circuit boards (PCBs). Miner. Process. Extr. Metall. Rev. 32, 90–104.

Lovinger, A.J., 1982. Poly (vinylidene fluoride), in: Developments in Crystalline Polymers—1. Springer, pp. 195–273.

Mäkinen, J., Bachér, J., Kaartinen, T., Wahlström, M., Salminen, J., 2015. The effect of flotation and parameters for bioleaching of printed circuit boards. Miner. Eng. 75, 26–31.

Mark, F.E., Lehner, T., 2000. Plastics recovery from waste electrical & electronic equipment in non-ferrous metal processes. Assoc. Plast. Manuf. Eur. Brussels, Belgium.

Marsden, J., House, I., 2006. The Chemistry of Gold Extraction. Society for Mining, Metallurgy, and Exploration.

McLaughlin, W., Adams, T.S., 1999. Li reclamation process. USA. Patent no. US5888463A.

Meng, F., McNeice, J., Zadeh, S.S., Ghahreman, A., 2019. Review of lithium production and recovery from minerals, brines, and lithium-ion batteries. Miner. Process. Extr. Metall. Rev. 1–19.

Meshram, P., Pandey, B.D., Mankhand, T.R., 2014. Extraction of lithium from primary and secondary sources by pre-treatment, leaching and separation: A comprehensive review. Hydrometallurgy 150, 192–208.

Mishra, D., Kim, D.J., Ralph, D.E., Ahn, J.G., Rhee, Y.H., 2008. Bioleaching of metals from spent lithium ion secondary batteries using Acidithiobacillus ferrooxidans. Waste Manag. 28, 333–338.

Mishra, D., Rhee, Y.-H., 2010. Current research trends of microbiological leaching for metal recovery from industrial wastes. Curr Res Technol Educ Top. Appl Microbiol Microb Biotechnol 2, 1289–1292.

Mostafavi, M., Mirazimi, S.M.J., Rashchi, F., Faraji, F., Mostoufi, N., 2018. Bioleaching and kinetic investigation of WPCBs by A. ferrooxidans, A. thiooxidans and their mixtures. J. Chem. Pet. Eng. 52, 81–91.

Mukherjee, T.K., 2019. Hydrometallurgy in Extraction Processes, Volume I. Routledge.

Nayaka, G.P., Pai, K. V, Santhosh, G., Manjanna, J., 2016. Recovery of cobalt as cobalt oxalate from spent lithium ion batteries by using glycine as leaching agent. J. Environ. Chem. Eng. 4, 2378–2383.

Nekouei, R.K., Pahlevani, F., Golmohammadzadeh, R., Assefi, M., Rajarao, R., Chen, Y.-H., Sahajwalla, V., 2019. Recovery of heavy metals from waste printed circuit boards: statistical optimization of leaching and residue characterization. Environ. Sci. Pollut. Res. 26, 24417–24429.

Neto, I.F.F., Sousa, C.A., Brito, M.S.C.A., Futuro, A.M., Soares, H.M.V.M., 2016. A simple and nearly-closed cycle process for recycling copper with high purity from end life printed circuit boards. Sep. Purif. Technol. 164, 19–27.

Oh, C.J., Lee, S.O., Yang, H.S., Ha, T.J., Kim, M.J., 2003. Selective leaching of valuable metals from waste printed circuit boards. J. Air Waste Manage. Assoc. 53, 897–902.

Pagnanelli, F., Moscardini, E., Altimari, P., Abo Atia, T., Toro, L., 2016. Cobalt products from real waste fractions of end of life lithium ion batteries. Waste Manag. 51, 214–221.

Pant, D., Joshi, D., Upreti, M.K., Kotnala, R.K., 2012. Chemical and biological extraction of metals present in e-waste: A hybrid technology. Waste Manag. 32, 979–990.

Peng, C., Liu, F., Wang, Z., Wilson, B.P., Lundström, M., 2019. Selective extraction of lithium (Li) and preparation of battery grade lithium carbonate (Li2CO$_3$) from spent Li-ion batteries in nitrate system. J. Power Sources 415, 179–188.

Petrovic, S., 2021. Basic Electrochemistry Concepts, in: Electrochemistry Crash Course for Engineers. Springer, pp. 3–10.

Petter, P.M.H., Veit, H.M., Bernardes, A.M., 2014. Evaluation of gold and silver leaching from printed circuit board of cellphones. Waste Manag. 34, 475–482.

Pham, V.A., Ting, Y.P., 2009. Gold bioleaching of electronic waste by cyanogenic bacteria and its enhancement with bio-oxidation, in: Advanced Materials Research. Trans Tech Publ, pp. 661–664.

Porvali, A., Ojanen, S., Wilson, B.P., Serna-Guerrero, R., Lundström, M., 2020. Nickel Metal Hydride Battery Waste: Mechano-hydrometallurgical Experimental Study on Recycling Aspects. J. Sustain. Metall. 6, 78–90.

Priya, A., Hait, S., 2017. Comparative assessment of metallurgical recovery of metals from electronic waste with special emphasis on bioleaching. Environ. Sci. Pollut. Res. 24, 6989–7008.

Pryor, M.R., 2012. Mineral processing. Springer Science & Business Media.

Pu, L., Yang, D., Guo, Y., 2012. Analyzing the main elements in waste TFT-LCD panel using inductively coupled plasma atomic emission spectrometry. Environ. Pollut. Control 76–78.

QIU, K., WU, Q., ZHAN, Z., 2009. Vacuum pyrolysis characteristics of waste printed circuit boards epoxy resin and analysis of liquid products. J. Cent. South Univ. Sci. Technol. 40, 1209–1215.

Quinet, P., Proost, J., Van Lierde, A., 2005. Recovery of precious metals from electronic scrap by hydrometallurgical processing routes. Mining, Metall. Explor. 22, 17–22.

Ray, H.S., 2006. Introduction to Melts: Molten Salts, Slags and Glasses. Allied Publishers.

Reddy, M. V, Sharma, N., Adams, S., Rao, R.P., Peterson, V.K., Chowdari, B.V.R., 2015. Evaluation of undoped and M-doped TiO2, where M= Sn, Fe, Ni/Nb, Zr, V, and Mn, for lithium-ion battery applications prepared by the molten-salt method. Rsc Adv. 5, 29535–29544.

Reeve, S., Eduljee, G., 2020. Chapter 4 An Overview of Electronic Waste Management in the UK, in: Electronic Waste Management (2). The Royal Society of Chemistry, pp. 101–136.

Ren, G., Xiao, S., Xie, M., Bing, P.A.N., Jian, C., Wang, F., Xing, X.I.A., 2017. Recovery of valuable metals from spent lithium ion batteries by smelting reduction process based on $FeO–SiO2–Al_2O_3$ slag system. Trans. Nonferrous Met. Soc. China 27, 450–456.

Robinson, B.H., 2009. E-waste: an assessment of global production and environmental impacts. Sci. Total Environ. 408, 183–191.

Roshanfar, M., Golmohammadzadeh, R., Rashchi, F., 2019. An environmentally friendly method for recovery of lithium and cobalt from spent lithium-ion batteries using gluconic and lactic acids. J. Environ. Chem. Eng. 7, 102794.

Ross, G.J., Watts, J.F., Hill, M.P., Morrissey, P., 2001. Surface modification of poly (vinylidene fluoride) by alkaline treatment Part 2. Process modification by the use of phase transfer catalysts. Polymer (Guildf). 42, 403–413.

Sahajwalla, V., Hossain, R., 2020a. The science of microrecycling: a review of selective synthesis of materials from electronic waste. Mater. Today Sustain. 9, 100040.

Sahajwalla, V., Hossain, R., 2020b. The science of microrecycling: a review of selective synthesis of materials from electronic waste. Mater. Today Sustain. 9, 100040.

Salbidegoitia, J.A., Fuentes-Ordóñez, E.G., González-Marcos, M.P., González-Velasco, J.R., Bhaskar, T., Kamo, T., 2015. Steam gasification of printed circuit board from e-waste: effect of coexisting nickel to hydrogen production. Fuel Process. Technol. 133, 69–74.

Seh-Bardan, B.J., Othman, R., Wahid, S.A., Husin, A., Sadegh-Zadeh, F., 2012. Bioleaching of heavy metals from mine tailings by Aspergillus fumigatus. Bioremediat. J. 16, 57–65.

Shen, Y., Chen, X., Ge, X., Chen, M., 2018. Thermochemical treatment of non-metallic residues from waste printed circuit board: Pyrolysis vs. combustion. J. Clean. Prod. 176, 1045–1053.

Sheng, P.P., Etsell, T.H., 2007. Recovery of gold from computer circuit board scrap using aqua regia. Waste Manag. Res. 25, 380–383.

Shuey, S.A., Taylor, P., 2005. Review of pyrometallurgical treatment of electronic scrap. Min. Eng. 57, 67–70.

Sloop, S.E., 2014. Reintroduction of lithium into recycled battery materials.

Smith, C., 2015. The Economics of E-Waste and the Cost to the Environment. Nat. Resour. Env't 30, 38.

Sodhi, M.S., Reimer, B., 2001. Models for recycling electronics end-of-life products. OR-Spektrum 23, 97–115.

Sojka, R., 1998. Innovative recycling technologies for rechargeable batteries, in: 4th International Battery Recycling Congress. Anais. Hamburg, Alemanha. pp. 1–3.

Song, Q., Liu, Y., Zhang, L., Xu, Z., 2020. Selective Electrochemical Extraction of Copper from Multi-metal E-waste Leaching Solution and Its Enhanced Recovery Mechanism. J. Hazard. Mater. 124799.

Srichandan, H., Mohapatra, R.K., Parhi, P.K., Mishra, S., 2019. Bioleaching approach for extraction of metal values from secondary solid wastes: A critical review. Hydrometallurgy 189, 105122.

Sum, E.Y.L., 1991. The recovery of metals from electronic scrap. Jom 43, 53–61.

Sun, L., Qiu, K., 2012. Organic oxalate as leachant and precipitant for the recovery of valuable metals from spent lithium-ion batteries. Waste Manag. 32, 1575–1582.

Tan, Q., Li, J., 2015. Recycling metals from wastes: a novel application of mechanochemistry. Environ. Sci. Technol. 49, 5849–5861.

Tansel, B., 2017. From electronic consumer products to e-wastes: Global outlook, waste quantities, recycling challenges. Environ. Int. 98, 35–45.

Tedjar, F., Foudraz, J.C., 2010. Method for the mixed recycling of lithium-based anode batteries and cells.

Terakado, O., Ohhashi, R., Hirasawa, M., 2013. Bromine fixation by metal oxide in pyrolysis of printed circuit board containing brominated flame retardant. J. Anal. Appl. Pyrolysis 103, 216–221.

Terakado, O., Ohhashi, R., Hirasawa, M., 2011. Thermal degradation study of tetrabromobisphenol A under the presence metal oxide: comparison of bromine fixation ability. J. Anal. Appl. Pyrolysis 91, 303–309.

Tesfaye, F., Lindberg, D., Hamuyuni, J., 2017. Valuable metals and energy recovery from electronic waste streams, in: Energy Technology 2017. Springer, pp. 103–116.

Thakur, P., Kumar, S., 2020. Metallurgical processes unveil the unexplored "sleeping mines" e-waste: a review. Environ. Sci. Pollut. Res. 1–12.

Tressaud, A., 2006. Fluorine and the environment: atmospheric chemistry, emissions & lithosphere. Elsevier.

Trumpy, B.H., Millis, N.F., 1963. Nutritional requirements of an Aspergillus niger mutant for citric acid production. J. Gen. Microbiol. 30, 381–393.

Tuncuk, A., Stazi, V., Akcil, A., Yazici, E.Y., Deveci, H., 2012. Aqueous metal recovery techniques from e-scrap: Hydrometallurgy in recycling. Miner. Eng. 25, 28–37.

Veldhuizen, H., Sippel, B., 1994. Mining discarded electronics. Ind. Environ.(Switzerland) 17, 7–11.

Venkatesan, P., Vander Hoogerstraete, T., Hennebel, T., Binnemans, K., Sietsma, J., Yang, Y., 2018. Selective electrochemical extraction of REEs from NdFeB magnet waste at room temperature. Green Chem. 20, 1065–1073.

Verma, H.R., Singh, K.K., Basha, S.M., 2018. Effect of milling parameters on the concentration of copper content of hammer-milled waste PCBs: a case study. J. Sustain. Metall. 4, 187–193.

Wang, H., Zhang, S., Li, B., Pan, D., Wu, Y., Zuo, T., 2017. Recovery of waste printed circuit boards through pyrometallurgical processing: A review. Resour. Conserv. Recycl. 126, 209–218.

Wang, J., Guo, J., Xu, Z., 2016. An environmentally friendly technology of disassembling electronic components from waste printed circuit boards. Waste Manag. 53, 218–224.

Wang, J., Lv, J., Zhang, M., Tang, M., Lu, Q., Qin, Y., Lu, Y., Yu, B., 2020. Recycling lithium cobalt oxide from its spent batteries: An electrochemical approach combining extraction and synthesis. J. Hazard. Mater. 124211.

Wang, M., Tan, Q., Liu, L., Li, J., 2019. Efficient separation of aluminum foil and cathode materials from spent lithium-ion batteries using a low-temperature molten salt. ACS Sustain. Chem. Eng. 7, 8287–8294.

Wang, R., Xu, Z., 2014. Recycling of non-metallic fractions from waste electrical and electronic equipment (WEEE): a review. Waste Manag. 34, 1455–1469.

Wang, W., Chen, W., Liu, H., 2019. Hydrometallurgical preparation of lithium carbonate from lithium-rich electrolyte. Hydrometallurgy 185, 88–92.

Wang, X., Lu, X., Zhang, S., 2013. Study on the waste liquid crystal display treatment: Focus on the resource recovery. J. Hazard. Mater. 244, 342–347.

Widmer, R., Oswald-Krapf, H., Sinha-Khetriwal, D., Schnellmann, M., Böni, H., 2005. Global perspectives on e-waste. Environ. Impact Assess. Rev. 25, 436–458.

Wu, W., Liu, X., Zhang, X., Li, X., Qiu, Y., Zhu, M., Tan, W., 2019. Mechanism underlying the bioleaching process of LiCoO2 by sulfur-oxidizing and iron-oxidizing bacteria. J. Biosci. Bioeng. 128, 344–354.

Wu, Y., Yin, X., Zhang, Q., Wang, W., Mu, X., 2014. The recycling of rare earths from waste tricolor phosphors in fluorescent lamps: A review of processes and technologies. Resour. Conserv. Recycl. 88, 21–31.

Wu, Z., Yuan, W., Li, J., Wang, X., Liu, L., Wang, J., 2017. A critical review on the recycling of copper and precious metals from waste printed circuit boards using hydrometallurgy. Front. Environ. Sci. Eng. 11, 8.

Xakalashe, B.S., Mintek, R., Seongjun, K., Cui, J., 2012. An overview of recycling of electronic waste Part 2. Chem. Technol.

Xia, M.C., Wang, Y.P., Peng, T.J., Shen, L., Yu, R.L., Liu, Y.D., Chen, M., Li, J.K., Wu, X.L., Zeng, W.M., 2017. Recycling of metals from pretreated waste printed circuit boards effectively in stirred tank reactor by a moderately thermophilic culture. J. Biosci. Bioeng.

Xiao, J., Li, J., Xu, Z., 2017. Novel approach for in situ recovery of lithium carbonate from spent lithium ion batteries using vacuum metallurgy. Environ. Sci. Technol. 51, 11960–11966.

Xiao, R., Zhang, Y., Yuan, Z., 2016. Environmental impacts of reclamation and recycling processes of refrigerators using life cycle assessment (LCA) methods. J. Clean. Prod. 131, 52–59.

Xin, B., Zhang, D., Zhang, X., Xia, Y., Wu, F., Chen, S., Li, L., 2009. Bioleaching mechanism of Co and Li from spent lithium-ion battery by the mixed culture of acidophilic sulfur-oxidizing and iron-oxidizing bacteria. Bioresour. Technol. 100, 6163–6169.

Xin, Y., Guo, X., Chen, S., Wang, J., Wu, F., Xin, B., 2016. Bioleaching of valuable metals Li, Co, Ni and Mn from spent electric vehicle Li-ion batteries for the purpose of recovery. J. Clean. Prod. 116, 249–258.

Xue, Y., Wang, Y., 2020. Green electrochemical redox mediation for valuable metal extraction and recycling from industrial waste. Green Chem.

Yang, C., Li, J., Tan, Q., Liu, L., Dong, Q., 2017. Green process of metal recycling: coprocessing waste printed circuit boards and spent tin stripping solution. ACS Sustain. Chem. Eng. 5, 3524–3534.

Yang, H., Liu, J., Yang, J., 2011. Leaching copper from shredded particles of waste printed circuit boards. J. Hazard. Mater. 187, 393–400.

Yang, W.-S., Lee, J.-S., Park, S.-W., Kang, J.-J., Alam, T., Seo, Y.-C., 2016. Gasification applicability study of polyurethane solid refuse fuel fabricated from electric waste by measuring syngas and nitrogenous pollutant gases. J. Mater. Cycles Waste Manag. 18, 509–516.

Yang, Y., Zheng, X., Cao, H., Zhao, C., Lin, X., Ning, P., Zhang, Y., Jin, W., Sun, Z., 2017. A closed-loop process for selective metal recovery from spent lithium iron phosphate batteries through mechanochemical activation. ACS Sustain. Chem. Eng. 5, 9972–9980.

Yazici, E.Y., Deveci, H., 2013. Extraction of metals from waste printed circuit boards (WPCBs) in H2SO$_4$–CuSO$_4$–NaCl solutions. Hydrometallurgy 139, 30–38.

Yousef, S., Tatariants, M., Bendikiene, R., Denafas, G., 2017. Mechanical and thermal characterizations of non-metallic components recycled from waste printed circuit boards. J. Clean. Prod. 167, 271–280.

Yu, J., He, Y., Ge, Z., Li, H., Xie, W., Wang, S., 2018. A promising physical method for recovery of LiCoO2 and graphite from spent lithium-ion batteries: Grinding flotation. Sep. Purif. Technol. 190, 45–52.

Zeng, G., Luo, S., Deng, X., Li, L., Au, C., 2013. Influence of silver ions on bioleaching of cobalt from spent lithium batteries. Miner. Eng. 49, 40–44.

Zeng, X., Song, Q., Li, J., Yuan, W., Duan, H., Liu, L., 2015. Solving e-waste problem using an integrated mobile recycling plant. J. Clean. Prod. 90, 55–59.

Zhang, L., Xu, Z., 2016a. A Review of Current Progress of Recycling Technologies for Metals from Waste Electrical and Electronic Equipment. J. Clean. Prod. 127, 19–36.

Zhang, L., Xu, Z., 2016b. A review of current progress of recycling technologies for metals from waste electrical and electronic equipment. J. Clean. Prod. 127, 19–36.

Zhang, S., Forssberg, E., 1998. Optimization of electrodynamic separation for metals recovery from electronic scrap. Resour. Conserv. Recycl. 22, 143–162.

Zhang, S., Yoshikawa, K., Nakagome, H., Kamo, T., 2013. Kinetics of the steam gasification of a phenolic circuit board in the presence of carbonates. Appl. Energy 101, 815–821.

Zhang, X., Zhang, C., Zheng, F., Ma, E., Wang, R., Bai, J., Yuan, W., Wang, J., 2020. Alkaline electrochemical leaching of Sn and Pb from the surface of waste printed circuit board and the stripping of gold by methanesulfonic acid. Environ. Prog. Sustain. Energy 39, e13324.

Zheng, Q., Shibazaki, K., Ogawa, T., Kishita, A., Hiraga, Y., Nakayasu, Y., Watanabe, M., 2020a. Continuous hydrothermal leaching of LiCoO2 cathode materials by using citric acid. React. Chem. Eng.

Zheng, Q., Watanabe, M., Iwatate, Y., Azuma, D., Shibazaki, K., Hiraga, Y., Kishita, A., Nakayasu, Y., 2020b. Hydrothermal leaching of ternary and binary lithium-ion battery cathode materials with citric acid and the kinetic study. J. Supercrit. Fluids 104990.

8 Biotechnological Management, Extraction and Recycling of Metals from E-Waste
The Present Scenario

Satarupa Dey and Biswaranjan Acharya*

CONTENTS

8.1 INTRODUCTION

Waste electric and electronic equipment and other devices such as household appliances, telecommunication equipment, medical device, lighting devices, automatic dispensers, electronic and electrical tools, leisure and sport equipment on the whole can be referred to as e-waste (Huang et al. 2014; Chen et al. 2015; Ilankoon et al. 2018). Apart from these, printed circuit boards (PCBs), cathode ray tubes, lead capacitors, electric cables activated glass and batteries are also included in the category of e-waste (Lambert et al. 2015). These e-wastes have become a global environmental

DOI: 10.1201/9781003095972-11

and health concern as it is being accumulated in exponential rate and currently over the 43.8 million tons (Mt) have been generated globally in 2015 (Breivik et al. 2016) which has grown to 49.8 Mt in the year 2018.

E-waste contains several valuable metals, toxic heavy metals and persistent organic pollutants (POPs) such as polycyclic aromatic hydrocarbons, polybrominated diphenyl ethers, polychlorinated dibenzo-p-dioxins and polychlorinated biphenyls dibenzofurans (Wu et al. 2016) which cause severe environmental impact due to leaching of toxic particles (Chen et al. 2016; Huang et al. 2016). This e-waste can also act as an alternative source of precious metals and the processes of metal extraction from them can result in positive environmental impact (Cucchiella et al. 2016).

Several industrialized and developing countries like China, Nigeria, Ghana, India and Pakistan have reported a record increase in their e-waste as no strict government regulation are implemented there. These countries are open to import of electronic waste from developed countries out of which 80 % is illegally done (Sthiannopkao and Wong 2013; UNEP 2005). Nearly 50-80% of the imported e-waste are either recycled or reused (Widmer et al. 2005) but they are mostly done informally in unorganized sector. Several countries have implemented "Extended producer responsibility (EPR) regulations" in the last decade as a measure to reduce electronic waste problem. According to EPR the producers of the electronic products are required to collect the sold products back once they have served their purpose. However, it has got several limitations such as technical issues, political, legislative issues and also lack of awareness in society and on the part of customers which acts as a hurdle for the process of recycling of e scraps. In most of the developing countries electronic wastes are recycled in an unauthorized workshop where it is sorted, fired, incinerated and washed with acid which generates highly toxic and acidic leachates which are harmful to human health. The details of the routes of e-waste and its treatment are represented in details in Figure 8.1. At present, both chemical and biological leaching have been practiced for treatment of e-waste and subsequent metal extraction. However, both the processes have their own limitations. Biological leaching can be considered as environment friendly and more economical in nature but it is time consuming. On the other hand, chemical leaching is much more rapid and efficient. Other techniques like nano remediation can also be an alternative for extraction of metals from e-waste. This chapter mainly deals with the issues related to all these metal extraction techniques and highlights their feasibility in treatment of e-waste for metal.

8.2 MANAGEMENT PROTOCOLS OF E-WASTE IN DEVELOPING COUNTRIES

Dumping of e-waste is emerging as a critical problem in developed, developing and underdeveloped countries. However, most developing and underdeveloped countries lack any protocol related to the proper handling and management of this e-waste, which has become a major nuisance due to its toxicity and other negative impact

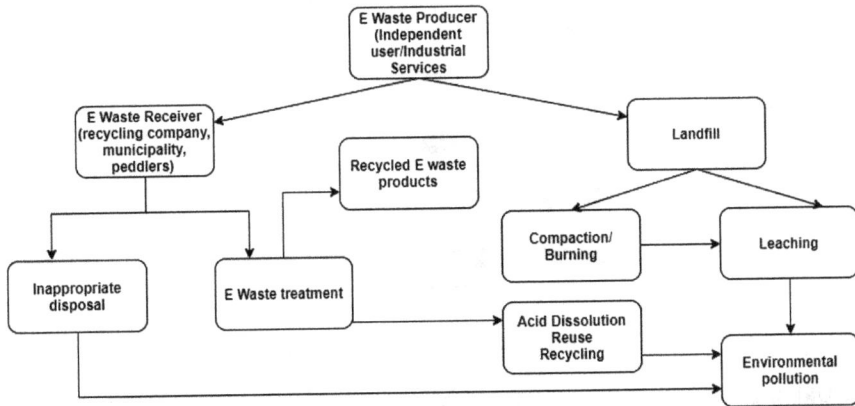

FIGURE 8.1   Routes of treatment of E-waste.

on environment. (Hammad et al. 2017). It is a reserve of several important precious metals and plastics, thus it is necessary to reutilize them and turn them to an asset. The European Union has set a roundabout budget where e-waste is considered as an asset and has stressed the need of reprocessing them. However, most of the developed nations do not reuse the e-waste due to stringent environmental protocols and most of this e-waste is exported to developing nations (Robinson 2009) which is again controlled and prohibited by the Basel Convention.

Many developing countries of Asia and Africa such as Cambodia, China, Malaysia, Nigeria, Pakistan and Vietnam have forbidden the import of e-waste, but countries like Ghana, Philippines and Thailand have allowed the entry of e-waste in exchange for exceptional endorsements (Jinhui et al. 2013). Many developing countries such as Bangladesh, China, Hongkong, India, Indonesia, Malaysia, Nigeria, Sri Lanka and Vietnam are important regions for dumping of e-waste, where the e-waste is disposed in open ground (Heart and Agamuthu 2012). In the absence of any formal waste management framework and proper organization most of the e-waste is processed through informal and unorganized processing sectors following primitive techniques leading to production of toxic fumes, highly acidic toxic leachates which contaminate both air and water. Most of the workers in this sector are ill payed and work without proper safety measures (Nasim et al. 2016, 2018) which jeopardizes their health. Such e-waste landfills are characterized by intense ecological contamination, toxic and lethal leachates produced by these e-waste flows into water source and contaminate streams, lakes, groundwater and soils.

In India the scenario of e-waste management is a serious issue due to informal recycling activities and there exists no policy to check the flow of e-waste. The business sector is the main e-waste generator which accounts for nearly 78 % of wastes in which the top ten states of India accounts for 70% are states like Maharashtra, Tamil Nadu, Andhra Pradesh, Uttar Pradesh, West Bengal, Delhi, Karnataka, Gujarat, Madhya Pradesh and Punjab. Although, there exist several small scale e-waste dismantling

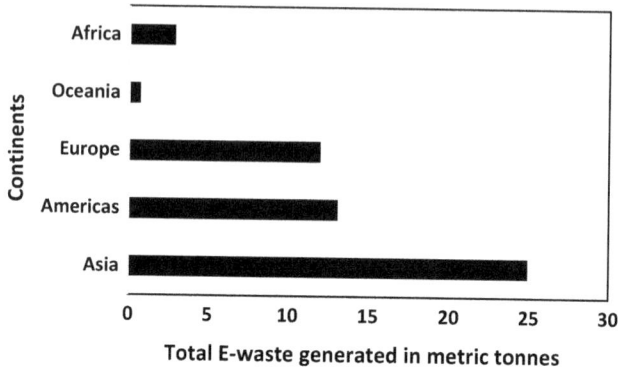

FIGURE 8.2 Total E-waste generated from the different continents. (Compiled following Forti et al. 2020.)

facilities in cities like Chennai and Bangalore, mostly e-waste recycling is done in an unorganized sector.

The amount of e-waste generated in different continents all around the world are presented in the Figure 8.2 (Forti et al. 2020).

8.3   HEALTH CONCERNS RELATED TO E-WASTE

E-wastes are known to have several health hazards on human beings which may be largely attributed to the presence of several toxic metals in them. Metals like cadmium, arsenic, chromium and nickel present in LED, rechargeable batteries, contact switches, old CRTs can develop several symptoms. Human exposure to such mixture results in various types of disorders such as calcium homeostasis, epigenetic modifications, endocrine disruption and may impair neurotransmission (Chen et al. 2011).

Cadmium compounds which occur extensively in infra-red detectors, semiconductor chips and surface mount devices chips are highly toxic and can bioaccumulate posing irreversible damage on human health. Cadmium is a redox – inactive metal and forms compound with sulphydryl group of proteins, is found to damage kidney, bones and adversely affect the neurodevelopment at the fetus stage (Pan et al. 2010; Patra et al. 2011). Arsenic, which is present in e-waste as gallium-arsenide mainly in Light Emitting Diode (LED) is known to cause irregular heart beat and bladder cancer. It produces reactive oxygen species, causes oxidative damage and produces bioactive molecules and may lead to renal diseases, effect reproductive health and causes liver and gastrointestinal disturbances (Jomova et al. 2011). Similarly Nickel present in Ni-Cd batteries can give rise to cancer and skin allergies (Padiyar 2011). Several compounds of Cr(VI) is extensively used in galvanized steel plates and production of metal housing which is highly toxic and carcinogenic in nature (Tsydenova and Bengtsson 2011). It also causes lung disorder, defects in neurodevelopment when exposed in high dose (Chen et al.2011). Flat screen displays contain Hg adversely effects the growth of both brain and central nervous system

(Langford and Ferner 1999). On the other hand, exposure to lead present largely in cathode ray tubes (CRT) and monitors can lead to impaired intellectual development due to damage in the nervous system in both children and adults (Poon 2008). Being a non-redox metal it also leads to headache, nausea and gastric ulcers (Monika 2010; Patra et al. 2011). Elements like lithium, selenium and americium found in medical equipment, photo drums of Xerox machines, smoke detector and lithium batteries can lead to cytogenic and chromosomal damage (Kelly and Dagle 1974) which leads to fatigue and gastrointestinal disorder (Patrick et al. 1999). Heavy metals present in e-waste results in several reproductive problems as it results in decreased sperm count and disrupts endocrine system of the human body (Balabanic et al. 2011). The leachates containing Pb, As, Se etc which leaches from e-waste after acid treatment may cause severe health hazards to the people who regularly work in these sites (Liu et al. 2008).

8.4   ENVIRONMENTAL HAZARDS ASSOCIATED WITH E-WASTE

E-waste is known to contaminate soil, water and air and has a profound impact on the flora as well as fauna. Contamination of heavy metal and percolation of acid may result in infertility of the soil due to metal accumulation and reduction in organic carbon content of soil. Water sources including river pond and ground water have also become highly contaminated with metals and e-waste incineration and cycling in an unorganized manner have led to the release of leachate to the water bodies turning it to be acidic in nature. The dumping of e-waste near water resources also increases the risk of heavy metal contamination from the leachates. This change in pH and increase in heavy metal content has resulted in adverse impact on the water flora and fauna. Moreover, if such contaminated water is used in irrigation and cultivation of vegetables it may result in bioaccumulation of toxic metals. The soot generated from the informal processing of e-waste are known to emit Cd, Cr, Ni, Pb and As which may lead to contamination of air which on inhalation may lead to serious health hazards (Grant et al. 2013; Wu et al. 2015).

8.5   PROCESSES OF RECOVERY OF METAL VALUES

It has been estimated that a total 41.8 million tons of e-waste have been generated by the year 2014 which was forecasted to be around 50 million tons in 2018 by Balde et al. 2015, of which most of the generated waste is exported to different countries of Asia and Africa (Takashima 1999). Formidable challenges are confronted by the developing countries in handling this waste as most of them do not have any proper management techniques and most of the waste is recycled manually. These handling and manual treatments pose considerable environmental threat and health hazards (Widmer et al. 2005; Babu et al. 2007). Thus, it has become increasingly important to reduce, reuse and recycle them in an organized way. According to Global E-Waste Monitor report in 2017, approximately 2 million tons (MT) of e-waste is generated annually in India among which 0.036 MT is processed and the rest 95% are recycled in unorganized sector in a crude manner.

The main constituents of e-waste include a wide array of elements which consist of around 50 % iron and steel elements followed by around 21% plastic, 13% non ferrous metals and rest include rubber, concrete and ceramics. Apart from this basic constituent e-waste also contains diverse hazardous elements like mercury, lead, cadmium, selenium arsenic and hexavalent chromium. They are also found to be a good source of copper and precious metals including gold, platinum and palladium which are found almost in the same amount as in natural ores (Hadi et al. 2015). Mostly these toxic metals and harmful organic compounds can create serious health hazards and toxic environments if not processed and handled properly (Chatterjee 2008). Thus, it has become essential to ameliorate the disposal techniques of e-waste in the most cost-effective and in an organized way. Both "Waste Electrical and Electronic Equipment (WEEE) directive" and "Restriction on the use of Hazardous Substances (RoHS)-directive" have instructed the manufacturers to not only take responsibility for management of e-waste, but also to restrict the use of heavy metals, polybrominated biphenyls, and polybrominated diphenyl ethers (PBDEs) in newly manufactured electrical and electronic gadgets. Apart from these several legislations and policies have been developed throughout the world on the management and processing of e-wastes and "Hazardous Materials Laws and Rules" was drafted in India with similar objectives (LaDou and Lovegrove 2008).

8.5.1   Metal Extraction from E-waste

Several technologies have been successfully deployed for recovering the metals, which are broadly classified into two major types of treatment, which include chemical and biological treatment. Chemical treatment may further belong to a Hydrometallurgical method where aqueous solution is used to extract metal values; the Pyrometallurgical method in which extreme heat is used to recover metals and the biological treatment includes Biohydrometallurgical processes where different types of microbes were used for bioleaching (Chauhan et al. 2018). In most cases the e-waste is first treated by the mechanical methods such as crushing, jigging and incineration to expose the metals present in the e-waste. The different types of treatment of e-waste are represented in Figure 8.3.

8.5.1.1   Chemical Leaching of Metal Values from E-waste

The entire process of recycling of e-waste is divided into three broad steps which are the following: i) collection, ii) preprocessing and iii) end processing. The preprocessing can be done following different processes including hydrometallurgy, pyrometallurgy, electrometallurgy and biometallurgy and through a combination of all these processes (Khaliq et al. 2014; Meskers et al. 2015). However, as most of the e-waste consists of a diverse range of materials and thus separation process of these metals is highly complex.

Pyrometallurgical methods have been employed widely for metal extraction from e-waste, despite having several disadvantages such as high energy requirement and being costly. It also generates a huge amount of heavy metal containing slag and it is extremely difficult to recover them thus the slag cannot be reutilized. The main

E- Waste

| Mechanical Methods
Incineration
Crushing
Jigging
Electrostatic Separation | Chemical methods
Extractive metallurgy
Pyro- metallurgical
processing
Hydrometallurgical
processing
Bio-metallurgical
processing | Biological methods
Bioleaching
Biosorption
Biomining
Biomineralisation
Bioprecipitation |

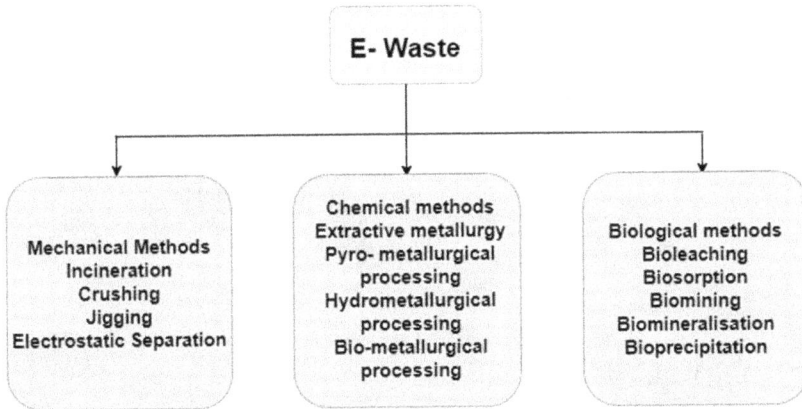

FIGURE 8.3   The different types of treatment in E-wastes.

processes include incineration and roasting in presence of selective gas at high temperature (Kaya 2016). However, the entire process results in release of halogenated and chlorinated compound which is considered as a major drawback of the process (Tsydenova and Bengtsson 2011). Moreover, other drawbacks of the process include high dioxin formation and release of toxic metals and a high amount of slag formation. An insight into the pyrometallurgical process is given in Figure 8.4.

On the other hand, in hydrometallurgical process extraction of valuable metals are done in acidic or alkaline solutions through leaching. In the subsequent step the extracted metals dissolved in the solution are recovered using different techniques such as precipitation, absorption, solvent extraction, ion exchange and electrowinning. As compared to pyrometallurgy it is considered to be more eco-friendly and can be done under controlled temperature (Birloaga et al. 2014). In case of e-waste, acids or caustic leaching is extensively practiced for selective dissolution of target metals, the dissolved metals are then further purified, refined and recovered through electro refining (Khaliq et al. 2014).

Different leaching methods that are used in hydrometallurgy include the use of acids (HCl, HNO_3, H_2SO_4), use of ligands, by using cyanide, thiosulfate and alkali (Guo et al. 2015; Bas et al. 2015; Jha et al. 2012; Rocchetti et al. 2013). Cyanide based treatment has been extensively used for recovery of gold, other techniques include alkaline solution or alkaline based metal cyanide solutions for extraction of metals (Bisceglie et al. 2017). However, in these processes the presence of air, pH, temperature, presence of anions or cations in the solid and liquid phase, stirring rate and surface area of contact is also an important factor for successful extraction. For gold extraction, compounds like KCN and 3 nitrobenzene sulfonic acid are extensively used (Bisceglie et al. 2017), on the other hand for copper, acid sulfate followed by cyanide treatment is practiced (Kamberovic et al. 2011). Oh et al. 2003 stated the use of different acids such as sulfuric, nitric and hydrochloric acid for metal recovery. Metals such as cobalt and lithium are removed by application of sodium hydroxide from lithium-ion batteries (Jadhav and Hocheng 2015). Apart from inorganic acid,

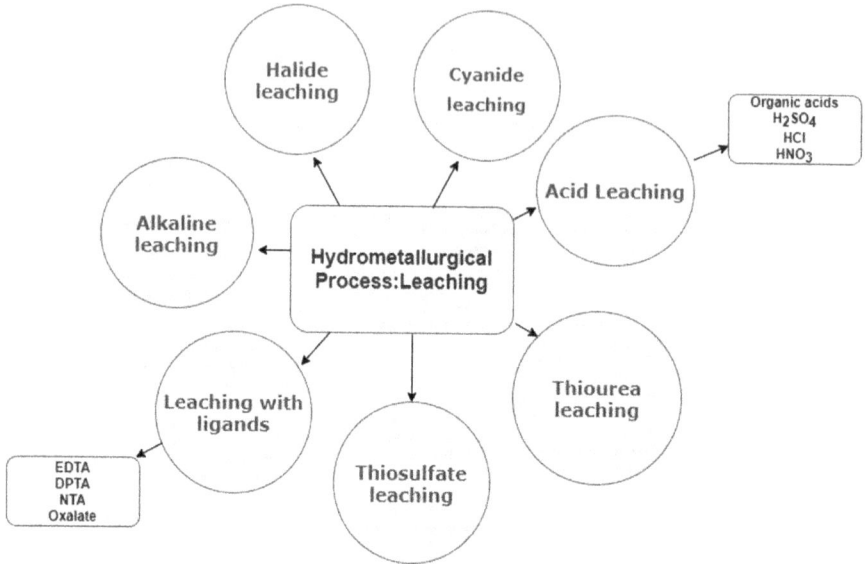

FIGURE 8.4   Different strategies undertaken in pyrometallurgical treatment of E-wastes.

organic acids including ascorbic acid, citric acid and acetic acid are also employed for removal of light metals. A combination of citric acid and hydrogen peroxides are also used for complete recovery of Li from lithium-ion battery (Li et al. 2010).

Thiosulfate leaching is extensively used for recovery of Au and Ag present in mobile phones. The major advantage of this method of leaching includes less interference with other cationic species and easy conversion to recoverable complexes. Ammonium and sodium thiosulfate are reported to recover precious metals such as gold, silver and platinum and forms stable anionic complexes without forming any oxides (Ubaldini et al. 2000), sometimes coleaching agents such as ammonia and thiourea are also used (Ha et al. 2010). The entire process is relatively nontoxic and non-corrosive compared to cyanide leaching as reaction between gold and ammoniacal thiosulfate results in formation of highly soluble and stable compound. Ethylenediaminetetraacetic acid (EDTA) and monoethanolamine acid (MEA) (Puente-Siller et al. 2013; 2017) are sometimes also used as additives. Apart from thiosulphate leaching gold can also be obtained by halide leaching using chloride, bromide and iodide which is characterized by improved redox potential and high solubility (Liddicoat and Dreisinger 2007). Similarly, thiourea leaching is a sulfur based organic agent which can be used to leach gold and silver from e-waste (Jing-ying et al. 2012; Veit et al. 2015). It gives quick rates of leaching compared to other leaching methods which was characterized by less interference of ions and is cost effective as well as environmentally friendly. The different types of hydrometallurgical are represented in Figure 8.5.

8.5.1.2   Biological Leaching of Metal Values from E-waste

The informal recycling of e-waste results in insufficient and incomplete recovery of heavy metals (Song and Li 2014) and generates a huge amount of persistent

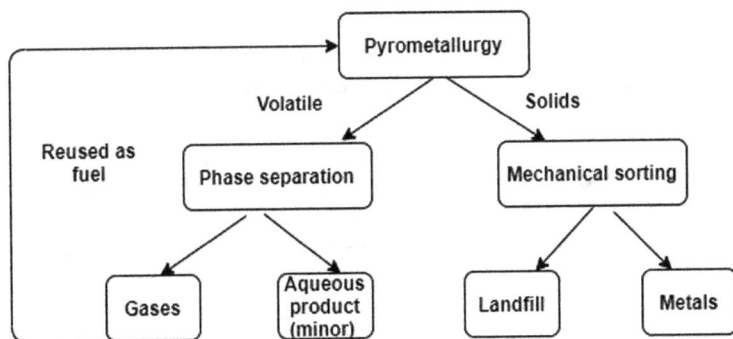

FIGURE 8.5 Different strategies of hydrometallurgical techniques used in E-waste treatment.

organic pollutants such as a various thermosetting plastics and heavy metals which is discharged in enormous quantities in soil (Olafisoye et al. 2013). The presence of POPs and heavy metal not only causes harm to ecosystem but also affects the soil microbial consortium. Liu et al. (2015) reported presence of microbes which were able to metabolize and remediate POPs. Different bioremediation strategies can be applied for management of e-waste which may be directed for treatment of both organic and inorganic parts. The thermosetting plastics present in e-waste are halogenated in nature and microbes that help in the process of dehalogenation can be used in their bioremediation. Compounds like PCB, PBDE are known to be highly persistent, toxic and tend to bioaccumulate and several complex environmental processes are required to degrade them (Luo et al. 2009). On the other hand, there also exist microbes which are capable of leaching of both metallic and non-metallic components (Pant et al. 2018) of e-waste. 16 S rRNA sequencing of the soil samples from e-waste sites showed the presence of bacteria *Solibacter, Nitrospira, Acidobacter, Chloroflexi* which belongs mostly to Bacteroidetes, Deltaproteobacteria and Firmicutes (Liu et al. 2018; Wu et al. 2017). Several PCB degradative bacteria have also been reported by different authors which efficiently degraded PCB (Luo et al. 2009).

Regarding extraction of metals from e-waste chemical methods are considered to be the most reliable method within short duration but it has several drawbacks such as requirement of huge amounts of chemical reagents, chances of secondary contamination and risk of handing. Apart from these there is always an associated problem such as difficulty in detaching the impurities from the metals. Microbial removal of metals from e-waste on the other hand is more specific and can recover metal even from low grade e-waste following relatively simple operational methodology (Priya and Hait 2017). However, the prerequisite of removal of metal by microbes includes removal of epoxy layer coatings to render the metals accessible to the bacteria. Microorganisms including bacteria, yeast and algae can undergo several processes such as bioleaching, biosorption, biomineralization, bioaccumulation, biooxidation and microbially mediated chemisorption of metals which can be also used in extraction of metals from E waste (Dixit et al. 2015).

5.1.3 Bioleaching

Bioleaching is the capability of microbes to convert solid metallic compounds to soluble and extractable form and their natural ability to oxidize and degrade inorganic and organic substances by acidolysis, complexolysis are the common mechanisms that are used and exploited for bioleaching (Bosshard et al. 1996). Several microbial groups such as autotrophic bacteria, heterotrophic bacteria and heterotrophic fungi have been reported to be used for bioleaching mechanisms (Karwowska et al. 2014). Chemolithotrophs such as sulfur and iron oxidizing bacteria are also used for extraction procedures (Clark and Norris 1996) and in the majority of these cases studied so far ambient temperature (40∘C or less), consortium of both sulfur and iron – oxidizing microbes are the primary requirement for successful functioning of the procedure (Rawlings et al. 2003). Acidophilic microorganisms such as *Acidithiobacillus ferrooxidans*, *Acidithiobacillus thiooxidans*, *Leptospirillum ferrooxidans*, and *Sulfolobus* sp. are very crucial for leaching of heavy metals from e-waste. All these bacteria and bacterial consortium can withstand acidic pH and help in the leaching process (Hong and Valix 2014). On the other hand, heterotrophic fungal genera such as *Penicillium* and *Aspergillus niger* are reported to help in metal bioleaching from e-waste (Mishra and Rhee 2010). The main mechanism comprises of oxidation and conversation of insoluble sulfides of copper, zinc and nickel to soluble sulfate form. For successful extraction of Au and Ag bacteria like *Chromobacterium violaceum*, *Acidithiobacillus* sp. *Ferromicrobium* sp., *Leptospirillum* sp. and *Acidiphilium* sp. and *Pseudomonas fluorescens* (Alan et al. 2005; Olson 2006; Pham and Ting 2009; Tay et al. 2013). Creamer et al. 2006 reported the use of *Desulfovibrio desulfuricans* for recovery of Au, whereas use of *Hymeniacidon heliophila* sponge was reported by Rozas et al. 2017 for extraction of copper. *Acidithiobacillus ferrooxidans* can also assist in extraction of both Au and Cu after proper pre-treatment of e-waste. A consortium of *Gallionella* sp., *Acidithiobacillus* sp. and *Leptospirillum* sp. was found to be effective for extraction of Cu from PCB. Apart from Copper and Gold, Zinc can also be extracted from their sulfide ore by bioleaching in both direct and indirect way. In direct leaching direct contact of bacteria is essential whereas in indirect leaching ferric ion is used as a mediator as stated in studies of Suzuki 2001 and Cui and Zhang 2008. *Thiobacillus ferrooxidans* is the most studied microorganism which is also used at high temperature and low pH, whereas *Sulfobacillus thermosulfidooxidans* is a thermophilic, acidophilic-chemolithotrophic bacteria is used for bioleaching of metals (Ilyas et al. 2007). Metal tolerant fungi such as *A. niger* and *P. simplicissimum* can be used to leach Au, Al, Cu, Pb, Zn as well as Ni (Madrigal-Arias et al. 2015; Brandl et al. 2001; Mulligan and Kamali 2003). Fungal species mainly interact with gold by production of cyanide and aurocyanide complexes which assists in biochemical leaching. However, bioleaching despite being a cost-effective method has several disadvantages such as complete recovery of metal may not be feasible and slowness of the process (Pant et al. 2012). The list of microbes playing a vital role in bioleaching of E-waste are tabulated in Table 8.1.

TABLE 8.1
List of Microbes Used for Extraction of Metals by Bioleaching

Metals Recovered	Microbes Used	References
Au	*Acidiphilium* *Acidithiobacillus* sp. *Chromobacterium violaceum* *Desulfovibrio desulfuricans* *Ferromicrobium* *Pseudomonas fluorescens* *Leptosprillum thiooxidans*	Alan et al. (2005);Creamer et al. (2006); Olson (2006);Pham and Ting (2009); Solisio et al. (2002)
Ag	*Acidiphilium* *Acidithiobacillus* sp. *Ferromicrobium* *Leptosprillum thiooxidans*	Alan et al. (2005); Olson (2006)
Cu	*Acidithiobacillus ferrooxidans* *Acidithiobacillus thiooxidans* *Gallionella* sp. *Leptospirillum* sp. *Sulfobacillus thermosulfidooxidans* *Sulfobacillus thermotolerans* *Thermoplasma acidophilum* *Aspergillus niger* *Penicillium simplicissimum*	Brandl et al. (2001); Chang-bin et al. (2007);Isildar et al. (2015); Johnson et al. (2008); Makinen et al. (2015); Mulligan and Kamali (2003);Suzuki (2001); Tetsuo and Atsushi (2001); Wang et al. (2009); Xiang et al. (2010)
Pb	*Acidithiobacillus thiooxidans* *Thiobacillus ferrooxidans* *Aspergillus niger* *Penicillium simplicissimum*	Johnson et al. (2008); Patel and Kasture (2014)
Zn	*Acidithiobacillus thiooxidans* *Thiobacillus ferrooxidans* *Sulfobacillus thermosulfidooxidans* *Thermoplasma acidophilum* *Aspergillus niger* *Penicillium simplicissimum*	Brandl et al. (2001); Ilyas et al. (2014); Konishi et al. (1992); Solisio et al. (2002); Wang et al. (2009)
As	*Acidithiobacillus thiooxidans* *Thiobacillus ferrooxidans* *Aspergillus fumigatus*	Patel and Kasture (2014)
Al	*Acidithiobacillus thiooxidans* *Acidithiobacillus ferrooxidans* *Bacillus circulans* *Bacillus mucilaginosus* *Sulfobacillus thermosulfidooxidans* *Thermoplasma acidophilum* *Aspergillus niger* *Penicillium simplicissimu*	Brandl et al. (2001); Groudev (1987); Ilyas et al. (2014)
Ni	*Acidithiobacillus thiooxidans* *Acidithiobacillus ferrooxidans* *Aspergillus niger* *Penicillium simplicissimum*	Brandl et al. (2001)

(continued)

TABLE 8.1 (Continued)
List of Microbes Used for Extraction of Metals by Bioleaching

Metals Recovered	Microbes Used	References
Cd	*Micrococcus roseus* *Sulfobacillus thermosulfidooxidans* *Thermoplasma acidophilum* *Aspergillus niger*	Ilyas et al. (2014); Patel and Kasture (2014); Ren et al. (2009),

8.5.1.4 Biosorption

Biosorption process is a physicochemical and metabolically independent process which can function as a cost effective alternative for recovery of precious metals from metal laden solutions (Fomina and Gadd 2014). Several biomaterials such as algae, fungi, actinomycetes, bacteria, yeast and biopolymer have been reported to bind metals to the different cell wall receptors (Das and Das 2013). This process has both advantages and disadvantages, the main advantages of the process are its cost effective nature along with reduced treatment time. Along with there being no requirement of nutrients, no toxicity and easy mathematical modelling. However, the main drawback includes early saturation of the biomaterial, a problem with reusability (Hansda and Kumar 2016).

Plant barks, sawdust, leaves, peat moss, rice husk and sugarcane bagasse have been reported to be a very efficient biosorbant which can be used in removal of metals (Michalak et al. 2013). Chitin was also used as a biomaterial for biorecovery of Au (Cortes et al. 2015). *Cladophora hutchinsiae*, a green algae and bacterium *Halomonas* BVR 1 have been reported to be used as removal of highly toxic chemical such as uranium and cadmium respectively (Bagda et al. 2017; Rajesh et al. 2014). The list of microbes used for biosorption is enlisted in Table 8.2.

Removal of Organic Pollutants by Bacteria

The organic component of the e-waste consists of different types of polymers such as Acrylonitrile Butadiene Styrene (ABS), Polyethylene (PE), Polypropylene (PP), Polytetrafluoroethylene (PTFE), Polycarbonate (PC) and High Impact Polystyrene (HIPS) (Stevens and Goosey 2008). PCB is the major pollutants that is available in e-waste are of great concern as they are persistant, toxic and tend to bioaccumulate (Zhang et al. 2010; Zheng et al. 2015). Microbial degradation of Printed Circuit Board (PCB) is considered as an eco-friendly and cost-effective option that can be utilized for e-waste management (Pieper et al. 2005). However, the process of degradation of PCB is extremely complex due to its structural nature differing mainly in the number and distribution of chlorines attached with the biphenyl ring. Potrawfke et al. (1998) first reported *Burkholderia xenovorans* LB 400 which was able to utilize PCB having two chlorine residues. Various bacteria such as *Paenibacillus*, *Pseudomonas* and *Stenotrophomonas* have been reported to utilize PCB as a sole carbon source

TABLE 8.2
List of Microorganisms Used for Biosorption of Metals from Electronic Wastes

Metals Recovered	Microbes Used	References
	Bacteria	
Gold	*Lactobacillus acidophilus*	Sheel and Pant 2018
Chromium	*Bacillus sphaericus*	Patel and Kasture 2014
Uranium	*Pseudomonas aeruginosa,*	Patel and Kasture 2014
	Myxococcus xanthus,	
Cadmium	*Acidithiobacillus ferrooxidans,*	He et al. 2012;
	Bacillus subtilis	Patel and Kasture 2014;
	Halomonas BVR 1	Rajesh et al. 2014
	Magnetospirillum gryphiswaldense	
	Pseudomonas putida X4	
	Pseudomonas aeruginosa	
Lead	*Streptoverticillium cinnamoneum*	Patel and Kasture 2014
	Fungi	
Cadmium	*Saccharomyces cerevisiae*	
Uranium	*Rhizopus arrhizus*	
	Algae	
Chromium	*Scenedesmus obliquus*	Bagda et al. 2017;
Copper	*Chlorella vulgaris*	Chojnacka et al. 2005;
	Synechocystis sp.	Gajendiran and Abraham
Uranium	*Cladophora hutchinsiae*	2015
Zinc	*Ecklonia maxima`*	

(Sakai et al. 2005; Furukawa et al. 2008). Apart from the conventional technology, addition of these degraders appears to be quite promising in removal of PCB from e-waste contaminated sites. However, performance of all these biological degraders is highly dependent on the environmental condition and the ability of these microbes to tolerate heavy metals which is the main limitation of the process of bioremediation (Hosokawa et al. 2009; Barbato et al. 2016). Bioaugmentation with indigenous microbes can be considered as a potential and viable alternative in the mitigation of soil contaminated with PCB. Although a major drawback of this protocol is lack of indigenous cultivable PCB degraders. Several studies identified PCB metabolizing bacteria via culture independent method of which Song et al. 2015 and Tang et al. 2013 reported the presence of different iron-dissimilatory iron reducing bacteria which was associated with PCB dehalogenation. The ability of degradation of PCB of a microbe largely depends on the number of chlorines present, strains such as *Alcaligenes eutrophus* H850, *Rhodococcus* sp RHAI and *Burkholderia xenovorans* LB400 were able to transform hexachlorobiphenyls (Seto et al. 1995). The degradation rate of PCB is however reported to decrease with the increase in number of chlorine substitution increases and five chlorine substitutions are almost recalcitrant to aerobic cometabolism (Adebusoye 2007).

8.5.2   PROSPECTS OF NANOREMEDIATION IN E-WASTE MANAGEMENT

Nanoremediation can also be considered as an ideal candidate for remediation of heavy metals from different types of industrial waste. The particles mainly have a size ranging from 10-100nm, and are characterized with high surface area, reactivity, adsorptivity, photocatalytic properties which assist in analytical detection and subsequent remediation. Zero valent iron particles of nanoscale range are extensively used in nanoremediation and can act as an ideal candidate to remediate heavy metals (Liu et al. 2011; Tratnyek and Johnson 2006; Karn et al. 2009). These zero valent iron particles have a metallic iron core which has electron donating power, whereas the surface iron hydroxide attract and absorb charged heavy metals and can function in removal and immobilization of oxyanions such as As(V), Cr(VI), Zn(II), Pb (II), Pd(II) and Ni(II) (Ponder et al. 2000). Aluminium nanoparticles can result in 97% removal of Ni(II) from metal laden solution and can be used as an alternative strategy for the treatment of e-waste (Sharma et al. 2008). Similarly, zeolite aided iron nanoparticles can also be used for adsorption of heavy metals from aqueous samples (Li et al. 2018). Nagarajah et al. (2017) reported, nanomagnetite coated by silica and MgO (MTM) was capable of removing Pb, Cd, Cu mainly by substitution, followed by precipitation mechanism. Despite several advantages the basic technology is still not much evolved for large scale application and also the main drawback is that they tend to self-aggregate and associate with suspended particles and bioaccumulate in the food chain (Karn et al. 2009; Kotnala 2009).

8.5.3   PROSPECTS OF HYBRID TECHNOLOGY FOR METAL EXTRACTION FROM E-WASTE

Although bioleaching has been considered as an economic and environmentally safe approach for the treatment of e-waste, however, in this process prolonged and full recovery of metals is not possible (Ren et al. 2009). Contrarily, chemical leaching is less time consuming and productive but produces harmful slag. Employment of a combination of both the techniques can be considered as an improved and efficient method for recovery of metals. Metals like Zn, Cd, Cu and Pb can be extracted using EDTA with *A. ferrooxidans*, *Shewanella putrefaciens* or bacterial strain DSM 9103 as stated in reports of Cheikh et al. (2010) and Satroutdinov et al. (2000). Both ammonium citrate and oxalate along with a combination of *Aspergillus niger* and *Penicillium* sp. was used for extraction of heavy metals including Cd, Pb and Cu, similarly, Au and Cu can be efficiently removed by *A. ferrooxidans*, *Sulfobolus* spp (Wasay et al. 1998; Kohr 1995). A fungal genus *Phanerochaete chrysosporium* was pretreated with sulfide and was reported to successfully extract Au (Sarpong et al. 2011). Biodegradable ligands such as Ethylene diamine di-succinic acid, Diethylene triamine penta-acetate, and Nitrilotriacetic acid have been already proved to be quite efficient in removal of Cd, Pb, Cu, Zn and Fe fromores (Hong et al. 2000).

8.6   CONCLUSION

E-waste is a source of precious metal which can act as an alternative source of mineral and both chemical as well as biological leaching can be used effectively for extraction of metals from them. However, both these methods have both merit and demerits, biological method is more eco-friendly option for e-waste treatment but the process is time consuming. On the other hand, a chemical process which requires high temperature and pressure as compared to a biological process is much faster and economically feasible. However, a hybrid technology involving both chemical and biological is considered as a sustainable and more feasible option, also the efficiency can be increased in combination with the use of nano particles. Thus, a combination of different procedures can be suggested to control e-waste pollution.

REFERENCES

Adebusoye, S.A., F.W. Picardal, M.O.Ilori, O.O. Amund, C. Fuqua, and N. Grindle. 2007. Growth on dichlorobiphenyls with chlorine substitution on each ring by bacteria isolated from contaminated African soils. *Applied Microbiology Biotechnology* 74:484–492.

Alan, N., B. J. De Klerk, D.D. William, and B.Petrus. 2005. Recovery of precious metal from sulphide minerals by bioleaching. US Patent No. 6860919 (B1).

Babu, R., A.K. Parande, and A.C.Basha. 2007. Electrical and electronic waste: a global environmental problem. *Waste Management & Research* 25: 307–318.

Bagda, E., M. Tuzen, and A. Sarı. 2017. Equilibrium, thermodynamic and kinetic investigations for biosorption of uranium with green algae (*Cladophora hutchinsiae*). *Journal of Environmental Radioactivity* 175:7–14.

Balabanič, D., M. Rupnik, and A.K. Klemenčič. 2011. Negative impact of endocrine-disrupting compounds on human reproductive health. *Reproduction, fertility, and development* 23(3): 403–416. https://doi.org/10.1071/RD09300

Balde, C.P., F. Wang, R.Kuehr, and J.Huisman. 2015. The Global E-waste Monitor 2014. Quantities Flows and Resources. United Nations University, IAS e SCYCLE, Bonn, Germany, pp. 1-41. Institute for the Advanced Study of Sustainability. http://i.unu.edu/media/ias.unu.edu-en/news/7916/Global-E-waste-Monitor-2014-small.pdf.

Barbato, M., F. Mapelli, M. Magagnini, B. Chouaia, M. Armeni, R. Marasco, E. Crotti, D. Daffonchio, and S. Borin.2016. Hydrocarbon pollutants shape bacterial community assembly of harbor sediments. *Marine Pollution Bulletin* 104 (1–2): 211–220.

Bas, A.D., E. Koc, Y.E. Yazici, and H. Deveci. 2015. Treatment of copper-rich gold ore by cyanide leaching, ammonia pretreatment and ammoniacal cyanide leaching. *Transactions of Nonferrous Metals Society of China (Engl. Ed.)* 25(2): 597–607. https://doi.org/10.1016/S1003-6326(15)63642-1.

Birloaga, I., V. Coman, B. Kopacek, and F. Veglio`. 2014. An advanced study on the hydrometallurgical processing of waste computer printed circuit boards to extract their valuable content of metals. *Waste Management* 34(12): 2581–2586. https://doi.org/10.1016/j.wasman.2014.08.028.

Bisceglie, F., D. Civati, B. Bonati, and F.D. Faraci. 2017. Reduction of potassium cyanide usage in a consolidated industrial process for gold recovery from wastes and scraps. *Journal of Cleaner Product* 142: 1810–1818. https://doi.org/10.1016/j.jclepro.2016.11.103.

Bosshard, P.P., R. Bachofen, and H. Brandl. 1996. Metal leaching of fly ash from municipal waste incineration by *Aspergillus niger*. *Environmental Science and Technology* 30:3066–3070.

Brandl, H., R. Bosshard, and M.Wegmann. 2001. Computer-munching microbes: metal leaching from electronic scrap by bacteria and fungi. *Hydrometallurgy* 59 (2–3): 319–326.

Breivik, K., J.M. Armitage, F. Wania, A.J. Sweetman, and K.C. Jones. 2016. Tracking the Global Distribution of Persistent Organic Pollutants Accounting for E-Waste Exports to Developing Regions. *Environmental Science & Technology* 50 (2): 798-805. DOI: 10.1021/acs.est.5b04226

Chatterjee, P. 2008. Health costs of recycling. *British Medical Journal* 337: 376–377.

Chang-bin, W., Z. Wei-min, Z. Hong-bo, F. Bo, H. Ju-fang, Q. Guan-zhou, and W. Dian-zuo. 2007. Bioleaching of chalcopyrite by mixed culture of moderately thermophilic microorganisms. *Journal of Central South University of Technology* 14(4): 474.

Chauhan, G., P.R. Jadhao, K.K. Pant, and K.D.P. Nigam. 2018. Novel technologies and conventional processes for recovery of metals from waste electrical and electronic equipment: Challenges & opportunities – A review. **Error! Hyperlink reference not valid.** 6:1288-1304.

Chen, M., O.A. Ogunseitan, J. Wang, H. Chen, B. Wang, and S. Chen. 2016. Evolution of electronic waste toxicity: trends in innovation and regulation. *Environment International* 89–90: 147–154.

Chen, J., H.C. Zhou, C.Wang, C.Q. Zhu, and N.F.Y. Tama. 2015. Short-term enhancement effect of nitrogen addition on microbial degradation and plant uptake of polybrominated diphenyl ethers (PBDEs) in contaminated mangrove soil. *Journal of Hazardous Material* 300: 84-92.

Chen, A., K.N. Dietrich, X. Huo, and S.M. Ho. 2011. Developmental neurotoxicants in Ewaste: an emerging health concern. *Environmental Health Perspectives* 119 (4): 431–433.

Cheikh, M., J.P. Magnin, N. Gondrexon, J. Willisn, and A. Hassen. 2010. Zinc and lead leaching from contaminated industrial waste sludges using coupled processes. *Environmental Technology* 31(14): 1577–1585.

Chojnacka, K., A. Chojnacki, and H. Gorecka. 2005. Biosorption of Cr3+, Cd2+ and Cu2+ ions by blue– green algae *Spirulina* sp.: kinetics, equilibrium and the mechanism of the process. *Chemosphere* 59:75–84.

Clark, D.A., and P.R. Norris. 1996. *Acidimicrobiumferrooxidans* Gen. Nov., Sp. Nov.: mixed-culture ferrous iron oxidation with *Sulfobacillus* species. *Microbiology* 142:785–790.

Cortes, L.N., E.H. Tanabe, D.A. Bertuol, and G.L. Dotto. 2015. Biosorption of gold from computer microprocessor leachate solutions using chitin. *Waste Management* 45:272–279.

Creamer, N.J., V.S. Baxter-Plant, J. Henderson, M. Potter, and L.E. Macaskie. 2006. Palladium and gold removal and recovery from precious metal solutions and electronic scrap leachates by *Desulfovibriodesulfuricans*. *Biotechnology Letters* 28(18):1475–1484.

Cucchiella, F., I. D'Adamo, P. Rosa, and S. Terzi. 2016. Scrap automotive electronics: a minireview of current management practices. *Waste Management Research* 34:3–10.

Cui, J., and L. Zhang. 2008. Metallurgical recovery of metals from electronic waste: a review. *Journal of Hazardous Materials* 158: 228–256.

Das, N., and D. Das. 2013. Recovery of rare earth metals through biosorption: an overview. *Journal of Rare Earths* 31:933–943.

Dixit, R., D. Malaviya, K. Pandiyan, U.B. Singh, A. Sahu, R. Shukla, and D. Paul. 2015. Bioremediation of heavy metals from soil and aquatic environment: an overview of principles and criteria of fundamental processes. *Sustainability* 7:2189–2212.

Fomina, M., and G.M. Gadd. 2014. Biosorption: current perspectives on concept, definition and application. *Bioresource Technology* 160:3–14.

Forti, V., P.B. Cornelis, R. Kuehr, and B. Garam. 2020. The Global E-waste Monitor 2020 Quantities,flows, and the circular economy potential. UNU/UNITAR and ITU.

Furukawa, K. and H. Fujihara. 2008. Microbial degradation of polychlorinated biphenyls: Biochemical and molecular features. *Journal of Bioscience Bioengineering* 105 (5):433–449.

Gajendiran, A., and J. Abraham. 2015. Mycoadsorption of mercury isolated from mercury contaminated site. *Pollution Research Journal* 34:535–538.

Grant, K. et al. 2013. Health consequences of exposure to e-waste: a systematic review. *The Lancet Global Health* 1(6):350–361. https://doi.org/10.1016/S2214-109X(13)70101-3

Groudev, S.N. 1987. Use of heterotrophic microorganisms in mineral biotechnology. *Acta Biotechnology* 7 (4): 299–306.

Guo, X., J. Liu, H. Qin, Y. Liu, Q. Tian, and D. Li. 2015. Recovery of metal values from waste printed circuit boards using an alkali fusion-leaching-separation process. *Hydrometallurgy* 156: 199–205. https://doi.org/10.1016/j.hydromet.2015.06.011.

Ha, V.H., J.c. Lee, J. Jeong, H.T. Hai, and M.K. Jha. 2010. Thiosulfate leaching of gold from waste mobile phones. *Journal of Hazardous Material* 178(1–3):1115–1119.https://doi.org/10.1016/j.jhazmat.2010.01.099.

Hadi, P., M. Xu, C.S.K. Lin, C.W. Hui, and G.McKay. 2015. Waste printed circuit board recycling techniques and product utilization. *Journal of Hazardous Material* 283:234–243. https://doi.org/10.1016/j. jhazmat.2014.09.032.

Hammad, H.M., W. Farhad, F. Abbas,S. Fahad, S. Saeed, W.Nasim, and H.F. Bakhat. 2017. Maize plant nitrogen uptake dynamics at limited irrigation water and nitrogen. *Environmental Science Pollution Research* 24:2549–2557.

Hansda, A., and V. Kumar. 2016. A comparative review towards potential of microbial cells for heavy metal removal with emphasis on biosorption and bioaccumulation. *World Journal of Microbiology and Biotechnology* 32:170.

He, X., W. Chen, and Q. Huang. 2012. Surface display of monkey metallothionein α tandem repeats and EGFP fusion protein on *Pseudomonas putida* X4 for biosorption and detection of cadmium. *Applied Microbiology and Biotechnology* 95:1605–1613.

Heart, S., and P. Agamuthu. 2012. E-waste: a problem or an opportunity? Review of issues, challenges and solutions in Asian countries. *Waste Management Research* 30:1113–1129.

Hong, Y., and M. Valix. 2014. Bioleaching of electronic waste using acidophilic sulfur oxidising bacteria. *Journal of Cleaner Product* 65:465–472.

Hong, K.J., S. Tokunaga, and T. Kajiuchi. 2000. Extraction of heavy metals from msw incinerator fly ashes by chelating agents. *Journal of Hazardous Materials* 75(1):57–73.

Hosokawa, R., M.Nagai, M. Morikawa, and H. Okuyama, 2009. Autochthonous bioaugmentation and its possible application to oil spills. *World Journal of Microbiology and Biotechnology*25 (9): 1519–1528.

Huang, C., L. Bao, P. Luo, Z. Wang, S. Li, and E.Y. Zeng. 2016. Potential health risk for residents around a typical e-waste recycling zone via inhalation of size-fractionated particle-bound heavy metals. *Journal of Hazardous Material* 317:449–456.

Huang, J., M. Chen, H. Chen, S. Chen, and Q. Sun.2014. Leaching behavior of copper from waste printed circuit boards with Bronsted acidic ionic liquid. *Waste Management* 34:483–488.

Ilankoon, I.M.S.K.,Y. Ghorbani, M.N. Chong, G. Herath, T. Moyo, and J. Petersen. 2018. E-waste in the international context – A review of trade flows, regulations, hazards, waste management strategies and technologies for value recovery. *Waste Management* 82:258–275.

Ilyas, S., J.C. Lee, and B.S. Kim. 2014.Bioremoval of heavy metals from recycling industry electronic waste by a consortium of moderate thermophiles: process development and optimization. *Journal of Cleaner Product* 70:194–202.

Ilyas, S., M.A. Anwar, S.B. Niazi, and M.A. Ghauri. 2007. Bioleaching of metals from electronic scrap by moderately thermophilic acidophilic bacteria. *Hydrometallurgy* 88 (1–4):180–188.

Isildar, A., J. van de Vossenberg, E.R. Rene, E.D. van Hullebusch, and P.N.L. Lens. 2015. Two-step bioleaching of copper and gold from discarded printed circuit boards (PCB). *Waste Management* 57:149–157.

Jadhav, U., and H. Hocheng. 2015. Hydrometallurgical recovery of metals from large printed circuit board pieces. *Scientific Reports* 5 (101):1–10. https://doi.org/10.1038/srep14574.

Jha, M.K., P.K. Choubey, A.K. Jha, A. Kumari, J.C. Lee, V. Kumar, and J. Jeong. 2012. Leaching studies for tin recovery from waste e-scrap. *Waste Management* 32(10): 1919–1925. https://doi.org/ 10.1016/j.wasman.2012.05.006.

Jing-ying, L., X. Xiu-li, and L. Wen-quan. 2012. Thiourea leaching gold and silver from the printed circuit boards of waste mobile phones. *Waste Management* 32(6):1209–1212. https://doi.org/ 10.1016/j.wasman.2012.01.026.

Jinhui, L., N.L.N. Brenda, L.Lili, Z.Nana,Y. Keli, and Z. Lixia. 2013. Regional or global WEEE recycling. Where to go? *Waste Management* 3:923–934.

Johnson, D.B., N. Okibe, K. Wakeman, and L.Yajie. 2008. Effect of temperature on the bioleaching of chalcopyrite concentrates containing different concentrations of silver. *Hydrometallurgy* 94(1–4):42–47.

Jomova, K., Z. Jenisova, M. Feszterova, S. Baros, J. Liska, D. Hudecova, C.J. Rhodes, and M. Valko. 2011. Arsenic: toxicity, oxidative stress and human disease. *Journal of Applied Toxicology* 31(2):95–107.

Kamberovic, Z., M. Korac, and M. Ranitovic. 2011. Hydrometallurgical process for extraction of metals from eectronic waste. Part II. Development of the processes for the recovery of copper from Printed Circuit Boards (PCB). *Metalurgiya—MJoM* 17 (3):139–149.

Karn, B., T. Kuiken, and M. Otto. 2009. Nanotechnology and in situ remediation: a review of the benefits and potential risks. *Environmental Health Perspectives* 117 (12):1813–1831.

Karwowska, E., D. Andrzejewska-Morzuch, M. Lebkowska, A. Tabernacka, M. Wojtkowska, A.Telepko, and A. Konarzewska. 2014. Bioleaching of metals from printed circuit boards supported with surfactant-producing bacteria. *Journal of Hazardous Material* 264:203–210.

Kaya, M. 2016. Recovery of metals and nonmetals from electronic waste by physical and chemical recycling processes. *Waste Management* 57:64–90. https://doi.org/10.1016/j.wasman.2016.08.004.

Khaliq, A., M. Rhamdhani, G. Brooks, and S. Masood. 2014. Metal extraction processes for electronic waste and existing industrial routes: a review and Australian perspective. *Resources* 3 (1):152–179. https://doi.org/10.3390/resources3010152.

Kelly, S., and A. Dagle. 1974. Cytogenetic damage in americium poisoning. *New York State Journal of Medicine* 74 (9):1597–1598.

Kohr, W.J. 1995. Method for rendering refractory sulphide ores more susceptible to bio-oxidation, US Patent No. 5431717

Konishi, Y., H. Kubo, and S. Asai. 1992. Bioleaching of zinc sulfide concentrate by Thiobacillus ferrooxidans. *Biotechnology and Bioengineering* 39 (1):66–74.

Kotnala, R.K. 2009. New Nanotechniques, Ethical Issues of Nanotechnology, Nova Science Publishers, New York, USA (Chapter 7) (ISBN 978-1-60692-516-4).

LaDou, J., and S. Lovegrove. 2008. Export of electronics equipment waste. *International Journal of Occupational Environmental Health* 14:1–10.

Lambert, F., S. Gaydardzhiev, G. Leonard, G. Lewis, and P.-F. Bareel. 2015. Copper leaching from waste electric cables by biohydrometallurgy. *Mineral Engineering* 76:38–46.

Langford, L.J.,and R.E. Ferner 1999. Toxicity of mercury. *Journal of Human Hypertension* 13: 651–656.

Li, Z., L. Wang, J. Meng, X. Liu, J. Xu, F. Wang, and P. Brookes. 2018. Zeolite-supported nanoscale zero-valent iron: new findings on simultaneous adsorption of Cd (II), Pb(II), and As (III) in aqueous solution and soil. *Journal of Hazardous Material* 344:1–11.

Li, L., J. Ge, F. Wu, R. Chen, S. Chen, and B.Wu. 2010. Recovery of cobalt and lithium from spent lithium ion batteries using organic citric acid as leachant. *Journal of Hazardous Material*176(1-3): 288–293. DOI: 10.1016/j.jhazmat.2009.11.026.

Liddicoat, J., and D. Dreisinger. 2007. Chloride leaching of chalcopyrite. *Hydrometallurgy* 89 (3–4): 323–331. https://doi.org/10.1016/j.hydromet.2007.08.004.

Liu, J., X. Chen, H.Y. Shu, X.R. Lin, Q.X. Zhou, T. Bramryd, and L.N. Huang. 2018. Microbial community structure and function in sediments from e-waste contaminated rivers at Guiyu area of China. *Environmental Pollution* 235:171–179.

Liu, J., X.X. He, X.R. Lin, W.C. Chen, Q.X. Zhou, W.S. Shu, and L.N. Huang. 2015. Ecological effects of combined pollution associated with e-waste recycling on the composition and diversity of soil microbial communities. *Environmental Science and Technology* 49:6438–6447.

Liu, Y., G. Su, B. Zhang, G. Jiang, and B. Yan. 2011. Nanoparticle-based strategies for detection and remediation of environmental pollutants. *Analyst* 136:872–877.

Liu, Q., J. Cao, K.Q. Li, X.H., Miao, G. Li, F.Y. Fan, and Y.C. Zhao. 2008. Chromosomal aberrations and DNA damage in human populations exposed to the processingof electronics waste. *Environmental Science and Pollution Research* 16 (3): 329–338.

Luo, X.J., X.L. Zhang, J. Liu, J.P.Wu, Y.Luo, S.J.Chen, B.X. Mai, and Z.Y.Yang. 2009. Persistent halogenated compounds in waterbirds from an e-waste recycling region in South China. *Environmental Science and Technology* 43:306–311.

Madrigal-Arias,J.E.,R.Argumedo-Delira,A.Alarcón,M.Mendoza-López,O.García-Barradas, J.S. Cruz- Sánchez, and M.Jiménez-Fernández. 2015. Bioleaching of gold, copper and nickel from waste cellular phone PCBs and computer goldfinger motherboards by two *Aspergillus niger* strains. *Brazilian Journal of Microbiology* 46:707–713.

Makinen, J., J. Bacher, T.Kaartinen, M. Wahlstrom, and J.Salminen. 2015. The effect of flotation and parameters for bioleaching of printed circuit boards. *Mineral Engineering* 75:26–31.

Monika, J.K. 2010. E-waste management: as a challenge to public health in India. *Indian Journal of Community Medicine* 35 (3):382–385.

Michalak, I., K. Chojnacka, and A.Witek-Krowiak. 2013. State of the art for the biosorption process—a review. *Applied Biochemistry and Biotechnology* 170:1389–1416.

Mishra, D., and Y.H. Rhee. 2010. Current research trends of microbiological leaching for metal recovery from industrial wastes. Current Research, Technology and Education Topics. *Applied Microbiology and Microbial Biotechnology* 2:1289–1292.

Mulligan, C.N., and M. Kamali. 2003. Bioleaching of copper and other metals from lowgrade oxidized mining ores by *Aspergillus niger*. *Journal of Chemical Technology and Biotechnology* 78 (5):497–503.

Nagarajah, R., K.T. Wong, G.Lee, K.H.Chu, Y.Yoon, N.C. Kim, and M. Jang. 2017. Synthesis of a unique nanostructured magnesium oxide coated magnetite cluster composite and its application for the removal of selected heavy metals. *Separation and Purification Technology* 174:290–300.

Nasim, W., A. Ahmad, A. Amin, M. Tariq, M. Awais, M. Saqib, K. Jabran, G.M. Shah, S.R. Sultana, H.M. Hammad, and M.I.A. Rehmani. 2018. Radiation efficiency and nitrogen fertilizer impacts on sunflower crop in contrasting environments of Punjab-Pakistan. *Environmental Science Pollution Research* 25: 1822–1836.

Nasim, W., H. Belhouchette, M.Tariq, S.Fahad, H.M. Hammad, M. Mubeen, M.F.H. Munis, H.J.Chaudhary, I.Khan, F.Mahmood, and T. Abbas. 2016. Correlation studies on nitrogen for sunflower crop across the agroclimatic variability. *Environmental Science Pollution Research* 23:3658–3670.

Padiyar, N. 2011. Nickel allergy-is it a Cause of concern in everyday dental practice. *International Journal of Contemporary Dentist* 12 (1):80–81.

Pan, J., J.A. Plant, N. Voulvoulis, C.J. Oates, and C. Ihlenfeld. 2010. Cadmium levels in Europe: implications for human health. *Environment Geochemistry and Health* 32 (1):1–12.

Pant, D., A. Giri, and V.Dhiman. 2018. Bioremediation techniques for E-waste Management. In: Waste bioremediation. Springer, Singapore. 105–125.

Pant, D., D. Joshi, M.K.Upreti, and R.K.Kotnala. 2012. Chemical and biological extraction of metals present in E-waste: a hybrid technology. *Waste Management* 32:79–990.

Patel,S., and A. Kasture. 2014. E (electronic) waste management using biological systems-over-view. *International Journal of Current Microbiology and Applied Science* 3:495–504.

Patra, R.C., A.K. Rautray, and D. Swarup. 2011. Oxidative stress in lead and cadmium toxicity and its amelioration. *Research Veterinary Medicine International* 9: 1–2.

Patrick, N.S., D. Pharm, D. and F.C. Lee. 1999. Elevated lithium level: a case and brief over-view of lithium poisoning. *Psychosomatic Medicine* 61: 564–565.

Pham, V.A., and Y.P. Ting. 2009. Gold bioleaching of electronic waste by cyanogenic bacteria and its enhancement with bio-oxidation.*Advanced Materials Research* 1:71–73.

Pieper, D. H. 2005. Aerobic degradation of polychlorinated biphenyls. *Applied Microbiology and Biotechnology* 67 (2):170–191.

Poon, C.S. 2008. Management of CRT glass from discarded computer monitors and TV sets. *Waste Management* 28:1499.

Ponder, S.M., J. Darab, and T. Mallouk. 2000. Remediation of Cr(VI) and Pb(II) aqueous solutions using supported, nanoscale zero-valent iron. *Environmental Science and Technology* 34:2564–2569.

Potrawfke, T., T.H. Löhnert, K.N. Timmis, and R.M. Wittich. 1998. Mineralization of low-chlorinated biphenyls by *Burkholderia* sp. strain LB400 and by a two-membered con-sortium upon directed interspecies transfer of chlorocatechol pathway genes. *Applied Microbiology and Biotechnology* 50: 440–446.

Priya, A., and S. Hait. 2017. Comparative assessment of metallurgical recovery of metals from electronic waste with special emphasis on bioleaching. *Environmental Science Pollution Research* 24: 6989–7008.

Puente-Siller, D.M., J.C. Fuentes-Aceituno, and F.Nava-Alonso. 2017. An analysis of the effi-ciency and sustainability of the thiosulfate-copper-ammonia-monoethanolamine system for the recovery of silver as an alternative to cyanidation. *Hydrometallurgy* 169: 16–25. https://doi.org/ 10.1016/j.hydromet.2016.12.003.

Puente-Siller, D.M., J.C. Fuentes-Aceituno, and F.Nava-Alonso. 2013. A kinetic-thermo-dynamic study of silver leaching in thiosulfate-copper-ammonia-EDTA solutions. *Hydrometallurgy* 134–135: 124–131. https://doi.org/10.1016/j.hydromet.2013.02.010.

Oh, C.J., S.O. Lee, H.S. Yang, T.J. Ha, and M.J.Kim. 2003. Selective leaching of valuable metals from waste printed circuit boards. *Journal of Air and Waste Management Association* 1995. https://doi.org/10.1080/10473289.2003.10466230.

Olafisoye, O.B., T. Adefioye, and O.A.Osibote. 2013. Heavy metals contamination of water, soil, and plants around an electronic waste dumpsite. *Polish Journal of Environmental Studies* 22:1431–1439.

Olson, G.J. 2006. Microbial oxidation of gold ores and gold bioleaching. *FEMS Microbiology Letters* 119(1–2): 1–6.

Rajesh, V., A.S.K. Kumar, and N. Rajesh. 2014. Biosorption of cadmium using a novel bacterium isolated from an electronic industry effluent. *Chemical Engineering Journal* 235:176–185.

Rawlings, D.E., D. Dew, and du, C. Plessis. 2003. Biomineralization of metal-containing ores and concentrates. *Trends in Biotechnology* 21(1): 38–44.

Ren, W., P. Li, Y.Geng, and X. Li. 2009. Biological leaching of heavy metals from a contaminated soil by *Aspergillus niger*. *Journal of Hazardous Materials* 167 (1–3):164–169.

Robinson, B.H. 2009. E-waste: an assessment of global production and environmental impacts. *Science of the Total Environment* 408:183–191.

Rocchetti, L., V. Fonti, F. Veglio`, and F.Beolchini. 2013. An environmentally friendly process for the recovery of valuable metals from spent refinery catalysts. *Waste Management Research* 31 (6): 568–576. https://doi.org/10.1177/0734242X13476364.

Rozas, E.E., M.A. Mendes, C.A.Nascimento, D.C. Espinosa, R.Oliveira, G. Oliveira, and M.R. Custodio. 2017. Bioleaching of electronic waste using bacteria isolated from the marine sponge *Hymeniacidonheliophila* (Porifera). *Journal of Hazardous Material* 329:120–130.

Sakai, M., S.Ezaki, N. Suzuki, and R. Kurane. 2005. Isolation and characterization of a novel polychlorinated biphenyl-degrading bacterium, *Paenibacillus*sp KBC101. *Applied Microbiology and Biotechnology* 68 (1):111–116.

Sarpong, G.O., K.O. Asare, and M. Tien. 2011. Fungal pretreatment of sulfides in refractory gold ores. *Minerals Engineering* 24(6): 499–504.

Satroutdinov, A.D., E.G. Dedyukhina, T.I. Chistyakova, M.Witschel, I.G. Minkevich, V.K. Eroshin, and T. Egli. 2000. Degradation of metal–EDTA complexes by resting cells of the bacterial strain DSM 9103. *Environmental Science and Technology* 34(9): 1715–1720.

Seto, M., K.Kimbara, M. Shimura, T. Hatta, M. Fukuda, and K. Yano. 1995. A Novel Transformation of Polychlorinated Biphenyls by *Rhodococcus* sp. Strain RHA1. *Applied and environmental microbiology* 61(9):3353–3358. https://doi.org/10.1128/AEM.61.9.3353-3358.1995

Sharma, Y.C., V.S. Srivastava, N. Upadhyay, and C.H.Weng. 2008. Alumina nanoparticles for the removal of Ni(II) from aqueous solutions. *Industrial and Engineering Chemistry Research* 47 (21): 8095–8100.

Sheel, A., and D. Pant. 2018. Recovery of gold from electronic waste using chemical assisted microbial biosorption (hybrid) technique. *Bioresource Technology* 247:1189–1192.

Solisio, C., A. Lodi, and F.Veglio. 2002. Bioleaching of zinc and aluminium from industrial waste sludges by means of *Thiobacillus ferrooxidans*. *Waste Management* 22 (6): 667–675.

Song, M.K., C.L. Luo, F.B. Li, L.F. Jiang, Y. Wang, D.Y. Zhang, and G. Zhang. 2015. Anaerobic degradation of Polychlorinated Biphenyls (PCBs) and Polychlorinated Biphenyls Ethers (PBDEs), and microbial community dynamics of electronic waste-contaminated soil. *Science of the Total Environment* 502: 426–433.

Song, Q., and J.Li. 2014.Environmental effects of heavy metals derived from the e-waste recycling activities in China: a systematic review *Waste Management* 34 (12):2587–2594.

Stevens, G.C., and M.Goosey. 2008. in Electronic Waste Management, DOI: 10.1039/9781847559197-00040 40-74.

Sthiannopkao, S., and M.H. Wong. 2013. Handling e-waste in developed and developing countries: Initiatives, practices, and consequences. *Science of The Total Environment* 463–464: 1147–1153.https://www.sciencedirect.com/science/journal/00489697

Suzuki, I. 2001. Microbial leaching of metals from sulfide minerals. *Biotechnology Advances* 19 (2): 119.

Takashima, M. 1999. Method for recovering aluminum from materials containing metallic aluminum. US Patent No. 5 855, 644.

Tang, X.J., J.N.Qiao, C. Chen, L.T. Chen, C.N. Yu, C.F. Shen, and Y.X.Chen. 2013. Bacterial Communities of Polychlorinated Biphenyls Polluted Soil Around an E-waste Recycling Workshop. *Soil Sediment Contamination* 22 (5):562–573.

Tay, S.B., G. Natarajan, M.N. bin Abdul Rahim, H.T.Tan, M.C.M.Chung,Y.P.Ting, and W.S.Yew. 2013. Enhancing gold recovery from electronic waste via lixiviant metabolic engineering in *Chromobacteriumviolaceum*. *Scientific Report* 3:2236.

Tetsuo, I., and S. Atsushi. 2001. Process for leaching copper from copper sulfide using bacteria. United States Patent: 6168766.

Tratnyek, P.G., and R.L. Johnson. 2006. Nanotechnologies for environmental cleanup. *Nanotoday* 1 (2).

Tsydenova, O., and M. Bengtsson. 2011. Chemical hazards associated with treatment of waste electrical and electronic equipment. *Waste Management*, 31(1): 45–58. https://doi.org/10.1016/j.wasman.2010.08.014.

Ubaldini, S., F.Veglio`, P.Fornari, and C. Abbruzzese. 2000. Process flow-sheet for gold and antimony recovery from stibnite. *Hydrometallurgy* 57(3): 187–199. https://doi.org/10.1016/S0304-386X(00)00107-9.

UNEP. 2005. UN concerned over E-waste in Asia. (www.incommunicado.info/node/215).

Veit, H.M., A.M. Bernardes, A.C. Kasper, N.C.Juchneski, C.O. de Calgaro, and E.H.Tanabe. 2015.Veit, H.M., & Moura Bernardes, A. (Eds.), *Electronic Waste*. Springer International Publishing, Cham. https://doi.org/10.1007/978-3-319-15714-6.

Wang, J., J. Bai, J. Xu, and B. Liang. 2009. Bioleaching of metals from printed wire boards by *Acidithiobacillusferrooxidans* and *Acidithiobacillusthiooxidans* and their mixture. *Journal of Hazardous Materials* 172 (2–3): 1100–1105.

Wasay, S.A., S.F. Barrington, and S.Tokunaga. 1998. Remediation of soils polluted by heavy metals using salts of organic acids and chelating agents. *Environmental Technology* 19(4): 369–379.

Widmer, R., H. Oswald-Krapf, D. Sinha-Khetriwal, M.Schnellmann, and H. Bon. 2005. Global perspectives on E-waste. *Environmental Impact Assessment Review* 25(5): 436–458.

Wu, W., C. Dong, J. Wu, X. Liu, Y. Wu, X. Chen, andS.Yu. 2017. Ecological effects of soil properties and metal concentrations on the composition and diversity of microbial communities associated with land use patterns in an electronic waste recycling region. *Science of the Total Environment* 601:57–65.

Wu, Y., Y. Li, D. Kang, J. Wang, Y. Zhang, D. Du, D., et al. 2016. Tetrabromobisphenol A and heavy metal exposure via dust ingestion in an e-waste recycling region in Southeast China. *Science of the Total Environment* 541: 356–364.

Wu, Q. et al (2015). Heavy metal contamination of soil and water in the vicinity of an abandoned e-waste recycling site: implications for dissemination of heavy metals. *Science of the Total Environment* 506–507:217–225. https://doi.org/10.1016/j.scitotenv.2014.10.121

Xiang, Y., P.Wu, N.Zhu, T.Zhang, W.Liu, J.Wu, and P.Li. 2010. Bioleaching of copper from waste printed circuit boards by bacterial consortium enriched from acid mine drainage. *Journal of Hazardous Material* 184:812–818.

Zhang, J.Q.,Y.S. Jiang, J. Zhou, B. Wu, Y. Liang, Z.Q. Peng, D.K. Fang, B. Liu, H. Huang, Y. He, C.L. Wang, and F.N. Lu. 2010. Elevated Body Burdens of PBDEs, Dioxins, and PCBs on Thyroid Hormone Homeostasis at an Electronic Waste Recycling Site in China. *Environmental Science and Technology* 44 (10): 3956–3962.

Zheng, X. B., X.J.Luo, Y.H. Zeng, J.P. Wu, and B.X. Mai. 2015. Chiral Polychlorinated Biphenyls (PCBs) in Bioaccumulation, Maternal Transfer, and Embryo Development of Chicken. *Environment Science and Technology* 49(2): 785–791.

Part 4

Vision for the Future

Towards Resource Efficient E-waste Management

9 Pilot Production Experience of a Recycled Plastic Aggregate Manufactured Using Plastic From Waste of Electrical and Electronic Equipment

*Lucas E. Peisino, Ariel L. Cappelletti, Julián González Laría, Melina Gómez, Rosana Gaggino, Bárbara B. Raggiotti, and Jerónimo Kreiker**

CONTENTS

DOI: 10.1201/9781003095972-13

9.1   INTRODUCTION

Due to the large consumption and production of electrical and electronic equipment (EEE), waste of electrical and electronic equipment (WEEE) has grown notably in the last two decades. The constant replacement, either due to the quality of the products, the planned obsolescence or the need for increasingly innovative products, leads to an excessive production and as a consequence to a great pollution problem throughout the world (Pérez-Belis, Bovea, and Ibáñez-Forés 2015; Kiddee, Naidu, and Wong 2013).

Although most of the components of WEEE have a recycling and revaluation circuit (metals such as gold or copper, among others), WEEE plastics (WEEEP) are the only component that does not possess a correct final disposal, which generally leads to them being disposed of in open dumps, or in sanitary burials (Ilankoon et al. 2018). WEEE plastics have a very complex polymer composition, and while it is difficult to establish a generalized composition, in most waste devices the polymers acrilonitryle-butadiene-styrene (ABS) and high impact polyestyrene (HIPS) are found as major components, closed to 70 percent, and polycarbonate, polypropylene in smaller proportions. They also present other substances, such as brominated flame retardants (BFRs), heavy metals, antimony, among others, which give it a toxic potential that is very difficult to handle when finding a way to recycle and revalue this plastic waste (Signoret et al. 2020). So, it is very important to consider the safety of both workers and consumers and to take into account the environmental aspects and the life cycle of the elaborated products. Limiting the analysis to the recycling of the plastic fraction of WEEE, it is possible to identify different forms of recycling used in most plastics, including chemical, pelletized recycling, revaluation as fuel or crushed material for another purpose. The chemical recycling a big scale is not possible, because the different constituting polymer, because this type of recycling process needs a depolymerization process which is specific for each polymer and could have an undesirable effect in the other polymers. The pelletizing process involves the grinding, extrusion and conformed of pellets. With the WEEEP, it is very difficult because of the big difference in the melting point and decomposition temperature of different constitution polymers. During the heating process it may be the case that some polymer decomposes before other melts, so the mixture is unworkable. Another problem with the heating process is the large amount of hazardous compounds from additives or the same polymeric matrix that released during the heating process. The same problem of the release of hazardous substances is that which prevents the use of this waste as combustible material, which leads to a diminishing of the useful life of filters and the gas washers of the incineration plants. A promising technology for recycling WEEEP is by grinding it and then its incorporation as an aggregate in a composite material with cement or polymeric matrix. Despite the potential use as an aggregate, by its good mechanical and morphological behavior, the WEEEP shredded can not be directly used as an aggregate because of the contaminant compounds that could be leached from the composite material, when a stabilization strategy has been developed.

In our previous reports, we have demonstrated that it is possible to stabilize the contaminants present in WEEEP, the heavy metals (Peisino et al. 2019) and the BFRs (Gómez et al. 2020) through the use of a core-shell strategy approach in order to obtain a novel recycled plastic aggregate (RPA), where the WEEEP core is covered by a shell consisting in a cement and additives matrix. Heavy metals were proven to be stabilized for the physical barrier of the cement shell, while BFRs were contained from leachate through the use of activated charcoal as an additive for the cementitious mix, which was applied to form a triple layer shell around the WEEEP particles. Also, mechanical properties of RPA and building components manufactured based on RPA were characterized with satisfactory performances (Gómez et al. 2020).

Here we present the production of the RPA and RPA-based building components on a pilot scale, the experience in mounting the pilot plant, the preliminary results, future perspectives in order to implement this technology in the society through cooperatives and small businesses ventures, municipal solid waste management programs, NGOs, etc.

9.1.1   WEEEP Employed as Aggregate for Cement Mixtures

In order to recycle or reuse plastic from WEEE to produce building components, a few reports informed the utilization of recycled plastics as aggregates or fiber to manufacture mortars or concrete mixtures (Hannawi, Kamali-Bernard, and Prince 2010; Bágeľ and Matiašovský 2010; Lakshmi and Nagan 2010). As a general rule, this research is driven in order to diminish the waste amount and to improve the recycling policies or by analysis of the physical and mechanical properties of the mortars or concrete developed. Moreover, in these articles the effectproduced for the partial replacement of the natural aggregates (coarse or fine aggregates) by the WEEEP (Wang and Meyer 2012; Senthil Kumar and Baskar 2015b, [c] 2015) is widely analyzed. The main properties affected by the addition of the WEEE RPAs or PF are compressive strength (CS), flexural strength, tensile resistance, water absorption, thermal conductivity and density among others (Popovici et al. 2015; Farooq et al. 2019; Nowek 2016; Kurup and Kumar 2017). The decrease of the weight of mortar and concrete mixtures which contain WEEEP, has resulted in the proposal from authors to use them as lightweight building materials (Senthil Kumar and Baskar 2015a).

To the present, just a few studies that consider the effect of the cement matrix over thesepollutants were reported. The leaching of metals in concrete elaborated with WEEE plastic was considered by two reports (Senthil Kumar, Gandhimathi, and Baskar 2016; Peisino et al. 2019). On the other and, the leaching of BFRs in mortar elaborated with WEEEP was analized by one report (Gómez et al. 2020).

9.1.2 ABOUT THE TECHNOLOGY

9.1.2.1 Research Gap and Motivation

As we have shown in previous sections, WEEEP is generated in large quantities and its safe reuse and recycling is problematic. It is important for the health of operators, consumers and the environment to develop a safe technology for recycling this type of hazardous waste. The social responsibility of the scientific and technological researchers is vital in the development of harmless, low-cost and viable technologies for recycling waste and its proposed uses. For this reason, we have developed a technology into a safe and economical recycling paradigm. In this way, in previous reports we have presented a core-shell stabilization strategy for recycling WEEEP; in this chapter we present its implementation in a pilot scale to provide a potential solution through this responsible and safe technology for a large scale problem like hazardous plastic waste. The study results presented here could be interesting for industry and society.

9.1.2.2 Core-shell Recycled Plastic Aggregate (RPA) from WEEE Plastics

The experimental methodology for the manufacturing of the core-shell RPA was extensively described in our previous reports (Peisino et al. 2019; Gómez et al. 2020). A schematic description is represented in Figure 9.1.

The shell of RPAs was prepared employing masonry cement. Some of the additives that we have investigated for the production of RPA were polish stoneware tile residue (Kreiker et al. 2018), bentonite, activated charcoal (AC) and river sand and fine river sand in different dosages.

It was found that the use of masonry cement and AC in a three layer core-shell array was the composition with the best results, both for the stabilization of pollutants (heavy metals and BFRs) and for mechanical properties.

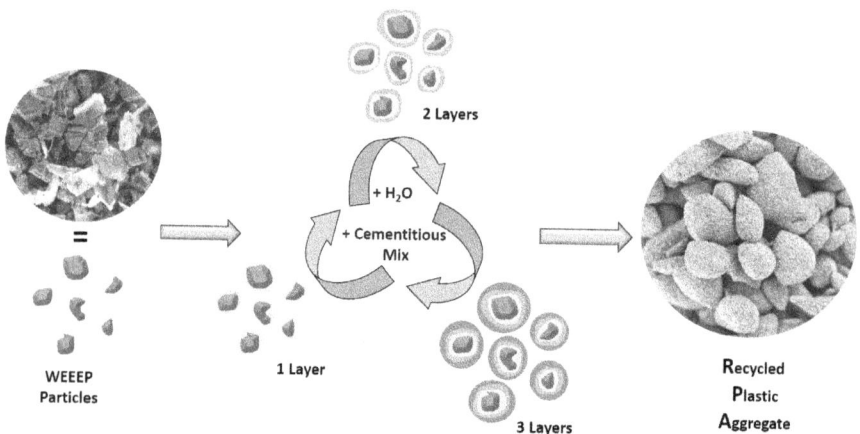

FIGURE 9.1   Schematic representation of the RPA manufacturing procedure.

In summary, a brief description of the procedure is explained as follows:

- Shredded WEEEP particles were placed in a mixer. The particles were wet by spraying water in order to increase the adhesion of cementitious shell.
- Next, using the mixer machine, the cementitious mixture (cement+additives) was added in small portions and the mixing was improved by hand enveloping movements. Water was added in necessary quantities in order to obtain a homogeneous cement shell coverage. This step was repeated in order to obtain RPA with a shell consisting of three layers of cementitious material. The setting time for each layer addition was overnight. After setting time, particles were sorted with a 4.8 mm sieve (4 mesh) before starting the coverage of the particles with a new layer.
- Finally, RPAs were cured with water immersion for 14 days at 20°C.

For inorganic pollutants stabilization evaluation an aqua regia (a mixture of nitric acid and hydrochloric acid in a molar ratio of 1:3) digestion was performed at 80°C in order to dissolve and quantify (by flame excited atomic absorption analysis) the concentration of metals and heavy metals present in the WEEEP sample. The cement shell proved to act as a physical barrier which can effectively stabilize the heavy metals present in WEEEP when TCLP (USEPA 1992) leaching condition was employed (Peisino et al. 2019).

Organic contaminants (BFRs) were extracted from RPAs through chemical extraction and quantified by Gas Chromatography. The focus of the analysis was centered on tetrabromobisphenol-A (TBBPA) because it is the most widely used BFR and with the highest concentration found in WEEEP (Gómez et al. 2020).

Taking into account the phenol nature of TBBPA (Figure 9.2), it was expected that the leaching of TBBPA was promoted by the cement shell, which is of a marked basic character, so the incorporation of an additive in order to counteract this effect became necessary. Activated carbon employment proved to be an effective additive for the stabilization of the organic contaminant present in the WEEEP particles. A dosage of 1.1 %w/w of the activated carbons assayed has been shown to be optimum in order to prevent the leaching of TBBPA.

From the technical point of view, the proposed RPA showed good physical properties that allow its use as an alternative aggregate for cement mortars or a concrete mixture. AC presence has a negative effect over compression resistance observed at 28 and 60 days of setting. Nevertheless, at 120 days compressive resistance was similar to the value for samples prepared without activated carbon. This indicates that the usage of AC as an additive does not compromise the desired mechanical properties for building components (Gómez et al. 2020).

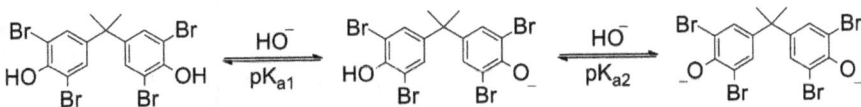

FIGURE 9.2 Acid-base equilibrium of TBBPA.

9.2   PILOT PRODUCTION

In our research institute, a pilot plant production for processing 100 kg per day of WEEEP was set up. The machinery employed for this scale of production is economical and commercially available. In the following sections a detailed description of this pilot production will be presented, both the RPA production and cement mortar building components prepared with the RPA.

9.2.1   RPA PRODUCTION

For the production of RPA, a simple methodology was implemented, which is graphically describe in a flowchart (Figure 9.3). The first step is to collect the WEEEP, and shredded to particle size of 40-50 mm, then shredded to 3-4 mm in a second stage. Once WEEEP particles of 3-4 mm were obtained, the next step in the manufacturing procedure is to cover them with the (masonry cement + additives) shell, for this purpose WEEEP must be firstly placed in the modified concrete mixer machine and moistened while mixing. The cementitious mixture is added slowly to the plastic particles in order to allow a uniform coverage. Next, the particles with a cement coating are left to settle overnight in a convenient container. Next day, particles must be sorted with a 4.8 mm sieve. This entire procedure to cover the particles is carried out repeatedly until a product with three layers of cementitious shell is obtained (see Figure 9.3). The final product with triple cementitious layers must be left to settle for 14 days at room temperature.

After 14 days of settling the obtained RPA are ready to be used in building component manufacturing.

9.2.1.1   WEEEP Shredder

The first step in a pilot plant to process the WEEEP is to reduce the size of the plastic casings. For this we propose the shear shredder of 5.5 kW is employed (see picture

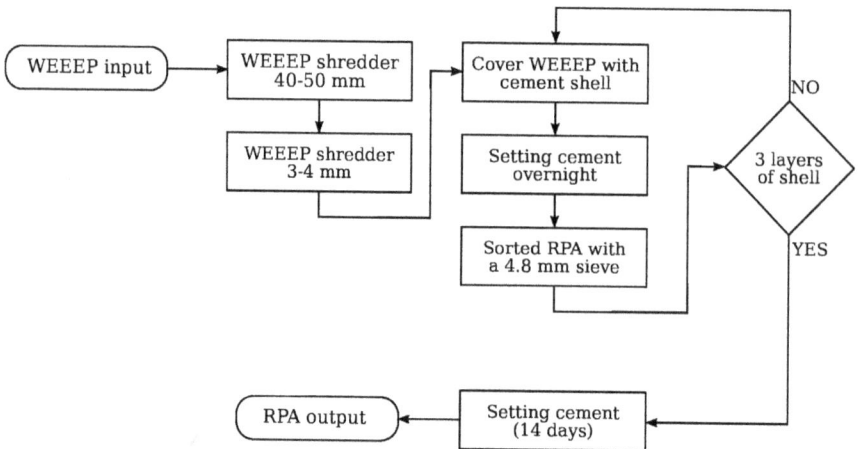

FIGURE 9.3   Flow diagram of RPA manufacturing procedure.

Figure 9.4a). As described in Table 9.1, this machine converts a plastic cabin in 40-50 mm WEEEP particles in a rate of 13 kg per hour, so to process 100 kg of plastic requires 8 hours. Moreover, the energy consumption in this time is 158.4 MJ.

Next, the 40-50 mm WEEEP particles were reduced to a size of 3-4 mm. For this purpose the chipper shredder of 3.7 kW is employed (see picture of Figure 9.4b). As described in Table 9.1, this machine converts a plastic of 40-50 mm size into WEEEP particles of 3-4 mm in a rate of 25 kg per hour, so to process 100 kg of plastic it takes 4 hours. Moreover, the energy consumption in this time is 53.3 MJ.

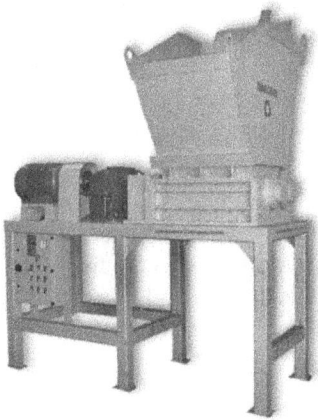
(a)

(b)

(c)

(d)

FIGURE 9.4 Machinery used in the recycling of WEEEP to produce the RPA. (a) shear shredder, (b) chipper shredder, (c) electromechanical concrete mixer and (d) finished RPA and 4 mesh sieve.

9.2.1.2 WEEEP Cover (Shell Growing)

The cover plastic with a cement/additives mixture is a distinguishing step of this recycling technology of WEEEP. As explained in a previous section and shown in Figure 9.3, the cover of WEEEP with a cement and additives mixture is a key step in this process. To achieve the shell over WEEEP at productive scale, we propose to employ a standard electromechanical concrete mixer of 0.4 m and 2.2 kW (see picture of Figure 9.4c). As in entry 3 of Table 9.1, this machine cover a plastic with shell cement in a rate of 100 kg per hour and the energy consumption in one hour of production is 7.9 MJ.

9.2.1.3 Setting Cement and RPA Sieving

Once the shell was made, the RPA is placed in a floor for 16 hours (overnight). Next day, the cement shell has increased its mechanical resistance by effect of the early cement setting. So, the RPA is desegregated by passing by sieve of 4.8 mm. This operation is made manually with a wide shovel. This step takes a lot of time and it is a labor intensive step, however the electrical energy consumption is zero.

9.2.1.4 RPA Curing

Looking at Figure 9.3, the steps of make shell, setting overnight cement and sieve must be repeated two more times to yield a three layers shell. After the final sieving, the RPA is stored in a pile and is watered daily to maintain a high humidity level. The RPA is stored in this condition by 14 days for a correct cement shell curing (see Figure 9.5).

(a)   (b)

FIGURE 9.5 Pictures of the plastic to be recycled (a) WEEEP and the finished product obtained after the recycling process (b) recycled plastic aggregate.

TABLE 9.1
Steps in RPA Production and its Production Parameters for a Processing of 100 kg of WEEEP

Entry	Step	t (h)	E (MJ)	Operator[a]	t_{global} (h)[b]	Yield (kg/h)	Yield (kg/MJ)
1	WEEEP shredder at 40-50 mm	8	158.4[c]	A	8	13	0.6
2	WEEEP shredder at 3-4 mm	4	53.3[d]	A	12	25	1.9
3	Cover WEEEP with cement shell (1st layer)	1	7.9[e]	B	13	100	12.7
4	Setting cement overnight	16	0[f]	B	13 (+ 1 day)	6	—
5	Sorted RPA with 4.8 mm sieve	1	0[f]	B	14 (+ 1 day)	33	—
6	Cover WEEEP with cement shell (2nd layer)	1	7.9[e]	B	15 (+ 2 days)	100	12.7
7	Setting cement overnight	16	0[f]	B	15 (+ 2 days)	6	—
8	Sorted RPA with 4.8 mm sieve	1	0[f]	B	16 (+ 2 days)	33	—
9	Cover WEEEP with cement shell (3rd layer)	1	7.9[e]	B	17 (+ 2 days)	100	12.7
10	Setting cement overnight	16	0[f]	B	17 (+ 3 days)	6	—
11	Sorted RPA with 4.8 mm sieve	1	0[f]	B	18 (+ 3 days)	33	—
12	Setting cement (14 days)	336	0[f]	B	17 days	—	—

[a] The process is carried out by two people: operator A takes care of milling activities and operator B takes the masonry activities.
[b] Summation of each process time, assuming a step by step linear production.
[c] Shredder machine power = 5.5 kW and a efficient factor = 1.
[d] Shredder machine power = 3.7 kW and a efficient factor = 1.
[e] Mixer machine power = 2.2 kW and a efficient factor = 1.
[f] Non electrical machine is employed in this step.

In Figure 9.6 is presented the layout plan of a pilot scale recycling plant of WEEEP. The total area of 200 m² is employed for the manufacturing process. Is clearly shown the flow of the material, which enters the left side. The first process is shredder WEEEP (at left of the plan) and require around of 50 m², because the high noise produced by the shredder machines this area could be soundproofed. Next, the make shell zone is linearly proposed towards the back of the establishment. The area necessary for this step is the biggest one, about 100 m², because the temporary storage and sieve activities are carried out directly on the floor. Before that final RPA comes out of the recycling pilot plant, this is storage on the floor and saturated with water for 14 days, the area for this step is represented by a 50 m² place inside the barn. However, the curing process is possible to be done outside.

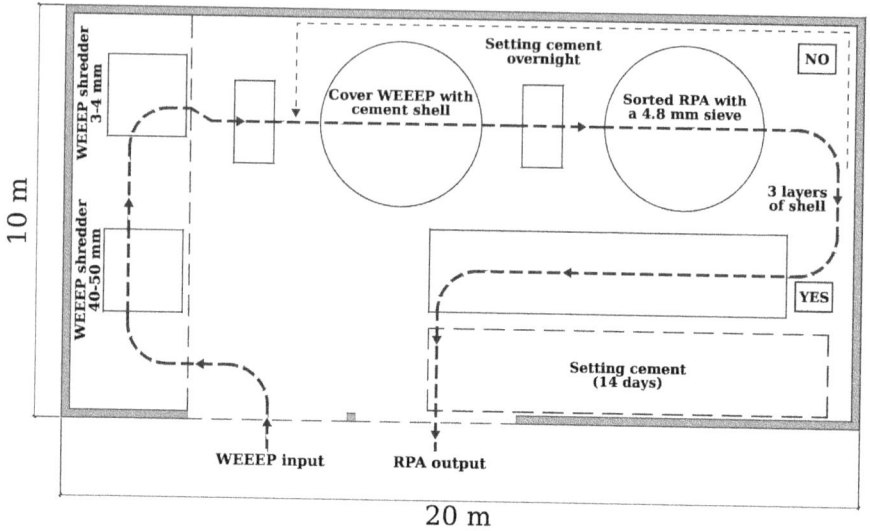

FIGURE 9.6   Space flow diagram of RPA manufacturing procedure (dimensions in m).

9.2.1.5   Labor and Time Considerations

The manufacture of the RPA is carried out in a pilot production plant, in which the capacity for process WEEEP is 100 kg per day. The labor is achieved by two operators, one to do shredder activities and one to do the cover WEEEP and sieve RPA.

Figure 9.7 presents an eight hour work-day diagram. As shown for a typical day, the shredder should be used to reduce the casing of electrical and electronic apparatus. This is the only step which must be done every day to maintain a processing level of 100 kg of WEEEP per day. It is possible to observe which takes eighth days reach the top of production capacity of the pilot plant. Next, a brief analysis of the production times of *batch 1* will be carried out (see Figure 9.7).

The first day of production, 100 kg of casing of WEEEP are reduced to a size of 40-50 mm (step A in Figure 9.7). This labor is carried out by one operator, this is the exclusive activity for this person. Also, the training for this work is the basic for operating the machinery and for safety at work. In the second day, the 40-50 mm WEEEP particles are reduced even more by chipper shredder machine at 3-4 mm particles (step B in Figure 9.7). Because in this operation the operator has to put crushed material in the machine and nothing else, the activity is carried out by the same worker which operates the shear shredder machine. So, a single worker is in charge of carrying out the two shredding tasks.

Once the plastic material is conditioned, on the third day, the stabilization shell for RPA begins. This task is achieved by a worker that must have knowledge of masonry and cement work. For the production of shell for 100 kg of WEEEP the time used is 1 hour, this step is carried out in four little batchs with 25 kg of WEEEP particles that are covered in 15 minutes each time. Plastic particles are wet and next the cementitious mixture is added slowly; when the shell is complete and has the correct consistency,

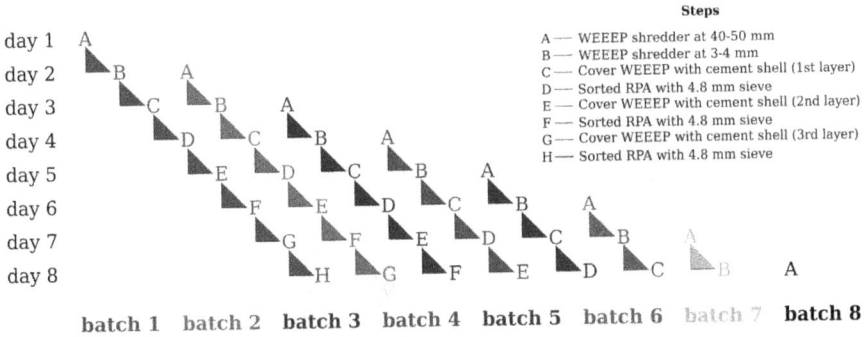

FIGURE 9.7 Timeline of the RPA pilot scale manufacture, complete manufacture of a one batch of RPA take eight days.

the RPA is overturned to a wheelbarrow, transported and deposited in curing place on the floor (step C in Figure 9.7). Next day, the plastic with one layer of shell is manually sieved to desegregate particles; this task is carried out with a shovel and a sieve of 4.8 mm (step D in Figure 9.7). On the fifth day, the RPA is again subjected to the cover process to yield the second layer of cement shell (step E in Figure 9.7). Next day, the plastic with one layer of shell is manually sieve to desegregate particles; this task is carried out with a shovel and a sieve of 4.8 mm (step F in Figure 9.7). The third layer of cement shell is made after seven days of WEEEP process begin (step G in Figure 9.7). On the eighth day the last sort for desegregate RPA particles is achieved and finally the stored RPA in a definitive curing place by 14 days before leaving the pilot production plant (step H in Figure 9.7).

Thus, it is possible to observe that the residence time of **batch 1** (or any batch) in the active production process is eight days. If the production takes place 20 days per month, so this pilot recycling plant is able to process 2000 kg of contaminated plastic from the waste of electrical and electronic equipment.

9.2.1.6 Supplies

For the proposed RPA manufacture technology, three main supplies materials are required:

- plastic from WEEE;
- masonry cement;
- activated carbon.

The first material listed corresponds to a secondary raw material, which means that a recycled material is used avoiding the excessive consumption of natural resources (see Figure 9.5a). The generation of WEEEP occurs mainly in urban settlements like metropolis and small cities, due to household and industrial wastes, however in small towns and rural areas this waste also is produced. This is an advantage, because this material is located in high amounts in relative small areas. This high density of

RPA composition (%w/w)

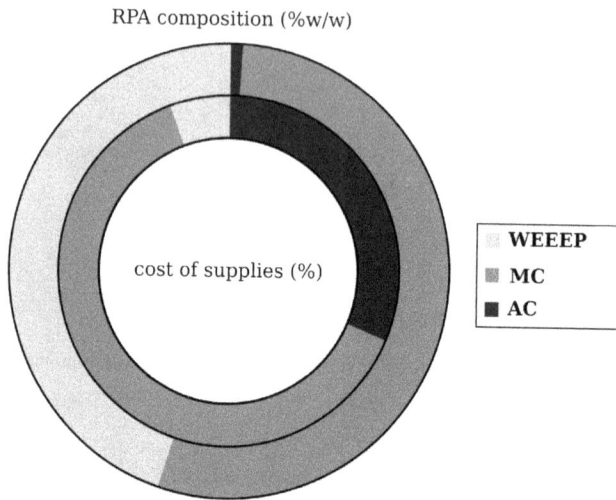

FIGURE 9.8 Proportion of each component needed for the manufacture of the RPA and the relative cost of each supply. The outside ring indicate the percentage by weight (%w/w) of components and the inside ring represent the cost percentage of the each components.

WEEEP impacts directly in the cost of transport of it, which is reduced if the recycled factory is located into the city. As can be observed in

Figure 9.8, this material represents the 45 % of the total mass of RPA. It is important to notice that the RPA is composed almost of a 50 % of recycled plastic that in the majority of cases has no proper final disposal, hence, this fact represents a big step in the direction of a circular manufacturing economy.

The second material employed is masonry cement (MC) type MC 12.5 (BSI 2011). The composition of this MC is at least 40 % by weight of Portland cement clinker, pulverized limestone in relative high percentage (i.e. 50–60 %), a low proportion of gypsum (i.e. 5–6 %) and less than 1 % by weight of organic additives. Even though the use of cement has a negative impact on the environment, due to the high energy demanding process by which it is obtained, the positive is that the proportion of cement used for the obtention of RPAs is the same used in the vast majority of conventional mortars. The use of masonry cement is well founded from the point of view of its workability, not least the fact that it is a material with easy access both logistically and economically.

Lastly, the third component listed as a supply for the manufacturing of RPA is the activated carbon. AC is a fairly common material that can be found in any city, although its cost is high, the proportion in which it is used is negligible (1 %), balancing the cost equation. AC function is very important, because it acts as a stabilizing agent for the organic pollutants present in the WEEEP matrix, and therefore, its employment is vital (Gómez et al. 2020). The physico-chemical properties of AC can vary, depending on the method of activation, place of origin, state of conservation, among others, for this reason we developed a simple and

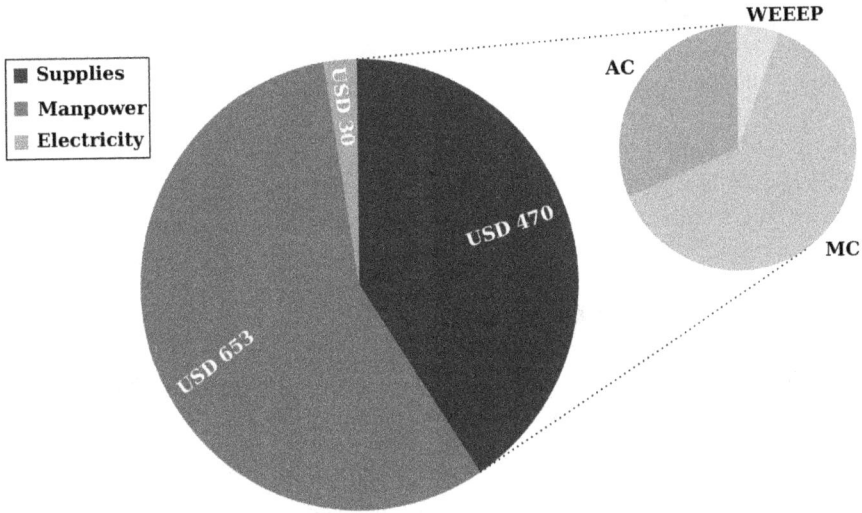

FIGURE 9.9   Cost contributions in the manufacture of the RPA. Indicated values for recycling 2000 kg of WEEEP.

robust study which allows us to predict the performance of different AC (brands, supplier, batches, etc.) based on the Iodine Number (IN) determination of them. In this way, with a simple chemical titration, which does not require expensive glassware or equipment, it can be predicted whether the available CA will fulfill its function correctly with a dosage of 1 %, or if it is necessary to correct the proportion of the additive.

9.2.1.7   Cost of Production

In Figure 9.9 a graphical representation of the main contributors to the cost of manufacture of the core-shell RPA is presented. On the one hand, when analyzing the cost of production of the RPA, it can be depicted that the biggest contributor to the total cost is manpower, which in our criteria it is a positive factor, since an enterprise that adopts this technology would generate a significant number of new jobs.

On the other hand, electricity consumption is almost negligible in comparison with manpower and supplies cost, this is advantageous from two points of view, the most obvious one is the economic point of view because electricity is an expensive resource (at least in Argentina), and from an environmental perspective a limited energy consumption is always welcome.

Considering the second largest contributor to the cost of production of the RPA (supplies), their cost is considerable and unfortunately difficult of been reduced due to the use of relatively low-cost materials, with the exception of activated carbon which is an expensive supply, but it is used in a minuscule proportion (1 %), see Figure 9.8.

It is expected that recycled material based building components are going to be more expensive than analogs produced with conventional materials, because sand went through a natural generation process and the cost associated with its used is mainly the cost of transport of it, instead of WEEE, which has to be transported, dismantled and shredded. However, if the material has an improved insulation behaviour, a smaller amount of building components may be needed, balancing the final costs.

9.2.1.8 Machinery Investment

The necessary hardware to start up the pilot production plant with a capacity of recycling 2000 kg/month of plastic from WEEE was detailed in a previous section. In this section we present the investment necessary in this machinery:

- shear shredder of 5.5 kW (USD 29000)
- chipper shredder of 3.7 kW (USD 4000)
- electromechanical concrete mixer of 0.4 m^3 and 2.2 kW (USD 2000)
- buckets; wide shovels; 4.8 mm sieve (4 mesh) of 4 m^2; garden hose; among others (USD 200)

Looking above, the total investment in necessary hardware for mounting the recycling plant of 100 kg per day of WEEEP is around USD 35000. In the diagram production here proposed, the capacity of the recycling plant is 2000 kg of WEEEP per month. However, if increase shifts to two or three instead of one, the capacity of the plant this same machinery investment could be increased to recycle 4000 and 6000 kg of WEEEP per month respectively.

The global average generation of WEEE per inhabitant in 2016 was 8.1 kg, the lowest value found in Niger (0.4 kg) and maximum was attributed to Norway (28.5 kg) (Mihai et al. 2019). The plastic fraction of WEEE is around the 20–30 %, so the global production per inhabitant of WEEEP cloud be estimated in 1.6–2.4 kg per year (Dimitrakakis et al. 2009; Wäger et al. 2012). Taking this into account, the pilot recycling plant of the WEEE plastic fraction presented here would be able to treat this waste of the a small city (i.e. 15000 people).

From a sustainable perspective, it is important to highlight that a circular economy alternative has been introduced by this proposed technology for WEEE plastics recycling, which nowadays are incinerated or disposal. Probably it will be an economical and social sustainable technology because this low tech approach could be implement both in developed and in developing countries. On the other hand, the RPA technology process proposes the use of MC in which the active component with the higher incorporated energy is Portland cement clinker; said component is approximately 40 %w/w of MC, so this component only represents 25 % of the total weight of recycled plastic aggregate obtained by recycling WEEEP. The other components are materials that have a very low incorporated energy and in the case of WEEE plastics are secondary raw materials; thus granting sustainability from the environmental point of view with respect to other traditional construction materials.

9.2.2 BUILDING COMPONENTS PRODUCTION

Building components (BC) for housing, based on cement, are proposed as a direct application of the RPA developed. The BC here presented were designed and manufactured under Argentine standards, considering several technical aspects as the granulometric curve of the RPA, compression, flexion and tensile strength, to mention a few. CIRSOC 201 regulation Norm have three categories for masonry: ceramic solid brick, ceramic hollow block and concrete hollow block, where compression strength must be greater than 5.0 MPa for the first one, and greater than 13.0 MPa for the remaining two (according to IRAM 12566-1 standard) (Parmigiani 2005).

Fine aggregate granulometry must be less than 3 mm, and the volume ratio of fine aggregate to cement/lime must be around 2.25–3%, according to CIRSOC 201.

For the use of a non-traditional construction material or system to be legal in Argentina, it is necessary to obtain the Certificate of Technical Aptitude (CAT, for its acronym in Spanish), granted by the Ministry of Housing. The future perspective of the development of the RPA is to transfer the technology know-how to local cooperatives or small companies for the application in housing construction, for this reason, all the requirements for the procurement of the CAT certificate. Different building components produced with RPA-containing mortars are discussed in the following sections.

9.2.2.1 Bricks

This component was manufactured as a traditional brick employed in Argentina, its dimensions are 25.0 x 12.5 x 5.5 cm as can be seen in Figure 9.10. Bricks were prepared using a mortar mixture with a volume ratio RPA:cement of 6:1, which the only aggregate employed was RPA. The mean particle size was 3 mm and the apparent density was 770 kg/m^3.

Because of the mortar used has a plastic core, the resulting brick will be lighter and will have better thermal insulation properties than conventional fired clay bricks. RPA bricks have a density of 960 kg/m^3 and a thermal conductivity of 0.332 W/K m, showing good properties as an isolation material. Further studies upon the thermal behavior are being carrying out.

(a)   (b)

FIGURE 9.10 Standard solid bricks made with core-shell RPA. (a) close view of one brick and (b) view of a pallet with an important production of bricks.

FIGURE 9.11   Standard hollow block made with core-shell RPA.

The compressive strength measured for this brick was (4.2 ± 0.4) MPa, an interesting alternative for masonry with non-structural performance requirements.

An example of a wall prepared using RPA bricks can be observed in Figure 9.12d, this image corresponds to the first wall of an prototype of house made using building component prepared exclusively with RPA as aggregate for mortars.

9.2.2.2   Hollow Blocks

The design and morphology of this block was made considering the standard concrete hollow block (Figure 9.11). The dimensions of this block are 19 x 39 x 19 cm, with a three cavities array, according to IRAM 11561-1 norm.

9.2.2.3   Structural Window Frame

Finally, precast structural window frame was developed. In this special case, the aggregate employed for its manufacture was core-shell RPA of two different mean particle size, 3 mm and 8 mm, due to the necessity of a fine and coarse aggregate. RPA was produced using the same methodology describe for RPA, but in this case using bigger shredded particles of WEEEP. The volume proportion used was of 3:1:1 (RPA:RPA:cement).

Besides, the cementitious mixture containing the RPA (3 and 8 mm) as aggregate was reinforced with steel bars (4.2 mm). The width of this window is 40 cm and the height can be variable according to the need, which can be observed in Figure 9.12e.

In addition, this window frame is compatible with a wide variety of construction systems such as traditional brick masonry, concrete block masonry and precast concrete boards.

9.2.2.4   Housing Prototype

A little prototype was built according to an affordable initial housing design widely used by CEVE. The dimensions of this prototype are 3 x 6 m, which represent a little office of 18 square meters (see Figure 9.12a). The dwelling design consists of

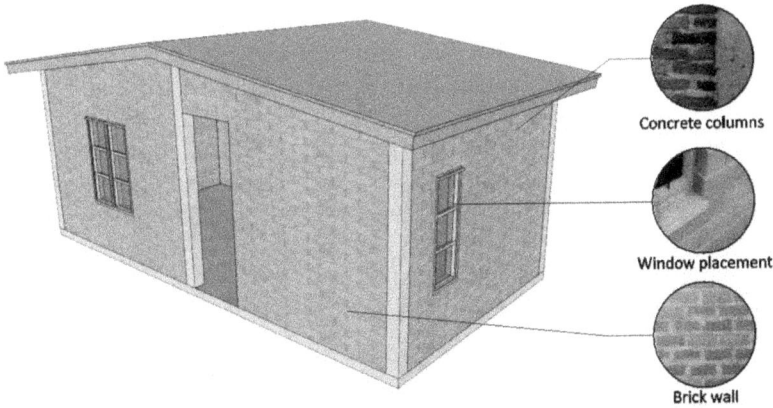

(a) 3D model of the build 18 m² housing prototype.

(b) Facade of the prototype under construction.

(c) Close picture of the prototype.

(d) Building of prototype wall with RPA bricks.

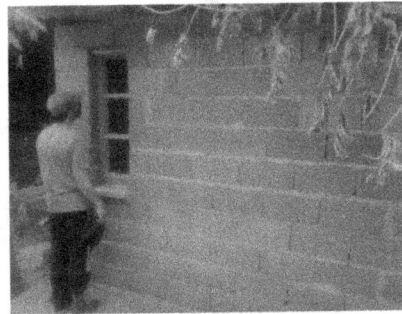

(e) Structural window frame made with core-shell RPA incorporated to wall.

FIGURE 9.12 Pictures of the build housing prototype in CEVE.

a dining space and a bedroom with a small balcony. This design is also used as a productive measure, since it represents the amount of bricks that can be produced with the recycled plastic aggregate that is obtained in a certain period. For this case study, were employed around 2000 bricks identical to the described above and 100 core-shell RPA mortar blocks. To produce this quantity of RPA it is necessary to process 1400 kg of WEEEP by the technological method described in an earlier section. So, to obtain the RPA necessary to build this 18 m² housing prototype require 14 days of production.

Therefore, the design of a production plan can be carried out in accordance with the amount of waste that can be obtained and also the amount of bricks and subsequently housing units that can be built from these. For the pilot scale production presented in this work, will be possible to obtain the aggregate material (RPA) to produce the bricks necessary for build one small house of approximately 25 m² in one month. This is equivalent to a 5.7 m³ of core-shell RPA, with which 3000 bricks can be manufactured.

As seen in pictures of Figure 9.12 of the prototype construction, different traditional masonry techniques were applied in order to evaluate the components performance, consisting in walls construction, concrete columns fill and windows installation. The results of the analysis shown that masonry techniques were positively carried out and no major obstacles were found.

9.3 PRODUCTIVE VENTURE

The first step to project a RPA production venture consists of carrying out a thorough and exhaustive analysis of the legislation that applies on the integral management of the electric and electronic waste in the country where the project is going to be located, since if there is no specific regulation about WEEE management, they are generally treated as special or hazardous waste, and general regulations apply to them along with other waste from specific treatment. This is currently the case in Argentina, where the integral management of WEEE is carried out under Argentine National Laws No. 24,051 and No. 23,992 on hazardous waste treatment (Ley 24051 1992). This regulatory framework requires that the plant where the process of disassembling and sorting the material streams from WEEE takes place performs under strict protocols to establish the traceability of the equipment entering for treatment and the final destination through a certificate of final disposal of materials that could not enter the recycling and marketing circuit, which has been happening with the WEEEP. In Argentina, several cases of informal treatment and outside the WEEE regulations have been detected, with the purpose of marketing platelets, cables and metals. In these plants, the WEEEP is also crushed and commercialization of it has been detected for use as an aggregate to cement (Bengoa 2015). This type of undertaking brings a huge risk for operators and the general population, for which it should not be carried out without due compliance with the regulatory framework, since not only the use of specific hygiene and safety equipment is omitted, also pollutant emission levels are not controlled in the stripping line, not all materials entering and leaving are examined or effluent control is not performed (Zhuang 2019).

As it is really difficult for a municipality or a cooperative of recycling, within its municipal solid waste (MSW) recycling activity, to comply with the requirements of a WEEE recycling plant for comprehensive management under regulations, the best alternative is the association with companies in the sector for the provision of WEEE plastic in the granulometry that is convenient for the recycling company and eventually reduced the particle size to the appropriate granulometry with a grinding mill at the RPA production site.

This dependence on a registered supplier to manage WEEE has been one of the biggest drawbacks to recycling WEEE plastics, due to the fact that it is not possible to influence the heterogeneity of the plastic that is available as raw material, since there is no separation according to the type of equipment discarded and everything is shredded together. A greater separation by type of plastic, supposes a much higher cost that makes the undertaking unsustainable. But this is not the case for the production of the RPA, since the robustness of the process allows incorporating different types of WEEEP, and only the granulometry of the residue has an impact on the properties of the final product.

This kind of limitation when treating the waste inevitably leads us to promote the association and/or entailment between a WEEE treatment plant that complies with the regulatory requirements for waste management with an RPA production entrepreneurship in the area of the work cooperatives. On the other hand, having overcome the inconveniences of WEEE management and the provision of plastic as a secondary raw material for the venture, it is necessary to address the cost of production as a limitation that is difficult to overcome if the problem is not considered in a comprehensive manner. The cost for produce is USD 192 per cubic meter of RPA, which when compared with the cost of one cubic meter of a natural aggregate, which can be obtained in the informal market at USD 25 gives us a very significant difference if RPA is intended to fully replace natural aggregate as an economically sustainable alternative. For this reason, it is necessary to frame the RPA production into a company's MSW program or within a recycling subsidy program, since in economic terms it is otherwise unsustainable.

This type of economic analysis for products that involve the use of a recycled material, are quite similar. In many cases, the final products have a cost significantly higher than traditional products, since it involves a rather expensive step of conditioning and/or production, compared to a non-renewable material or a "commodity". Considering this, it is essential that societies understand the need to propose strategies to promote the production that make ventures sustainable employing, for example:

- Reallocate transport-final disposal budget items directly as a production subsidy
- Consider a percentage within the marketing cost of the product for waste treatment
- Incorporate a percentage amount of recycled materials as a requirement to access public works tenders
- Establish a fund to promote recycling to support these ventures

Any strategy that contributes to economic sustainability will allow, at this stage, a sustained production of a sustainable and safe product from a waste that, in nowadays, does not find a viable recovery destination.

9.4   CONCLUSION

During the present work, it was demonstrated that through a simple but effective tech-nology, it is feasible to recycle a mixture of plastics as complex as WEEE plastics, even if they contain toxic substances such as heavy metals and BFRs. Through the use of a common additive such as activated carbon, which can be found in any country around the globe, it was possible to neutralize the release of contaminants present in the plastic matrix.

The building components developed based on the core-shell RPA have the advan-tage of having good mechanical properties, and also being lighter and with better heat insulation properties than their analogues produced with conventional materials.

From an economic perspective, investment in machinery for the assembly of a small plant of WEEEP recycling is relatively low (approximately USD 35,000) as well as the conditioning of the space for the production of RPA because it does not need special infrastructure requirements. This amount of investment can be carried out without major inconveniences by both governmental entities (e.g. local governments, municipalities, provinces) as well as small and medium-sized companies, cooperatives and NGOs.

The bottleneck of the process, from the temporal point of view, is the reduction in size of the WEEEP wastes by shredding. In fact, the production limit is given by the crushing capacity of the shear mill. The other stages of process of production of RPA, although they represent more labor, they are less time consuming. The produc-tion pilot plant here proposed / presented is capable of recycling 2000 kg of WEEEP per month in operations of 8 hours and 20 working days. However, it is possible to increase production through a simple restructuring of schedules and tasks with the addition of double or triple work shifts.

From an environmental perspective, it is important to highlight that a circular economy alternative is been introduced by the proposed technology for recycling of WEEE plastics, which nowadays are energetically revalued (incineration) in some places, and in most of the world, they are finally disposed by buried them, causing major environmental and health issues. Besides, the RPA technology process proposes the use of MC as an active component and the aggregate with a very low incorporated energy, in this case WEEE plastics are also secondary raw materials.

Concerning the cost of production of RPA based building components and as a personal opinion of the authors, although the materials produced with RPA are a little more expensive than their analogues made with traditional materials, it is a cost that today we must be willing to face, since as users and co-participants in environmental pollution we are all responsible for trying to improve our habitat and not to continue the indiscriminate consumption of the natural resources of our planet.

9.5   APPENDIX 1

List of abbreviations:

- ABS: Acrylonitrile-butadiene-styrene
- AC: Activated charcoal
- BC: Building Components

- BFR: Brominated flame retardant
- CEVE: Centro Experimental de la Vivienda Económica
- HIPS: High impact polystyrene
- IRAM: Instituto Argentino de Normalización y Certificación
- RPA: Recycled plastic aggregate
- TBBPA: Tetrabromobisphenol-A
- WEEE: Waste from Electrical and Electronic Equipment
- WEEEP: Plastic Waste from Electrical and Electronic Equipment
- MSW: Municipal Solid Waste

ACKNOWLEDGMENTS

Authors would like to thank to the Institutions in which the research was carried out. Centro Experimental de la Vivienda Económica (CEVE), dependent of the Consejo Nacional de Investigaciones Científicas y Técnicas (CONICET) and the Asociación Vivienda Económica (AVE). The Centro de Investigación, Desarrollo y Transferencia de Materiales y Calidad (CINTEMAC) belonging to the Universidad Tecnológica Nacional, Facultad Regional Córdoba (UTN-FRC). Also the authors are grateful to the Ministry of Science and Technology of Córdoba for the financial support. M.G., A.L.C. and J.G.L gratefully thank the fellowship granted from CONICET.

REFERENCES

Bágel', L, and P Matiašovský. 2010. "Surface Pretreatment-a Way to Effective Utilization of Waste Plastics as Concrete Aggregate. Review and First Experiences." In *Proceedings: CESB 2010 Prague – Central Europe towards Sustainable Building "From Theory to Practice,"* 1–10. www.scopus.com/inward/record.uri?eid=2-s2.0-84902476014&partnerID=40&md5=effcc2e4d2cf30aabd9a8272b3f2d47e.

Bengoa, Nicolás. 2015. "Se Construirán Adoquines Con Plástico Reciclado En Chajarí." *TalCual Chajarí*, December 16. www.talcualchajari.com.ar/17158/se-construiran-adoquines-con-plastico-reciclado-en-chajari/.

BSI. 2011. "BS:EN:413-1:2011 Masonry Cement Part 1: Composition, Specifications and Conformity Criteria." *BSI Standards Publication*. UK.

Dimitrakakis, E., A. Janz, B. Bilitewski, and E. Gidarakos. 2009. "Small WEEE: Determining Recyclables and Hazardous Substances in Plastics." *Journal of Hazardous Materials* 161 (2-3): 913–19. doi:10.1016/j.jhazmat.2008.04.054.

Farooq, Adil, A. M. Muneeb, T. Tauqeer, R. Mamoon, H. Waqas, M. Awais, and U. R. Mujeeb. 2019. "Impact on Concrete Properties Using E-Plastic Waste Fine Aggregates and Silica Fum." *Mineral Resources Management* 35 (2): 103–18. doi:10.24425/gsm.2019.128516.

Gómez, Melina, Lucas Ernesto Peisino, Jerónimo Kreiker, Rosana Gaggino, Ariel Leonardo Cappelletti, Sandra E Martín, Paula M Uberman, María Positieri, and Bárbara Belén Raggiotti. 2020. "Stabilization of Hazardous Compounds from WEEE Plastic: Development of a Novel Core-Shell Recycled Plastic Aggregate for Use in Building Materials." *Construction and Building Materials* 230 (January): 116977. doi:10.1016/j.conbuildmat.2019.116977.

Hannawi, Kinda, Siham Kamali-Bernard, and William Prince. 2010. "Physical and Mechanical Properties of Mortars Containing PET and PC Waste Aggregates." *Waste Management* 30 (11). Elsevier Ltd: 2312–20. doi:10.1016/j.wasman.2010.03.028.

Ilankoon, I.M.S.K., Yousef Ghorbani, Meng Nan Chong, Gamini Herath, Thandazile Moyo, and Jochen Petersen. 2018. "E-Waste in the International Context – A Review of Trade Flows, Regulations, Hazards, Waste Management Strategies and Technologies for Value Recovery." *Waste Management* 82 (December). Elsevier Ltd: 258–75. doi:10.1016/j.wasman.2018.10.018.

Kiddee, Peeranart, Ravi Naidu, and Ming H. Wong. 2013. "Electronic Waste Management Approaches: An Overview." *Waste Management* 33 (5). Elsevier Ltd: 1237–50. doi:10.1016/j.wasman.2013.01.006.

Kreiker, J., R. Gaggino, L.E. Peisino, and J. Gonzales Laria. 2018. "Residue of Manufacture of Porcelain Stoneware Tiles as Supplementary Material for Cement Mortars." *Journal of Materials and Environmental Sciences* 9 (1): 370–75. doi:10.26872/jmes.2018.9.1.40.

Kurup, Arjun Ramakrishna, and K. Senthil Kumar. 2017. "Novel Fibrous Concrete Mixture Made from Recycled PVC Fibers from Electronic Waste." *Journal of Hazardous, Toxic, and Radioactive Waste* 21 (2): 04016020. doi:10.1061/(ASCE)HZ.2153-5515.0000338.

Lakshmi, R., and S. Nagan. 2010. "Studies on Concrete Containing E Plastic Waste." *International Journal of Environmental Sciences* 1 (3): 270–81. www.ipublishing.co.in/jesvol1no12010/EIJES1026.pdf.

Ley24051. 1992. *Ley N° 24,051: Residuos Peligrosos – Generación, Manipulación, Transporte Y Tratamiento – Normas.* Argentina: Boletín Oficial de la República Argentina.

Mihai, Florin-Constantin, Maria-grazie Gnoni, Christia Meidiana, Chukwunonye Ezeah, and Valerio Elia. 2019. "Waste Electrical and Electronic Equipment (WEEE): Flows, Quantities, and Management—A Global Scenario." In *Electronic Waste Management and Treatment Technology*, 1–34. Elsevier. doi:10.1016/B978-0-12-816190-6.00001-7.

Nowek, Milena. 2016. "Performance of Sand-Lime Products Made with Plastic Waste." *E3S Web of Conferences* 10: 00066. doi:10.1051/e3sconf/20161000066.

Parmigiani, Marta S. 2005. *Reglamento Argentino De Estructuras De Hormigón.* Edited by INTI – CIRSOC. *Reglamento CIRSOC 201.* Julio 2005. Argentina: INTI – CIRSOC. doi:10.1007/s13398-014-0173-7.2.

Peisino, Lucas Ernesto, Melina Gómez, Jerónimo Kreiker, Rosana Gaggino, and Melina Angelelli. 2019. "Metal Leaching Analysis from a Core-Shell WEEE Plastic Synthetic Aggregate." *Sustainable Chemistry and Pharmacy* 12 (June). Elsevier: 100134. doi:10.1016/j.scp.2019.100134.

Pérez-Belis, V., M.D. Bovea, and V. Ibáñez-Forés. 2015. "An in-Depth Literature Review of the Waste Electrical and Electronic Equipment Context: Trends and Evolution." *Waste Management & Research* 33 (1): 3–29. doi:10.1177/0734242X14557382.

Popovici, Antoanela, Ofelia Corbu, Gabriela Emilia Popita, Cristina Rosu, Marian Proorocu, Andrei Victor Sandu, and Mohd Mustafa Al Bakri Abdullah. 2015. "Modern Mortars with Electronic Waste Scraps (glass and Plastic)." *Materiale Plastice* 52 (4): 588–92.

Senthil Kumar, K., and K. Baskar. 2015a. "Briefing: Shear Strength of Concrete with E-Waste Plastic." *Proceedings of the Institution of Civil Engineers – Construction Materials* 168 (2): 53–56. doi:10.1680/coma.14.00044.

Senthil Kumar, K., and K. Baskar. 2015b. "Development of Ecofriendly Concrete Incorporating Recycled High-Impact Polystyrene from Hazardous Electronic Waste." *Journal of Hazardous, Toxic, and Radioactive Waste* 19 (3): 04014042. doi:10.1061/(ASCE) HZ.2153-5515.0000265.

Kumar, Kaliyavaradhan Senthil, and Kaliyamoorthy Baskar. 2015c. "Recycling of E-Plastic Waste as a Construction Material in Developing Countries." *Journal of Material Cycles and Waste Management* 17 (4): 718–24. doi:10.1007/s10163-014-0303-5.

Senthil Kumar, K., R. Gandhimathi, and K. Baskar. 2016. "Assessment of Heavy Metals in Leachate of Concrete Made With E-Waste Plastic." *Advances in Civil Engineering Materials* 5 (1): 20160003. doi:10.1520/ACEM20160003.

Signoret, Charles, Marie Edo, Dominique Lafon, Anne-Sophie Caro-Bretelle, José-Marie Lopez-Cuesta, Patrick Ienny, and Didier Perrin. 2020. "Degradation of Styrenic Plastics During Recycling: Impact of Reprocessing Photodegraded Material on Aspect and Mechanical Properties." *Journal of Polymers and the Environment*, May. Springer US. doi:10.1007/s10924-020-01741-8.

USEPA. 1992. "Method 1311: Toxicity Characteristic Leaching Procedure, SW 846 Test Methods for Evaluation of Solid Wastes: Physical/Chemical Methods," no. July 1992: 1–35.

Wäger, Patrick A., Mathias Schluep, Esther Müller, and Rolf Gloor. 2012. "RoHS Regulated Substances in Mixed Plastics from Waste Electrical and Electronic Equipment." *Environmental Science & Technology* 46 (2): 628–35. doi:10.1021/es202518n.

Wang, Ru, and Christian Meyer. 2012. "Performance of Cement Mortar Made with Recycled High Impact Polystyrene." *Cement and Concrete Composites* 34 (9). Elsevier Ltd: 975–81. doi:10.1016/j.cemconcomp.2012.06.014.

Zhuang, Xuning. 2019. "Chemical Hazards Associated With Treatment of Waste Electrical and Electronic Equipment." In *Electronic Waste Management and Treatment Technology*, 311–34. Elsevier. doi:10.1016/B978-0-12-816190-6.00014-5.

10 Current Practices and Development of LCA Application in E-Waste Management Systems

Haikal Ismail and Marlia M. Hanafiah

CONTENTS

DOI: 10.1201/9781003095972-14

10.1 INTRODUCTION

The research on the waste of electrical and electronic equipment (WEEE or e-waste) has received exceptional interest from various stakeholders all over the world due to its volume significantly increasing over the years. According to recent UNU study (Forti et al. 2020), in 2019, the global e-waste generation was about 53.6 million tonnes (Mt) (7.3 kg per capita), and it is expected to increase up to 74.7 Mt or 9.0 kg per capita in 2030 (Figure 10.1 (a)). By region, it was recorded that the greatest amount of e-waste was generated from Asia with a total of 24.9 Mt or 47% from the total global e-waste generation in 2019, followed by Americas (13.1 Mt or 25%), Europe (12 Mt or 22%), Africa (2.9 Mt or 5%), and Oceania (0.7 Mt or 1%) (Figure 10.1 (b)). Furthermore, from six categories, the global e-waste generation from small equipment was topping the list with a total of 17.4 Mt (or 32%), whereas the least amount was lamps (or lighting equipment) with a total of 0.9 Mt (or 2%) (Figure 10.1 (c)).

With the quantity of global e-waste generation apparently increasing year by year, progress on sustainable e-waste management becoming indispensable (Gao et al. 2019). As a result, various studies on e-waste management were performed worldwide. The recent bibliometric studies found that the previous studies related to e-waste in past years was growing exponentially (Andrade et al. 2019; Zhang et al. 2019). According to Ismail and Hanafiah (2020), study on e-waste management system and practice that includes regulations and policies at international and in particular countries were among the primary research topics in recent years. The investigation on the current and future e-waste generation was another main important topic in this research area. Furthermore, several studies performed an evaluation of the chemical and physical composition of e-waste, and evaluation of various consumer aspects such as the e-waste disposal preferences and behaviour, and the knowledge and awareness regarding to improper e-waste disposal.

Generally, the waste management hierarchy that ranked variety of options from the most to the least preferred waste management options is applied as a guideline for decision-making process in waste management (Ewijk and Stegemann 2016; Andersson and Stage 2018; Ismail and Hanafiah 2017). In the literature, various waste management strategies were outlined within the waste management hierarchy, but generally, it consists of five common management strategies with the prevention was listed up as the top preferable waste management strategy, consecutively by re-use, recycling, recovery (including energy), and the least preferred option was land-fill (Figure 10.2), as developed by European Union for developing appropriate waste management policy and for decision making (Andersson and Stage 2018).

Compared to other form of waste, managing the e-waste is complex, as e-waste contains various types of valuable materials alongside the hazardous substances (Herat and Agamuthu 2012; Robinson 2009). Thus, the investigation of the complex management of e-waste required a suitable assessment tool and methodology.

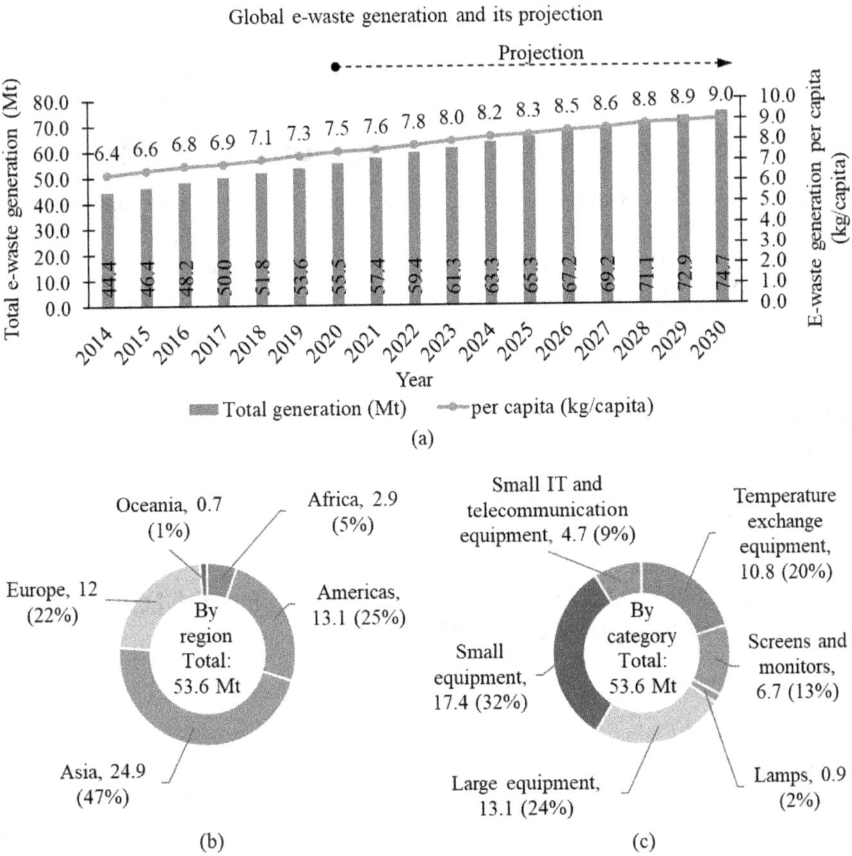

Global e-waste generation and its projection

FIGURE 10.1   (a) Global e-waste generation and its projection, (b) e-waste composition by region (in 2019), (c) e-waste composition by category (in 2019).

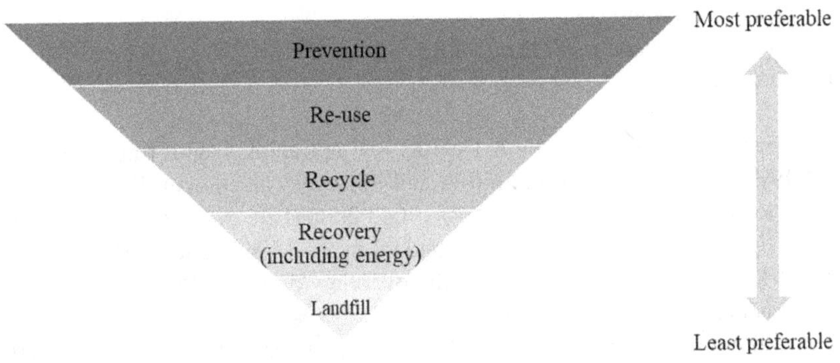

FIGURE 10.2   Waste management hierarchy concept.

The evaluation of the e-waste management performance can be conducted by using various assessment methods, *i.e.* material flow analysis (MFA), life cycle assessment (LCA), and environmental risk assessment (ERA) (Ismail and Hanafiah 2020).

10.2   AIMS AND MOTIVATION OF THE STUDY

The integration of LCA in e-waste management was explored in this chapter to give the overview on the recent progress of the topic discussed in this chapter. The selected LCA studies were mainly from the previous review study conducted by Ismail and Hanafiah (2019a). A brief introduction of LCA and the status of LCA application in e-waste management were provided in the following section, followed by an overview on LCA application in e-waste management. In this section, six aspects of selected LCA studies were analysed: (1) waste assessed, (2) system assessed, (3) modelling tool and software, (4) functional unit, (5) LCIA method, impact or damage categories, and (6) integration of LCA with other methods. Lastly, the summary of the existing LCA studies and recent update on LCA application in e-waste management were provided, followed by some future recommendations.

10.3   THE APPLICATION AND RECENT PROGRESS OF LCA IN E-WASTE MANAGEMENT

10.3.1   A BRIEF INTRODUCTION OF LCA

The internationally standardized LCA methodology was developed for assessing the environmental burdens of products (and services) from the acquisition of raw materials, to the production and product consumption, and until its final disposal (Harun et al. 2020; Aziz and Hanafiah 2020; Aziz et al. 2019; Ismail and Hanafiah 2019b). According to the International Organisation for Standardization (ISO) in the 14040 series (*i.e.,* ISO 14041:2006, ISO 14044:2006), the systemized LCA framework includes four steps (ISO 2006a; 2006b):

  i)   goal and scope definition,
  ii) life cycle inventory,
  iii) life cycle impact assessment, and
  iv) interpretation

The previous version of ISO 14040:1997 (Principles and framework), ISO 14041:1998 (Goal and scope definition and life cycle inventory), ISO 14042:2000 (Life cycle impact assessment), and ISO 14043:2000 (Life cycle interpretation) were replaced by these recent international standard in the 14040 series (ISO 1997; 1998; 2000a; 2000b). Figure 10.3 illustrated the overview on the LCA framework and its main elements.

The definition of goal and scope is the first step, where the goal of LCA states the intended application, the reason to perform an LCA study, and its intended audiences, whereas the scope of LCA states the breadth, depth and details of the study by using the following items or elements as the guidelines: the product system and its function,

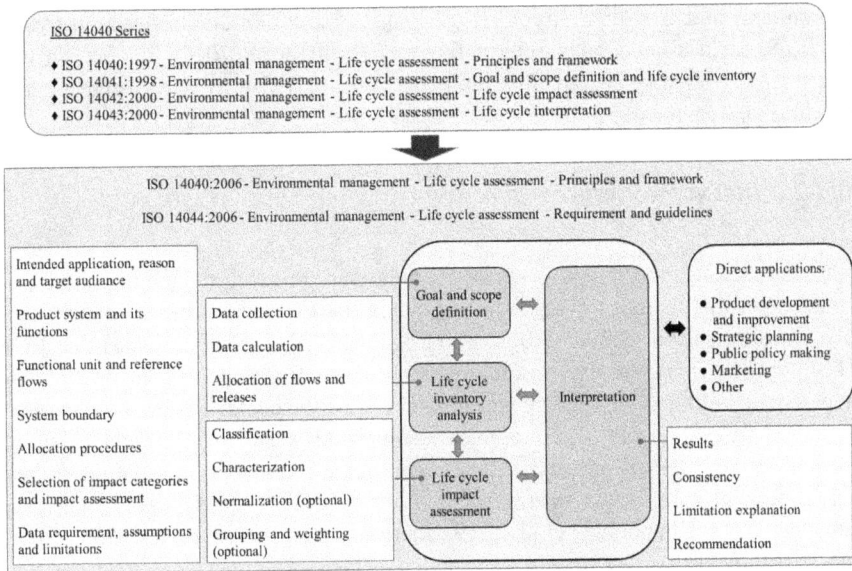

ISO 14040 Series

♦ ISO 14040:1997 - Environmental management - Life cycle assessment - Principles and framework
♦ ISO 14041:1998 - Environmental management - Life cycle assessment - Goal and scope definition and life cycle inventory
♦ ISO 14042:2000 - Environmental management - Life cycle assessment - Life cycle impact assessment
♦ ISO 14043:2000 - Environmental management - Life cycle assessment - Life cycle interpretation

ISO 14040:2006 - Environmental management - Life cycle assessment - Principles and framework

ISO 14044:2006 - Environmental management - Life cycle assessment - Requirement and guidelines

Intended application, reason and target audiance		Direct applications:
Product system and its functions	Data collection	Goal and scope definition
	Data calculation	• Product development and improvement
Functional unit and reference flows	Allocation of flows and releases	Life cycle inventory analysis
System boundary	Classification	
Allocation procedures	Characterization	Results
Selection of impact categories and impact assessment	Normalization (optional)	Life cycle impact assessment
Data requirement, assumptions and limitations	Grouping and weighting (optional)	Recommendation

FIGURE 10.3 Overview on LCA framework and its main elements.

the system boundary, the functional unit, the allocation procedures, the selection of method of impact assessment, and the data requirement, assumptions and limitation. The life cycle inventory (LCI) is the second step, where the data are collected for input and output processes of the product system. Then, calculation and validation of data are conducted, alongside the allocation of flow and releases. The life cycle impact assessment step translates the LCI results into understandable environmental impact of the product according to selected impact and/or damage categories. This phase involves with the selection of impact categories, category indicator and characterization model, before the classification (the allocation of LCI results) and the characterization (the assessment of category indicator results) are performed as mandatory elements, and it may have followed by grouping, normalization and weighting as optional elements in performing LCA study. The final phase in LCA is the interpretation. In this phase, it involves not only the presentation of the analysis results, but also includes the explanation of the key issues and limitations that were identified throughout the study, before the conclusion and recommendations for the study are made.

The information on the historical and development of LCA framework are available in various studies in the literature (Pryshlakivsky and Searcy 2013; Guinée et al. 2011; Rebitzer et al. 2004). In addition, a comprehensive description on theoretical and methodology guidelines for LCA based on the most recent international standards were published in many studies (Hauschild et al. 2018; Klopffer and Grahl 2014; Finnveden et al. 2009; Pennington et al. 2004). Furthermore, various guidelines and supporting documents regarding to the LCA, known as the International Reference Life Cycle Data System (ILCD) Handbook have been published by the Institute for

Environment and Sustainability for European Commission Joint Research Centre
(EC-JRC) are available for further references (Wolf et al. 2012; EC-JRC 2010a)
alongside a guideline specifically on integration of LCA in waste management
(Manfredi and Pant 2011).

10.3.2 Recent Progress of LCA Application in Solid Waste and E-Waste Management

By reviewing over 150 previous studies in assessing the performance of solid waste
management that includes e-waste management, the LCA was utilized in more than
40% of the reviewed studies (Allesch and Brunner 2014). This indicates that LCA
is becoming an indispensable tool in assessing the performance of solid waste management, including e-waste management. Previously, several review studies were
conducted to evaluate the LCA application in solid waste management (Laurent,
Clavreul, et al. 2014). Apart from the existing review studies on the LCA application in general solid waste management (Laurent, Bakas, et al. 2014; Yadav and
Samadder 2018; Khandelwal et al. 2019), there was also several review studies have
been performed to evaluate existing LCA studies in e-waste management (Ismail and
Hanafiah 2019a; Xue and Xu 2017; Rodriguez-Garcia and Weil 2016).

10.4 OVERVIEW ON APPLICATION OF LCA IN E-WASTE MANAGEMENT

10.4.1 Waste Assessed

Various e-waste types were evaluated in existing studies consists of three different
types of e-waste – product, component and residue (or mixture or other fraction)
(Ismail and Hanafiah 2019a). Based on the objective and scope, the existing studies
evaluated single type of e-waste or multiple types of e-waste across these three
different types of e-waste.

At product level, some studies evaluated single type or multiple types of e-waste
category. For an example, Unger et al. (2017) evaluated single type of e-waste categories (*i.e.,* small WEEE), whereas Biganzoli et al. (2015) evaluated multiple types of
e-waste categories, namely refrigerators and heaters (R1), large household appliances
(R2), monitors and TV (R3), small household appliances (R4), and lighting accessories and equipment (R5). Apart from that, some studies evaluated single type or
multiple types of e-waste product. For an example, Foelster et al. (2016) evaluated
single type of e-waste product (*i.e.,* refrigerator), whereas Menikpura et al. (2014)
evaluated multiple e-waste product types, namely refrigerator, washing machine
(WM), TV, and air-conditioner (AC). Unlike to these studies, some studies evaluated
single e-waste type from a mixture of various e-waste categories or various e-waste
products (mixed WEEE). For an example, Hischier et al. (2005) evaluated a mixture
of various e-waste categories, whereas Song et al. (2013b) evaluated a mixture of
various e-waste products.

Furthermore, some studies evaluated single type of e-waste component. For an
example, Xue et al. (2015) evaluated single type of e-waste component (*i.e.,* printed

wired board (PWB). Note that, none of the previous studies evaluated multiple types of e-waste component. In addition, Lu et al. (2017) evaluated the multiple types of e-waste at product- and component-level (*i.e.,* power supply unit (PSU) and refrigerator). Finally, some studies evaluated single type or multiple types of residue from e-waste (or mixture or other fraction from e-waste). For an example, Rocchetti et al. (2013) evaluated multiple types of residue from e-waste (*i.e.,* residues from fluorescent lamp, CRT, Li-ion accumulator, and PWB). In another study, Bigum et al. (2012) evaluated single type of e-waste mixture (*i.e.,* high-grade e-waste mixture). In addition, Alston and Arnold (2011) evaluated single type of other fraction from e-waste (*i.e.,* plastic from e-waste or WEEE plastic).

10.4.2 SYSTEM ASSESSED

Based on the objective and scope, the LCA was used to examine specific e-waste management option (single system), or various e-waste management options (multiple systems). For an example, Biswas and Rosano (2011) evaluated specific e-waste management option for compressor (*i.e.,* re-use by remanufacture). Similarly, Song et al. (2013b) evaluated specific e-waste management option for mixture of various e-waste product (mixed WEEE) (*i.e.,* recycling and recovery). In another study, Zink et al. (2014) evaluated various e-waste management options for smartphone (*i.e.,* reuse by refurbish and repurpose). Similarly, Rocchetti and Beolchini (2014) evaluated various management options for CRT monitor that includes the hazardous waste landfill, component recycling for production of new CRT, material recovery for production of new flat panel display (FPD), and material recovery with extended treatment for production of new FPD.

10.4.3 MODELLING TOOL AND SOFTWARE

Various types of modelling tools and software are available from various providers and developers to aid in performing LCA study in e-waste management. As presented in Table 10.1, these tools and software can be categorized as generic LCA software and dedicated waste LCA model or software. The generic LCA software such as SimaPro and GaBi was commonly used in the existing studies. While various existing dedicated waste LCA model or software such as EASEWASTE (Denmark), OWARE (Sweden), and LCA-IWM (the United Kingdom) were reviewed by Gentil et al. (2010), these model or software were less frequently used in the existing studies. Only Bigum et al. (2012) mentioned the used of EASEWASTE model for assessing the e-waste management in their study. Other than these tools and software, Menikpura et al. (2014) employed Microsoft Excel spreadsheet in their LCA studies.

10.4.4 FUNCTIONAL UNIT

Functional unit (F.U.) for examining the environmental impact of e-waste management using LCA was formulated differently by different study. As a result, it was classified into various categories with different meaning in previous review studies. In the review study by Xue and Xu (2017), the F.U. formulated for LCA application

TABLE 10.1
Various LCA Modelling Tools and Software

Software/Model	Provider/ Developer	Source	Remarks
SimaPro **(System for Integrated Environmental Assessment of Products)**	PRé Sustainability	https://simapro.com/	Generic LCA software
GaBi **(Ganzheitliche Bilanzierung)**	Sphere Solutions (formerly by PE International and thinkstep)	www.gabi-software.com/international/index/	Generic LCA software
Umberto	Ifu Hamburg	www.ifu.com/en/umberto/lca-software/	Generic LCA software
OpenLCA	Green Delta	www.openlca.org/greendelta/	Generic LCA software
CMLCA ©	Leiden University	www.cmlca.eu/	Generic LCA software
TEAM ™ **(Tool for Environmental Analysis and Management)**	Price Waterhouse and Cooper (PwC)	www.pwc.fr/dd	Generic LCA software
EASEWASTE **(Environmental Assessment of Solid Waste Systems and Technologies)**	Technical University of Denmark	Kirkeby et al. (2006)	Dedicated waste LCA model or software
WRATE **(Waste and Resources Assessment Tool for the Environment)**	Golder Associates and Environmental Resources Management (ERM)	www.wrate.co.uk/	Dedicated waste LCA model or software

in e-waste management was classified into quality-based F.U., quantity-based F.U., collection and treatment-based F.U., production-based F.U., and "Not described" for those LCA studies did not mention any F.U. in their studies. In another review study, Rodriguez-Garcia and Weil (2016) classified the F.U. into mass-based F.U., unit-based F.U., time and space-based F.U., operation-based F.U., and "Not stated" for those LCA studies did not specify any F.U. in their studies.

Compared to these review studies, however, the present study classified the F.U. formulated in previous LCA studies in e-waste management into "input-oriented F.U." and "output-oriented F.U.", where the former described the F.U. formulated based on the total amount of e-waste entering the product system and the latter described the F.U. formulated based on the amount of recovered materials leaving the product system. For an example, the F.U. formulated in the study by Song et al.

(2013b) (*i.e.,* 1 ton of WEEE) can be described as "input-oriented F.U.", and that formulated in the study by Rubin et al. (2014) (*i.e.,* 102g of copper recovery) can be described as "output-oriented F.U.". In addition, the "input-oriented F.U." was described either in mass of e-waste such as in the study by Song et al. (2013b) (as aforementioned), or in unit of e-waste Wang et al. (2014) (*i.e.,* 1 unit of LCD panel). The "input-oriented F.U." and the "output-oriented F.U." were formulated based on how the function of system under study was defined – either as the system for waste treatment or the system for production of secondary materials.

10.4.5   LCIA Methodology, Impact and/or Damage Categories Selected

Various LCIA methodologies that consists of various impact and/or damage categories have been developed. Generally, the LCIA methodologies can be classified into two major groups: problem-oriented approach (or midpoint) (*e.g.,* CML 2002) and damage-based approach (or endpoint) (*e.g.,* Eco-indicator 99). While earlier LCIA methodologies such as CML 2002 and Eco-indicator 99 were developed either based on problem-oriented approach or damage-oriented approach, several LCIA methodologies that combined both problem and damage-oriented approaches have been developed in recent years (*e.g.,* IMPACT 2002+, ReCiPe 2016). More information regarding to LCIA methodologies can be found in various studies elsewhere (Hauschild and Huijbregts 2015; EC-JRC 2010b).

The selection of impact and/or damage categories from various types of LCIA methodologies depend upon the objective and the scope of the existing studies. Some studies employed a single or multiple impact and/or damage categories from single LCIA methodology. For an example, Biswas and Rosano (2011) selected a single environmental impact category related to global warming potential (GWP)) by employing IPCC methodology. On the other hand, Xu et al. (2013) selected various impact categories by employing CML 2002 methodology, whereas Andreola et al. (2007) selected various damage categories by employing Eco-indicator 99. In another study, Compagno et al. (2014) and Wäger and Hischier (2015) have selected various impact and damage categories by employing IMPACT 2002+ methodology and ReCiPe 2016, respectively. Note that, some studies have selected various impact and damage categories from various LCIA methodologies. For an example, Alston and Arnold (2011) selected various impact and damage categories by employing CML 2002, Eco-indicator 99 and Eco-Scarcity 2006 methodologies.

10.4.6   Integration of LCA with Other Methods

By taking other aspects such as economic and social into consideration, some studies combined other methodologies and tools in their LCA studies in assessing the overall performance of the system under study. As presented in Table 10.2, the LCA studies were combined with technical cost modelling (TCM) (Xu et al. 2013), life cycle cost (LCC) (Mayers et al. 2005), and cost-benefit analysis (CBA) (Lu et al. 2006; Foelster et al. 2016) to include the economic aspect in evaluation of the system under study. Apart from that, Lu et al. (2017) employed sustainability LCA by combining the social (S-LCA) and the cost (LCC) aspects, alongside the environmental LCA study, whereas Song et al. (2013a) developed emergy-LCA index by combining the emergy analysis and LCA study in their analysis.

10.4.7 Summary on Recent Progress of LCA Application in E-waste Management

As described in a previous section, the previous studies assessed either a specific management option (single system), or various management options (multiple systems). Further analysis was performed by mapping various management options that were considered in the existing studies according to waste management hierarchy. As presented in Table 10.2, it was found that some studies evaluated a single system at specific waste management hierarchy order. For an example, the study by Biswas and Rosano (2011) evaluated a single management option for compressor (*i.e.,* re-use by remanufacture).

Similarly, Xue et al. (2015) evaluated a single management option for PWB (*i.e.,* recycling and recovery). On the other hand, some studies evaluated multiple systems at specific waste management hierarchy order. For an example, the study by Biswas et al. (2013) evaluated multiple management options for compressor (*i.e.,* re-use by remanufacture and repair). Furthermore, Zink et al. (2014) evaluated multiple management options for smartphone (*i.e.,* re-use by refurbish and repurpose). Unlike to these studies, some studies evaluated multiple systems across different waste management hierarchy order. For an example, the study by Lu et al. (2014) evaluated multiple management options across different waste management hierarchy order for mobile phone (*i.e.,* re-use of component, and recovery of materials). Similarly, Wäger et al. (2011) evaluated multiple management options across different waste management hierarchy order for mixture of various e-waste (or mixed WEEE) (*i.e.,* recycling and recovery, incineration with energy recovery, and disposal at landfill site). The main objectives of these studies are described in the following paragraphs.

10.4.7.1 Evaluation of Various Management and Disposal Options for E-Waste

One of the main objectives in previous LCA studies was to determine the best management option for e-waste. As described in a previous section, the existing studies have compared various management options, either at specific waste management hierarchy order (*e.g.,* (Biswas et al. 2013; Zink et al. 2014)), or across different waste management hierarchy order (*e.g.,* Lu et al. 2014; Wäger et al. 2011)).

Apart from that, some studies only evaluated a specific (or single) management option at specific waste management hierarchy (*e.g.,* (Biswas and Rosano 2011; Xue et al. 2015)). While some studies focused on comparing various e-waste management and disposal system, there were also studies only focused on comparison of various e-waste disposal options (*i.e.,* recycling, recovery, and landfill). For an example, Rocchetti and Beolchini (2014) performed LCA studies to compare the performance of various CRT monitor management and disposal options that include the hazardous waste landfill, the component recycling (or glass-to-glass recycling) for new CRT

TABLE 10.2
Summary of LCA Application in E-waste Management

References	E-waste assessed	System assessed					Other aspect/tool
		A	B	C	D	E	
Zink et al. (2014)	Smartphone	●					
Biswas and Rosano (2011)	Compressor	○					
Biswas et al. (2013)	Compressor	●					
Lu et al. (2017)	Power supply unit	●	●				
Lu et al. (2017)	Refrigerator	●	●				
Lu et al. (2014)	Mobile phone	●	●				Sustainability-LCA
Amato, Rocchetti, and Beolchini (2017)	LCD monitor			●	●	●	
Rocchetti and Beolchini (2014)	CRT monitor	●	●			●	
Xu et al. (2013)	CRT funnel glass	●	●			●	Economic (TCM)
Boyden et al. (2016)	Battery			●		●	
Niu et al. (2012)	CRT display			●	●		
Dodbiba et al. (2008)	WEEE plastic			●	●		
Hischier et al. (2005)	Mixed WEEE			●	●		
Wäger et al. (2011)	Mixed WEEE			●	●	●	
Eygen et al. (2016)	Laptop, Desktop			●		●	
Tran et al. (2018)	Battery			●	●		
Wäger and Hischier (2015)	WEEE plastic			●	●		
Alston and Arnold (2011)	WEEE plastic				●	●	
Apisitpuvakul et al. (2008)	Fluorescent lamp			●		●	
Alcántara-Concepción et al. (2016)	Computer			●		●	
Mayers et al. (2005)	Printer			●		●	Economic (LCC)
Lu et al. (2006)	Notebook		●	●	●	●	Economic (CBA)
Song et al. (2013b)	Mixed WEEE			○			
Song et al. (2013a)	Mixed WEEE			○			Emergy-LCA index
Menikpura et al. (2014)	TV, WM, AC, RF			○			
Biganzoli et al. (2015)	R1, R2, R3, R4, R5			○			
Noon et al. (2011)	CRT, LCD monitor			○			
Xue et al. (2015)	PWB			○			
Iannicelli-Zubiani et al. (2017)	PWB			○			
Abeliotis et al. (2017)	PWB			○			
Andreola et al. (2007)	CRT glass			○			
Bigum et al. (2012)	High-grade mixture			○			
Rocchetti et al. (2013)	Residues (FL, CRT, Li-ion, PWB)			○			

(continued)

TABLE 10.2 (Continued)
Summary of LCA Application in E-waste Management

References	E-waste assessed	A	B	C	D	E	Other aspect/tool
Campolina et al. (2017)	WEEE plastic			○			
Belboom et al. (2011)	Refrigerator and freezer			●			
Hong et al. (2015)	Mixed WEEE			●			
Foelster et al. (2016)	Refrigerator			●			Economic (CBA)
Bientinesi and Petarca (2009)	WEEE plastic			●			
Soo and Doolan (2014)	PWB			●			
Moraes et al. (2014)	Mobile phone			●			
Johansson and Björklund (2010)	Dishwasher			●			
Compagno et al. (2014)	CRT monitor			●			
Song et al. (2018)	CRT TV			●			
Ghodrat et al. (2017)	PWB			●			
Bian et al. (2016)	Mobile phone			●			
Arduin et al. (2017)	Tablet			●			
Dodbiba et al. (2012)	LCD monitor			●			
Wang et al. (2014)	LCD panel			●			
Rubin et al. (2014)	PWB			●			
Villares et al. (2016)	PWB			●			
Solé et al. (2012)	Electronic toys			○			
Unger et al. (2017)	Small WEEE			●			
Arduin et al. (2016)	Mobile phone charger			●			
Yao et al. (2018)	Mobile phone			●			
Barba-Gutiérrez et al. (2008)	TV, WM, RF, PC			○			
Xiao et al. (2016)	Refrigerator			○			
Gamberini et al. (2010)	Unspecified			○			

A: Prevention, B: Re-use, C: Recycling, D: Recovery/ Incineration, E: Landfill, ○: Single system, ●: Multiple systems

production, the recovery of materials for new flat panel display (FPD) production, and the recovery of materials with extended treatment (for further materials recovery) for new FPD production. On the contrary, Boyden et al. (2016) only focused on various disposal options by comparing the performance of landfill, pyro-metallurgy and hydro-metallurgy processes for lithium battery.

10.4.7.2 Evaluation of Burden or Benefit of Recycling and Recovery and Management Hotspots

Apart from the evaluation of different management and disposal options for e-waste, several LCA studies were conducted to evaluate the whole management chain for e-waste recycling and recovery, with the main objective was to investigate whether the e-waste recycling and recovery provide environmental benefit or burden. For instance, Menikpura et al. (2014) examined the performance of e-waste recycling and recovery system in Japan. Similarly, Biganzoli et al. (2015) and Song et al. (2013b) assessed the performance of e-waste recycling and recovery system in Italy and in the China, respectively.

Generally, the e-waste recycling and recovery contains two main sub-systems, namely the collection and transportation system, and the treatment and disposal system. Therefore, by examining the performance of e-waste recycling and recovery, some studies evaluated further the management chain in e-waste recycling and recovery to identify the hotspot in the management chain. However, the scope of these studies depends upon the methodological choices (*i.e.,* system boundary) and the availability of technical system, including data. For an example, compared to Menikpura et al. (2014) and Biganzoli et al. (2015), Song et al. (2013b) excluded the evaluation of management chain in e-waste recycling and recovery in their study. This is because the technical system of the collection and transportation system was not available in China. Furthermore, some studies only focusing on recycling and recovery for component or residue (including mixture and other fraction), where the prior activities related to collection and transportation system were excluded from their study such as the study by Rocchetti et al. (2013) in evaluation of recycling and recovery for various residues from e-waste.

10.4.7.3   Evaluation of Different Technologies in Recycling and Recovery

The evaluation on the different technologies in recycling and recovery for e-waste was another main objective of the LCA application in e-waste management. The performance of formal and informal e-waste recycling and recovery were compared in previous studies (Belboom et al. 2011; Hong et al. 2015; Foelster et al. 2016), there were also some studies compared the performance of e-waste recycling and recovery in different geographical areas (*i.e.,* country). For an example, Bientinesi and Petarca (2009) conducted LCA study to investigate the performance of different technologies in recycling and recovery for plastic from e-waste (WEEE plastic) in the Netherlands and Germany. Similarly, Soo and Doolan (2014) performed LCA study to compare the performance of printed wired board (PWB) recycling and recovery in Malaysia and in Australia. Furthermore, Bian et al. (2016) compared the performance of e-waste recycling and recovery that represent the technological capabilities in developing and developed countries, including the integrated system based on these countries. Apart from that, Arduin et al. (2017) compared the performance of e-waste recycling and recovery with different capabilities – optimistic recycling (using the best available technology), conservative recycling (using the best referenced recycling channel), and pessimistic recycling (worst recycling scenario).

10.4.7.4   Evaluation of Different Approaches and Methods in Recycling and Recovery

Some studies concerned the environmental impact of different approaches and methods in e-waste recycling and recovery. For an example, Moraes et al. (2014) performed LCA study to evaluate the current practice of mobile phone in Brazil, in which the mobile phone in Brazil was recycled partially and the rest was sent to European countries) to the proposed fully mobile phone recycling and recovery approach in Brazil. Furthermore, Johansson and Björklund (2010) compared the proposed recycling approach for dishwasher that include the 'pre-step' at pre-treatment process to the current recycling and recovery practice, whereas Compagno et al. (2014) compared the proposed recycling and recovery approach with extended treatment for further materials recovery to current recycling and recovery practice for CRT TV.

Apart from the evaluation of the different approaches in recycling and recovery for e-waste, some studies focused on the assessment of different approaches and methods for e-waste recycling and recovery. While Dodbiba et al. (2012) compared two liberation methods for e-waste recycling and recovery of LCD monitor, namely conventional grinding and electrical disintegration, the study by Wang et al. (2014) compared the performance of the distillation method to the supercritical method for liquid crystal recovery. In another study, Rubin et al. (2014) and Villares et al. (2016) compared the performance for copper recovery from PWB using different solutions (*i.e.*, aqua regia and sulfuric acid) and different recovery processes (*i.e.*, bioleaching and pyro-metallurgy), respectively.

10.4.7.5 Evaluation of Different Recycling and Recovery Rates

A number of studies compared the performance of different recycling and recovery rates and its environmental impact. Some studies compared various recycling and recovery rates with different management options such as landfill. For an example, Mayers et al. (2005) and Apisitpuvakul et al. (2008) compared various recycling rates with landfill disposal for printer and fluorescent lamp, respectively. In addition, there were also some studies focused only on different recycling and recovery rates and its environmental impact. For an example, Unger et al. (2017) performed a LCA study to compare the performance of current recycling rate of small WEEE with the required recycling rate according to European WEEE Directive, whereas Arduin et al. (2016) compared the current and future collection and recycling rates of mobile phone.

10.4.7.6 Evaluation of Collection and Transportation in Recycling and Recovery

Some studies concerned the impact of e-waste collection and transportation on the environmental benefit of recycling and recovery. For an example, Barba-Gutiérrez et al. (2008) and Xiao et al. (2016) compared the implications of e-waste collection and transportation towards the overall benefit of e-waste recycling and recovery in the Netherlands and China, respectively. Apart from that, Gamberini et al. (2010) evaluated various collection and transportation network for e-waste management in Italy.

10.5 DISCUSSION

Sustainable management of the end-of-life (EOL) e-waste has been considered to be a major concern in recent years due to its increasing trend in the consumption and shorter life span. Proper e-waste management is crucial due to its hazardous impact on environment and human health. However, various challenges in implementing sustainable e-waste management still remain untapped, particularly from a life cycle

TABLE 10.3
Challenges and Limitations in Adopting Sustainable E-waste Management

No.	E-waste Generation	E-waste Collection	E-waste Treatment
1	Illegal importation of e-waste	Lack of collection scheme	Low technological capacities – formal recycling sector
2	Increasing e-waste consumption	Low collection rate – consumer behaviour and awareness	Primitive methods – informal recycling sector
3	Increasing e-waste manufacturing industries	Evolvement of informal recycling sector	High cost of e-waste treatment

perspective. Table 10.3 shows the challenges in implementing sustainable e-waste management. Many countries are still lacking in regulations and policies for a proper management of e-waste, where e-waste is normally handled by non-formal recycling sectors. One of the major obstacles is recycling the printed circuit boards (PCBs) from the electronic wastes. The PCBs contain valuable materials such as platinum, gold, silver, etc. and base metals such as aluminium, copper, iron, etc. Usually, mechanical shredding and separation processes are used, but considered as inefficient in terms of the recycling rate of e-waste. Thus, affects the capacity of recovery of these valuable metals. Furthermore, the producers or importers of e-waste products and materials are not interested to participate in the voluntary take back scheme of e-wastes, thus makes the implementation difficult.

The complexity in e-waste management requires a better understanding on the benefit and burden of the e-waste materials, operations and processes. To overcome this issue, comprehensive framework and guideline for a sustainable management of e-waste need to be developed and to support the transition from traditional linear economy to circular and green economy. Figure 10.4 presented the framework for sustainable e-waste management. Estimation of future e-waste generation and its energy and materials flow can help to identify research gaps, opportunities and recommendations for investigating a more complete and transparent environmental performance of managing e-waste sustainably. Incorporating life cycle framework and thinking into e-waste management strategy produces a basis for making informed decisions to support sustainable e-waste management practices. Application of LCA in e-waste management can support the United Nation Agenda 2030 for sustainable development goals (SDGs) through sustainable e-waste management strategy towards circular economy and strategic decision making process for a cleaner and green industry.

10.6 CONCLUSION AND OUTLOOK

This chapter provides benefits to agencies, policy makers and industries pertaining to the e-waste management sector. The management of e-waste is facing a serious challenge, as e-waste contains various types of valuable materials and hazardous

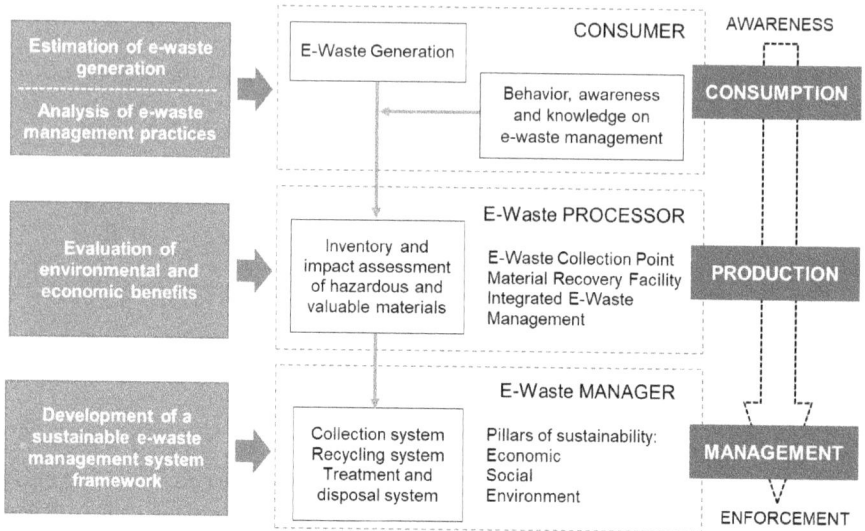

FIGURE 10.4 Framework for sustainable e-waste management.

substances, and therefore, it requires a proper treatment for disposal. With the complexity in e-waste management, various regulations and policies were introduced and implemented around the world. Currently, the best way to manage e-waste properly is still on-going research with various new techniques and approaches is still under development around the world. The study suggests that the LCA application in e-waste management could support the research agenda on establishment of sustainable e-waste management system, as LCA capable to provide a scientific evidence of any product system in capturing the environmental hotspot in the complex management chain of e-waste and identifying the improvement alternatives. Moreover, including LCA into e-waste management practice embraces the concept of waste-to-wealth and environmentally sound management of e-waste.

ACKNOWLEDGMENT

Marlia M. Hanafiah was financed by the National University of Malaysia (GUP-2020-034 & DIP-2019-001).

REFERENCES

Abeliotis, K., K. Lasaridi, C. Chroni, and A. Potouridis. 2017. Environmental Assessment of the Recovery of Printed Circuit Boards in Greece. In *5th International Conference on Sustainable Solid Waste Management*, 1–5.

Alcántara-Concepción, V., A. Gavilán-García, and I.C. Gavilán-García. 2016. Environmental Impacts at the End of Life of Computers and Their Management Alternatives in México. *Journal of Cleaner Production* 131: 615–628.

Allesch, A., and P.H. Brunner. 2014. Assessment Methods for Solid Waste Management: A Literature Review. *Waste Management & Research* 32, no. 6: 461–473.

Alston, S.M., and J.C. Arnold. 2011. Environmental Impact of Pyrolysis of Mixed WEEE Plastics Part 2: Life Cycle Assessment. *Environmental Science and Technology* 45, no. 21: 9386–9392.

Amato, A., L. Rocchetti, and F. Beolchini. 2017. Environmental Impact Assessment of Different End-of-Life LCD Management Strategies. *Waste Management* 59: 432–441.

Andersson, C., and J. Stage. 2018. Direct and Indirect Effects of Waste Management Policies on Household Waste Behaviour: The Case of Sweden. *Waste Management* 76: 19–27.

Andrade, D.F., J.P. Romanelli, and E.R. Pereira-Filho. 2019. Past and Emerging Topics Related to Electronic Waste Management: Top Countries, Trends, and Perspectives. *Environmental Science and Pollution Research* 26, no. 17: 17135–17151.

Andreola, F., L. Barbieri, A. Corradi, A.M. Ferrari, I. Lancellotti, and P. Neri. 2007. Recycling of EOL CRT Glass into Ceramic Glaze Formulations and its Environmental Impact by LCA Approach. *International Journal of Life Cycle Assessment* 12, no. 6: 448–454.

Apisitpuvakul, W., P. Piumsomboon, D.J. Watts, and W. Koetsinchai. 2008. LCA of Spent Fluorescent Lamps in Thailand at Various Rates of Recycling. *Journal of Cleaner Production* 16, no. 10: 1046–1061.

Arduin, R.H., C. Charbuillet, F. Berthoud, and N. Perry. 2016. What Are the Environmental Benefits of Increasing the WEEE Treatment in France? In *Electronics Goes Green 2016+, Berlin*, 1–7.

Arduin, R.H., C. Charbuillet, F. Berthoud, and N. Perry. 2017. Life Cycle Assessment of End-of-Life Scenarios: Tablet Case Study. In 16th International Waste Management and Landfill Symposium, Italy, 1–10.

Aziz, N.I.H.A., and M.M. Hanafiah. 2020. Life Cycle Analysis of Biogas Production from Anaerobic digestion of Palm Oil Mill Effluent. *Renewable Energy* 145: 847–857.

Aziz, N.I.H.A., M.M. Hanafiah, and S.H. Gheewala. 2019. A Review on Life Cycle Assessment of Biogas Production: Challenges and Future Perspectives in Malaysia. *Biomass and Bioenergy* 122: 361–374.

Barba-Gutiérrez, Y., B. Adenso-Díaz, and M. Hopp. 2008. An Analysis of Some Environmental Consequences of European Electrical and Electronic Waste Regulation. *Resources, Conservation and Recycling* 52, no. 3: 481–495.

Belboom, S., R. Renzoni, X. Deleu, J.-M. Digneffe, and A. Leonard. 2011. Electrical Waste Management Effects on Environment Using Life Cycle Assessment Methodology: The Fridge Case Study. In *SETAC EUROPE 17th LCA Case Study Symposium Sustainable Lifestyles*, 1–2.

Bian, J., H. Bai, W. Li, J. Yin, and H. Xu. 2016. Comparative Environmental Life Cycle Assessment of Waste Mobile Phone Recycling in China. *Journal of Cleaner Production* 131: 209–218.

Bientinesi, M., and L. Petarca. 2009. Comparative Environmental Analysis of Waste Brominated Plastic Thermal Treatments. *Waste Management* 29, no. 3: 1095–1102.

Biganzoli, L., A. Falbo, F. Forte, M. Grosso, and L. Rigamonti. 2015. Mass Balance and Life Cycle Assessment of the Waste Electrical and Electronic Equipment Management System Implemented in Lombardia Region (Italy). *Science of the Total Environment* 524–525: 361–375.

Bigum, M., L. Brogaard, and T.H. Christensen. 2012. Metal Recovery from High-Grade WEEE: A Life Cycle Assessment. *Journal of Hazardous Materials* 207–208: 8–14.

Biswas, W., and M. Rosano. 2011. A Life Cycle Greenhouse Gas Assessment of Remanufactured Refrigeration and Air Conditioning Compressors. *International Journal of Sustainable Manufacturing* 2, no. 2/3: 222.

Biswas, W.K., V. Duong, P. Frey, and M.N. Islam. 2013. A Comparison of Repaired, Remanufactured and New Compressors Used in Western Australian Small- and

Medium-Sized Enterprises in Terms of Global Warming. *Journal of Remanufacturing* 3, no. 1: 1–7.

Boyden, A., V.K. Soo, and M. Doolan. 2016. The Environmental Impacts of Recycling Portable Lithium-Ion Batteries. *Procedia CIRP* 48: 188–193.

Campolina, J.M., C.S.L. Sigrist, J.M.F. de Paiva, A.O. Nunes, and V.A. da S. Moris. 2017. A Study on the Environmental Aspects of WEEE Plastic Recycling in a Brazilian Company. *International Journal of Life Cycle Assessment* 22, no. 12: 1957–1968.

Compagno, L., C. Ingrao, A.G. Latora, and N. Trapani. 2014. Life Cycle Assessment of CRT Lead Recovery Process. *International Journal of Product Lifecycle Management* 7, no. 2/3: 201–214.

Dodbiba, G., H. Nagai, L.P. Wang, K. Okaya, and T. Fujita. 2012. Leaching of Indium from Obsolete Liquid Crystal Displays: Comparing Grinding with Electrical Disintegration in Context of LCA. *Waste Management* 32, no. 10: 1937–1944.

Dodbiba, G., K. Takahashi, J. Sadaki, and T. Fujita. 2008. The Recycling of Plastic Wastes from Discarded TV Sets: Comparing Energy Recovery with Mechanical Recycling in the Context of Life Cycle Assessment. *Journal of Cleaner Production* 16, no. 4: 458–470.

European Commission – Joint Research Centre – Institute for Environment and Sustainability (EC-JRC). 2010a. *International Reference Life Cycle Data System (ILCD) Handbook: General Guide for Life Cycle Assessment – Detailed Guidance.* Luxembourg. Publications Office of the European Union.

European Commission – Joint Research Centre – Institute for Environment and Sustainability (EC-JRC). 2010b. *International Reference Life Cycle Data System (ILCD) Handbook: Framework and Requirements for LCIA Models and Indicators.* Luxembourg. Publications Office of the European Union.

Ewijk, S. Van, and J.A. Stegemann. 2016. Limitations of the Waste Hierarchy for Achieving Absolute Reductions in Material Throughput. *Journal of Cleaner Production* 132: 122–128.

Eygen, E. Van, S. De Meester, H.P. Tran, and J. Dewulf. 2016. Resource Savings by Urban Mining: The Case of Desktop and Laptop Computers in Belgium. *Resources, Conservation and Recycling* 107: 53–64.

Finnveden, G., M.Z. Hauschild, T. Ekvall, J. Guinée, R. Heijungs, S. Hellweg, A. Koehler, D. Pennington, and S. Suh. 2009. Recent Developments in Life Cycle Assessment. *Journal of Environmental Management* 91, no. 1: 1–21.

Foelster, A.S., S. Andrew, L. Kroeger, P. Bohr, T. Dettmer, S. Boehme, and C. Herrmann. 2016. Electronics Recycling as an Energy Efficiency Measure – A Life Cycle Assessment (LCA) Study on Refrigerator Recycling in Brazil. *Journal of Cleaner Production* 129: 30–42.

Forti, V., C.P. Baldé, R. Kuehr, and G. Bel. 2020. *The Global E-Waste Monitor 2020: Quantities, Flows, and the Circular Economy Potential. United Nations University (UNU)/ United Nations Institute for Training and Research (UNITAR) – Co-Hosted SCYCLE Programme, International Telecommunication Union (ITU) & International Solid Waste Association (ISWA), Bonn/Geneva/Rotterdam.*

Gamberini, R., E. Gebennini, R. Manzini, and A. Ziveri. 2010. On the Integration of Planning and Environmental Impact Assessment for a WEEE Transportation Network – A Case Study. *Resources, Conservation and Recycling* 54, no. 11: 937–951.

Gao, Y., L. Ge, S. Shi, Y. Sun, M. Liu, B. Wang, Y. Shang, J. Wu, and J. Tian. 2019. Global Trends and Future Prospects of E-Waste Research: A Bibliometric Analysis. *Environmental Science and Pollution Research* 26, no. 17: 17809–17820.

Gentil, E.C., A. Damgaard, M. Hauschild, G. Finnveden, O. Eriksson, S. Thorneloe, P.O. Kaplan, et al. 2010. Models for Waste Life Cycle Assessment: Review of Technical Assumptions. *Waste Management* 30, no. 12: 2636–2648.

Ghodrat, M., M.A. Rhamdhani, G. Brooks, M. Rashidi, and B. Samali. 2017. A Thermodynamic-Based Life Cycle Assessment of Precious Metal Recycling out of Waste Printed Circuit Board through Secondary Copper Smelting. *Environmental Development* 24: 36–49.

Guinée, J., R. Heijungs, G. Huppes, A. Zamagni, P. Masoni, T. Ekvall, and T. Rydberg. 2011. Life Cycle Assessment: Past, Present and Future. *Environment Science And Technology* 45, no. 1: 90–96.

Harun, S.N., M.M. Hanafiah, and N.I.H.A. Aziz. 2020. An LCA-Based Environmental Performance of Rice Production for Developing a Sustainable Agri-Food System in Malaysia. *Environmental Management*. https://doi.org/10.1007/s00267-020-01365-7.

Hauschild, M.Z., and M.A.J. Huijbregts. 2015. *Life Cycle Impact Assessment*. Springer Netherlands.

Hauschild, M.Z., R.K. Rosenbaum, and S.I. Olsen. 2018. *Life Cycle Assessment: Theory and Practice*. Springer International Publishing AG.

Herat, S., and P. Agamuthu. 2012. E-Waste: A Problem or an Opportunity? Review of Issues, Challenges and Solutions in Asian Countries. *Waste Management and Research* 30, no. 11: 1113–1129.

Hischier, R., P. Wäger, and J. Gauglhofer. 2005. Does WEEE Recycling Make Sense from an Environmental Perspective? The Environmental Impacts of the Swiss Take-Back and Recycling Systems for Waste Electrical and Electronic Equipment (WEEE). *Environmental Impact Assessment Review* 25, no. 5: 525–539.

Hong, J., W. Shi, Y. Wang, W. Chen, and X. Li. 2015. Life Cycle Assessment of Electronic Waste Treatment. *Waste Management* 38, no. 1: 357–365.

Iannicelli-Zubiani, E.M., M.I. Giani, F. Recanati, G. Dotelli, S. Puricelli, and C. Cristiani. 2017. Environmental Impacts of a Hydrometallurgical Process for Electronic Waste Treatment: A Life Cycle Assessment Case Study. *Journal of Cleaner Production* 140: 1204–1216.

International Organization for Standardization (ISO). 1997. ISO 14040:1997 Environmental Management – Life Cycle Assessment – Principles and Framework.

International Organization for Standardization (ISO). 1998. ISO 14041:1998 Environmental Management – Life Cycle Assessment – Goal and Scope Definition and Inventory Analysis.

International Organization for Standardization (ISO). 2000a. ISO 14042:2000 Environmental Management – Life Cycle Assessment – Life Cycle Impact Assessment.

International Organization for Standardization (ISO). 2000b. ISO 14043:2000 Environmental Management – Life Cycle Assessment – Life Cycle Interpretation.

International Organization for Standardization (ISO). 2006a. ISO 14040:2006 Environmental Management – Life Cycle Assessment – Principles and Framework.

International Organization for Standardization (ISO). 2006b. ISO 14044:2006 Environmental Management – Life Cycle Assessment – Requirements and Guidelines.

Ismail, H., and M.M. Hanafiah. 2017. Management of End-of-Life Electrical and Electronic Products: The Challenges and the Potential Solutions for Management Enhancement in Developing Countries Context. *Acta Scientifica Malaysia* 1, no. 2: 5–8.

Ismail, H., and M.M. Hanafiah. 2019a. An Overview of LCA Application in WEEE Management: Current Practices, Progress and Challenges. *Journal of Cleaner Production* 232: 79–93.

Ismail, H., and M.M. Hanafiah. 2019b. Discovering Opportunities to Meet the Challenges of an Effective Waste Electrical and Electronic Equipment Recycling System in Malaysia. *Journal of Cleaner Production* 238: 117927.

Ismail, H., and M.M. Hanafiah. 2020. A Review of Sustainable E-Waste Generation and Management: Present and Future Perspectives. *Journal of Environmental Management* 264: 110495.

Johansson, J.G., and A.E. Björklund. 2010. Reducing Life Cycle Environmental Impacts of Waste Electrical and Electronic Equipment Recycling. *Journal of Industrial Ecology* 14, no. 2: 258–269.

Khandelwal, H., H. Dhar, A.K. Thalla, and S. Kumar. 2019. Application of Life Cycle Assessment in Municipal Solid Waste Management: A Worldwide Critical Review. *Journal of Cleaner Production* 209: 630–654.

Kirkeby, J.T., H. Birgisdottir, T.L. Hansen, T.H. Christensen, M. Hauschild, and G.S. Bhander. 2006. Environmental Assessment of Solid Waste Systems and Technologies: EASEWASTE. *Waste Management & Research* 24, no. 1: 3–15.

Klopffer, W., and B. Grahl. 2014. *Life Cycle Assessment (LCA): A Guide to Best Practice.* John Wiley, Weinheim, Germany.

Laurent, A., I. Bakas, J. Clavreul, A. Bernstad, M. Niero, E. Gentil, M.Z. Hauschild, and T.H. Christensen. 2014. Review of LCA Studies of Solid Waste Management Systems – Part I: Lessons Learned and Perspectives. *Waste Management* 34, no. 3: 573–588.

Laurent, A., J. Clavreul, A. Bernstad, I. Bakas, M. Niero, E. Gentil, T.H. Christensen, and M.Z. Hauschild. 2014. Review of LCA Studies of Solid Waste Management Systems – Part II: Methodological Guidance for a Better Practice. *Waste Management* 34, no. 3: 589–606.

Lu, B., B. Li, L. Wang, J. Yang, J. Liu, and X.V. Wang. 2014. Reusability Based on Life Cycle Sustainability Assessment: Case Study on WEEE. *Procedia CIRP* 15: 473–478.

Lu, B., X. Song, J. Yang, and D. Yang. 2017. Comparison on End-of-Life Strategies of WEEE in China Based on LCA. *Frontiers of Environmental Science and Engineering* 11, no. 5: 7.

Lu, L.T., I.K. Wernick, T.Y. Hsiao, Y.H. Yu, Y.M. Yang, and H.W. Ma. 2006. Balancing the Life Cycle Impacts of Notebook Computers: Taiwan's Experience. *Resources, Conservation and Recycling* 48, no. 1: 13–25.

Manfredi, S., and R. Pant. 2011. *Supporting Environmentally Sound Decisions for Waste Management: A Technical Guide to Life Cycle Thinking (LCT) and Life Cycle Assessment (LCA) for Waste Experts and LCA Practitioners.* Luxembourg: Publications Office of the European Union.

Mayers, C.K., C.M. France, and S.J. Cowell. 2005. Extended Producer Responsibility for Waste Electronics: An Example of Printer Recycling in the United Kingdom. *Journal of Industrial Ecology* 9, no. 3: 169–189.

Menikpura, S.N.M., A. Santo, and Y. Hotta. 2014. Assessing the Climate Co-Benefits from Waste Electrical and Electronic Equipment (WEEE) Recycling in Japan. *Journal of Cleaner Production* 74: 183–190.

Moraes, D. da G. e S.V.M. de, T.B. Rocha, and M.R. Ewald. 2014. Life Cycle Assessment of Cell Phones in Brazil Based on Two Reverse Logistics Scenarios. *Production* 24, no. 4: 735–741.

Niu, R., Z. Wang, Q. Song, and J. Li. 2012. LCA of Scrap CRT Display at Various Scenarios of Treatment. *Procedia Environmental Sciences* 16: 576–584.

Noon, M.S., S.J. Lee, and J.S. Cooper. 2011. A Life Cycle Assessment of End-of-Life Computer Monitor Management in the Seattle Metropolitan Region. *Resources, Conservation and Recycling* 57: 22–29.

Pennington, D.W., J. Potting, G. Finnveden, E. Lindeijer, O. Jolliet, T. Rydberg, and G. Rebitzer. 2004. Life Cycle Assessment Part 2: Current Impact Assessment Practice. *Environment International* 30, no. 5: 721–739.

Pryshlakivsky, J., and C. Searcy. 2013. Fifteen Years of ISO 14040: A Review. *Journal of Cleaner Production* 57: 115–123.

Rebitzer, G., T. Ekvall, R. Frischknecht, D. Hunkeler, G. Norris, T. Rydberg, W.-P. Schmidt, S. Suh, B.P. Wedema, and D.W. Pennington. 2004. Life Cycle Assessment Part 1: Framework, Goal and Scope, Inventory Analysis and Applications. *Environmental International* 30, no. 5: 701–720.

Robinson, B.H. 2009. E-Waste: An Assessment of Global Production and Environmental Impacts. *Science of the Total Environment* 408, no. 2: 183–191.

Rocchetti, L., and F. Beolchini. 2014. Environmental Burdens in the Management of End-of-Life Cathode Ray Tubes. *Waste Management* 34, no. 2: 468–474.

Rocchetti, L., F. Vegliò, B. Kopacek, and F. Beolchini. 2013. Environmental Impact Assessment of Hydrometallurgical Processes for Metal Recovery from WEEE Residues Using a Portable Prototype Plant. *Environmental Science and Technology* 47, no. 3: 1581–1588.

Rodriguez-Garcia, G., and M. Weil. 2016. Chapter 7 – Life Cycle Assessment in WEEE Recycling. In *WEEE Recycling: Research, Development, and Policies*, ed. A. Chagnes, G. Cote, C. Ekberg, M. Nilsson, T. Retegan, 177–207. Elsevier.

Rubin, R.S., M.A.S. De Castro, D. Brandão, V. Schalch, and A.R. Ometto. 2014. Utilization of Life Cycle Assessment Methodology to Compare Two Strategies for Recovery of Copper from Printed Circuit Board Scrap. *Journal of Cleaner Production* 64: 297–305.

Solé, M., J. Watson, R. Puig, and P. Fullana-I-Palmer. 2012. Proposal of a New Model to Improve the Collection of Small WEEE: A Pilot Project for the Recovery and Recycling of Toys. *Waste Management and Research* 30, no. 11: 1208–1212.

Song, Q., Z. Wang, and J. Li. 2013a. Sustainability Evaluation of E-Waste Treatment Based on Emergy Analysis and the LCA Method: A Case Study of a Trial Project in Macau. *Ecological Indicators* 30: 138–147.

Song, Q., Z. Wang, J. Li, and X. Zeng. 2013b. The Life Cycle Assessment of an E-Waste Treatment Enterprise in China. *Journal of Material Cycles and Waste Management* 15, no. 4: 469–475.

Song, X., C. Zhang, W. Yuan, and D. Yang. 2018. Life-Cycle Energy Use and GHG Emissions of Waste Television Treatment System in China. *Resources, Conservation and Recycling* 128: 470–478.

Soo, V.K., and M. Doolan. 2014. Recycling Mobile Phone Impact on Life Cycle Assessment. *Procedia CIRP* 15: 263–271.

Tran, H.P., T. Schaubroeck, P. Swart, L. Six, P. Coonen, and J. Dewulf. 2018. Recycling Portable Alkaline/ZnC Batteries for a Circular Economy: An Assessment of Natural Resource Consumption from a Life Cycle and Criticality Perspective. *Resources, Conservation and Recycling* 135: 265–278.

Unger, N., P. Beigl, G. Höggerl, and S. Salhofer. 2017. The Greenhouse Gas Benefit of Recycling Waste Electrical and Electronic Equipment above the Legal Minimum Requirement: An Austrian LCA Case Study. *Journal of Cleaner Production* 164: 1635–1644.

Villares, M., A. Işildar, A. M. Beltran, and J. Guinée. 2016. Applying an Ex-Ante Life Cycle Perspective to Metal Recovery from e-Waste Using Bioleaching. *Journal of Cleaner Production* 129: 315–328.

Wäger, P. A., R. Hischier, and M. Eugster. 2011. Environmental Impacts of the Swiss Collection and Recovery Systems for Waste Electrical and Electronic Equipment (WEEE): A Follow-Up. *Science of the Total Environment* 409, no. 10: 1746–1756.

Wäger, P.A., and R. Hischier. 2015. Life Cycle Assessment of Post-Consumer Plastics Production from Waste Electrical and Electronic Equipment (WEEE) Treatment Residues in a Central European Plastics Recycling Plant. *Science of the Total Environment* 529: 158–167.

Wang, Y., H. Hu, S. Qi, and G. Liu. 2014. Environmental Impacts Assessment of Liquid Crystal Extraction From Wasted LCD Panels. *Applied Mechanics and Materials* 496–500: 55–62.

Wolf, M.-A., R. Pant, K. Chomkhamsri, S. Sala, and D. Pennington. 2012. *The International Reference Life Cycle Data System (ILCD) Handbook: Towards More Sustainable Production and Consumption for a Resource-Efficient Europe.* Luxembourg: Publications Office of the European Union.

Xiao, R., Y. Zhang, and Z. Yuan. 2016. Environmental Impacts of Reclamation and Recycling Processes of Refrigerators Using Life Cycle Assessment (LCA) Methods. *Journal of Cleaner Production* 131: 52–59.

Xu, Q., M. Yu, A. Kendall, W. He, G. Li, and J.M. Schoenung. 2013. Environmental and Economic Evaluation of Cathode Ray Tube (CRT) Funnel Glass Waste Management Options in the United States. *Resources, Conservation and Recycling* 78: 92–104.

Xue, M., A. Kendall, Z. Xu, and J.M. Schoenung. 2015. Waste Management of Printed Wiring Boards: A Life Cycle Assessment of the Metals Recycling Chain from Liberation through Refining. *Environmental Science and Technology* 49, no. 2: 940–947.

Xue, M., and Z. Xu. 2017. Application of Life Cycle Assessment on Electronic Waste Management: A Review. *Environmental Management* 59, no. 4: 693–707.

Yadav, P., and S.R. Samadder. 2018. A Critical Review of the Life Cycle Assessment Studies on Solid Waste Management in Asian Countries. *Journal of Cleaner Production* 185: 492–515.

Yao, L., T. Liu, X. Chen, M. Mahdi, and J. Ni. 2018. An Integrated Method of Life-Cycle Assessment and System Dynamics for Waste Mobile Phone Management and Recycling in China. *Journal of Cleaner Production* 187: 852–862.

Zhang, L., Y. Geng, Y. Zhong, H. Dong, and Z. Liu. 2019. A Bibliometric Analysis on Waste Electrical and Electronic Equipment Research. *Environmental Science and Pollution Research* 26, no. 21: 21098–21108.

Zink, T., F. Maker, R. Geyer, R. Amirtharajah, and V. Akella. 2014. Comparative Life Cycle Assessment of Smartphone Reuse: Repurposing vs. Refurbishment. *International Journal of Life Cycle Assessment* 19, no. 5: 1099–1109.

11 Conclusion

Biswajit Debnath, Abhijit Das, Anil Potluri, and Siddhartha Bhattacharyya

Electronic waste (e-waste) has become an alarming issue worldwide, especially in developing countries due to improper treatment pathways employed by the informal sector. On the other hand, e-waste is a huge source of secondary raw materials. The term urban mining is synonymous with resource recovery from e-waste. Since e-waste is classified as hazardous waste, a lot of e-waste is shipped to low and medium-income countries such as India, China, Bangladesh, Vietnam, Africa, etc. China has banned the import of waste materials including plastic waste and electronic waste from other developed countries since 2018. As a result, routes of transboundary movement of e-waste have changed and pressure on countries such as Bangladesh, India, Vietnam has increased (Nair, 2021). Despite being a signatory of the Basel Convention, these kinds of illicit trade happen following the loopholes in the rules. On the other hand, this kind of trade becomes feedstock for the formal recycling units as the in-house collection scenario is dominated by the informal sector.

E-waste contains a wide variety of elements including common metals, rare earth metals, polymers, glass, glass fibre, rubber, concrete, and ceramics, etc. Metal recovery from e-waste is now technologically feasible, yet the sustainability of the business is a matter of concern as the electronics are becoming lighter (Balde, 2017). This is due to the percentage of metals in e-waste decreasing and the number of plastics increasing. The recovery and reuse of the plastic part of e-waste is comparatively a less discussed topic in the contemporary literature as well as the conferences. Utilization of this huge source is imperative to maintain business sustainability. Mechanical recycling is the most widely practiced for plastics recycling and extrusion is the most common method. The issues arise due to improper segregation as plastics containing halogens cannot be recycled in an environment-friendly way via extrusion. Additionally, to recycle a specific category of waste plastic, it is essential to avoid contamination with plastics containing Halogenated Flame Retardants (HFRs). While treatment of e-waste plastics is a concern, it is also important to recover metals from e-waste to extract the economic benefits (Forti, 2020). There are several other technologies for metal recovery from e-waste such as mechanical recycling, hydrometallurgical, pyro-metallurgical, and bio-metallurgical technologies. All these technologies complement urban mining. Utilization of the metals recovered from e-waste will not only enhance resource efficiency but also ensure a circular economy.

DOI: 10.1201/9781003095972-15

The term 'Circular Economy' (CE) is based on the very same concept of resource circulation and closing the loop. It is reasonable to link the concept of circular economy with the growing research and discussions on urban mining (Debnath et al. 2022).

To establish a sustainable and circular e-waste management system, timely intervention in technological aspects, policy modification, and environmental issues are required. E-waste is a complex material and the issues coming up are also very complex. While the hardware part of e-waste poses a health hazard to the informal e-waste workers, improper disposal of memory devices can pose security threats at different levels (Mao, 2020). With the advancement in reverse engineering technologies, such threats are becoming more prominent. While these issues exist in a hidden form at this moment, many other issues such as health issues due to improper handling of e-waste have already been proven. Despite the odds, emerging technologies in the field of information technology can very well be applied for better e-waste management. Intelligent and calculative strategies will prove to be triumphant towards paving a path, if not a vision towards a sustainable future.

The Sustainable Development Goals (SDGs), were adopted by all United Nations Member States in 2015 as a universal call to action to end poverty, protect the planet and ensure that all people enjoy peace and prosperity by 2030. These 17 SDGs recognize that action in one area will affect outcomes in others and that development must balance social, economic, and environmental sustainability. Achieving the SDGs requires the partnership of governments, the private sector, civil society, and citizens alike to make sure we leave a better planet for future generations. Recent advancements in the field of e-waste management can help to achieve some of the SDGs (Paben, 2021). Research outcomes in this field complementing the SDGs must be recognized and such approaches are praiseworthy. To achieve any big goal, it is advisable to follow the PDCA cycle i.e. the 'Plan-Do-Check-Act'. The timeline left to achieve the SDGs is only 10 years. As a checkpoint, we must look into the ever-evolving e-waste management sector, while understanding the paradigm shift towards a sustainable future.

Given the hazards induced by electronic wastes on the environment around the globe, it has become imperative for all nations to come forward and take a solemn resolution for evolving effective means for thwarting this ever-increasing danger to the environment. Although several initiatives have been taken up in this direction, yet most of these seem to be inadequate and ill-directed. A proper Government policy has this become a mandatory requirement to put things in place. To be specific the book covers the advancements in e-waste management in the past decade while capturing the proliferation in the technologies and tools for a sustainable future. This volume may come up as a whistle-blower in this regard and facilitate the policymakers to carve a fruitful plan to restore environmental sustainability from this menace.

REFERENCES

Nair, Abhijith. 2021. "E-Waste Management Market Size, Share and Industry Analysis | 2027"| 2027". www.alliedmarketresearch.com/e-waste-management-market. (Accessed 29 May 2021).

Baldé, Cornelis P., Vanessa Forti, Vanessa Gray, Ruediger Kuehr, and Paul Stegmann. *The global e-waste monitor 2017: Quantities, flows and resources.* United Nations University, International Telecommunication Union, and International Solid Waste Association, 2017.

Debnath, Biswajit, Ankita Das, and Abhijit Das. "Towards circular economy in e-waste management in India: Issues, challenges, and solutions." In Circular Economy and Sustainability, pp. 523–543. Elsevier, 2022.

Forti, Vanessa, Cornelis P. Balde, Ruediger Kuehr, and Garam Bel. "The Global E-waste Monitor 2020: Quantities, flows and the circular economy potential." (2020) United Nations University (UNU)/United Nations Institute for Training and Research (UNITAR) – co-hosted SCYCLE Programme, International Telecommunication Union (ITU) & International Solid Waste Association (ISWA), Bonn/Geneva/Rotterdam.

Mao, Shaohua, Weihua Gu, Jianfeng Bai, Bin Dong, Qing Huang, Jing Zhao, Xuning Zhuang, Chenglong Zhang, Wenyi Yuan, and Jingwei Wang. "Migration of heavy metal in electronic waste plastics during simulated recycling on a laboratory scale." *Chemosphere* 245 (2020): 125645.

Paben, Jared. 2021. "Copper Price Climbs To Recent Record – E-Scrap News". *E-Scrap News.* https://resource-recycling.com/e-scrap/2021/02/25/copper-price-climbs-to-recent-record/?utm_medium=email&utm_source=internal&utm_campaign=Feb+25+ESN. (Accessed 29 May 2021).

Index

For Product Safety Concerns and Information please contact our EU
representative GPSR@taylorandfrancis.com
Taylor & Francis Verlag GmbH, Kaufingerstraße 24, 80331 München, Germany

9 780367 559892